D1442577

ELECTRICITY, ELECTRONICS, and ELECTROMAGNETICS

Principles and Applications

ELECTRICITY, ELECTRONICS, and ELECTROMAGNETICS

Principles and Applications

ROBERT BOYLESTAD

Professor Queensborough College (City University of New York)

LOUIS NASHELSKY

Professor Queensborough College (City University of New York)

Prentice-Hall, Inc., Englewood Cliffs, New Jersey

Library of Congress Cataloging in Publication Data

BOYLESTAD, ROBERT L
    Electricity, electronics, and electromagnetics.

    Includes index.
    1. Electric engineering. 2. Electronics.
I. Nashelsky, Louis, joint author. II. Title.
TK146.B78   621.3   76–15000
ISBN 0–13–248310–6

PRENTICE-HALL INTERNATIONAL, INC., *London*
PRENTICE-HALL OF AUSTRALIA, PTY. LTD., *Sydney*
PRENTICE-HALL OF CANADA, LTD., *Toronto*
PRENTICE-HALL OF INDIA PRIVATE LIMITED, *New Delhi*
PRENTICE-HALL OF JAPAN, INC., *Tokyo*
PRENTICE-HALL OF SOUTHEAST ASIA (PTE.) LTD., *Singapore*

To our mothers,

Astrid Boylestad
Nellie Nashelsky

and in memory of our fathers

Anders Boylestad
Samuel Nashelsky

Contents

4 Magnetics   146

5 dc and ac Machinery   175

6 Basic Electronic Devices   214

7 Integrated Circuits (ICs)   258

Preface

The text covers the basic principles of all the major topics in electrical and electronic engineering and selected topics in electromechanics. The level of presentation is suitable for non-electrical students who require some background in electrical and electronic components, circuits, and applications. It can also be used satisfactorily by electrical students interested in selected topics, as a reference, or review of basic principles. The authors have attempted to concentrate on the most important concepts of each area covered, realizing that any in-depth treatment is best reserved for texts devoted to individual areas. Material is presented at a level that permits some application to practical problems without using lengthy derivations or deep mathematical development.

The material has been developed over a number of years of teaching and should reflect a level of reading and development suitable for student use. There is sufficient material for two semesters of coursework. In a one semester course the basic material should be covered followed by selected topics of particular interest.

The authors wish to thank Bill Gibson for his editorial help, Martha Gibson for her typing and Gene Panhorst for his diligent work in bringing the book through production.

ROBERT BOYLESTAD
Brookfield, Conn.

LOUIS NASHELSKY
Great Neck, N.Y.

1

Introduction

The areas of electricity and electronics have matured greatly over the last few decades until presently almost every person, household, or place in the country is dependent on them for their basic needs and comforts. We have long been dependent on electricity as a basic source of power. Both radio and television units can be considered a basic item in a great number of households in this country. The engineers of early 1900 would have been greatly impressed by the radio and TV industry of 1950. The engineers of 1950 can hardly believe the integrated circuit (IC) devices of 1975, and today's engineers will more than likely be astounded by the technological progress by the year 2000.

We are presently seeing only the start of the use of IC units in many commercial products. Smaller, less expensive, more sophisticated electronic units are already widely used in computers, calculators, radios, and cars, as examples. Electronic circuits will undoubtedly become a standard part of many products since these units offer considerable improvement, with the addition of very small, low-power, low-cost semiconductor devices and IC circuits. The widespread growth of the calculator is just one example of how electronic units are being quickly included in the commercial market. Where the slide rule was a standard calculating aid in engineering curricula for many years, the use of digital scientific calculators is making great inroads and will eventually become the accepted standard. The calculator circuit is a highly sophisticated electronic unit containing a great number of digital circuits.

The use of electronic circuitry has been steadily increasing in a great variety of areas and applications. This growth is partially due to the availability of low-cost, very small-sized, relatively complete electronic circuits referred to as large-scale integrated (LSI) circuits. This growth is well evident in the commercial marketplace as well as in the engineering profession. The large slide rule, hanging from the engineering student's belt, has been recently replaced by the more compact scientific calculator. Such electronic units as the HP-45 and SR-51 represent very sophisticated

1

electronic units, these units being virtually small, portable, fixed-program computers. It still astounds people that so much electronic circuitry can be contained in a device which is so compact. It is even more astounding to realize that within this small calculator there is essentially one small IC package, taking up less than one-tenth of the inside volume of the unit, and that within this LSI package is an even smaller chip on which the total circuitry exists. This LSI component contains tens of thousands of individual electronic components arranged in the various parts of a computer, including a fixed or read-only memory (ROM) and one or more erasable memory locations. This small electronic size is equally impressive when found in the growing area of digital watches. Again the single-chip LSI is the heart of a relatively complex total electronic package. Powered by a small battery, these electronic units contain circuitry to generate a reference clock (oscillator) signal, count these clock pulses, and display the ongoing time on a decimal display output. As the technology has improved the digital watch has been designed to also keep track of the date and display this information as well—and we are still only at the early stages of the IC revolution.

Radio has been an early indicator of the advances in electronics with small, portable radios or compact table models common. Linear integrated circuits are available for AM and FM radios—these IC units consisting of various amplifiers, detectors, and filtering circuits, essentially all the components required for these devices. Where economics will allow it the complete radio circuit can be built on a single chip of small size. Television circuitry is similarly becoming more dependent on IC units, and only the TV display tube is presently not a solid-state device. Basic to many of these circuits are the differential amplifier and operational amplifier circuits available individually as the building block of present electronic circuits, much like a single transistor was previously the primary electronic component.

The use of electronic components has been mainly used in engineering applications and special commercial products. The availability of small, low-cost electronics has made it practical to include electronic circuitry as part of many new areas of application, including measuring instruments, and in mechanical, civil, chemical, and biomedical engineering fields, among others. In each of these areas electronic circuitry has been added—both linear and digital IC—to provide more automatic operation, provide digital display, provide more centralized control, and add additional features not presently available. For example, digital thermometers now provide a faster, more easily visible reading in a small, compact, portable electronic measuring instrument. Similarly, in instrumentation, the digital display multimeter is fast replacing deflection dial meters. Oscilloscopes now have added display and control circuitry to provide automatic scale selection and digital readout.

To begin to better understand what is presently available in a variety of areas and to develop some framework in which to understand the future growth in this electronic area, the present text covers a range of subject matter from the initial chapters on fundamental analysis to the basic electronic areas and devices, concluding with a range of application areas to show how electronics is specialized for particular uses.

2

dc Networks

2.1 INTRODUCTION

The introduction to the elements of electricity and electronics will begin in the time-honored tradition with direct-current (dc) circuits. Systems of this type have fixed levels of electrical quantities that do not vary with time. The typical car battery is an excellent example of a dc quantity. If labeled as a 12-V battery, the terminal voltage will always be 12 V (ideally) and will not change with time. The electrical energy available at any home outlet is called an alternating-current (ac) voltage, which varies in a definite manner with time. This type of electrical quantity will be considered in detail in a later chapter. Since the quantities of interest in a dc network do not vary with time, it is considerably easier to introduce the basic laws of electrical systems with solely dc inputs. However, be assured that the similarities are so strong between the application of a theorem to a dc network as compared to an ac network that the analysis of ac systems will be made considerably easier by the knowledge gained from dc networks.

Since a broad range of subject matter must be covered in a minimum number of pages, there will be little time for derivations and lengthy discussions of any particular area. Further investigation must be left to the reader, who can use the wide variety of available references.

In the typical approach to this subject matter, instruments are usually considered in a separate chapter or text. Here they will be introduced the moment a measurable quantity is examined. In this way, the authors feel that the student using this type of text will be best prepared for practical application of the concepts discussed. There will, however, be little time to examine the guts of each instrument. The scales and proper use of the dials will be the common extent of the coverage of each instrument.

2.2 CURRENT

The first electrical quantity of major importance to be introduced is the *flow* variable: *current*. The rate of flow of charge through a conductor such as the wires in the walls of every home is a measure of the current in the conductor. The charge referred to above is the electrons in the wires that will move in a particular direction caused by an outside electrical supply of energy such as a battery (which will be discussed in detail in a later section). The greater the amount of charge that flows through a wire per unit time, the greater the current. In equation form,

$$\text{Current} = \frac{\text{Charge}}{\text{Time}}$$

or

$$I = \frac{Q}{t} \tag{2.1}$$

where I = current in amperes
Q = charge in coulombs
T = time in seconds

For conversion purposes,

$$1 \text{ coulomb (C)} = 6.24 \times 10^{18} \text{ electrons} \tag{2.2}$$

Note in Eq. (2.1) that current is measured in amperes (A) in honor of the French scientist Ampère. The equation states in words that if 6.24×10^{18} electrons (or 1 C of charge) pass through the imaginary surface of the wire of Fig. 2.1 in 1 s (all in one particular direction), the current is 1 A.

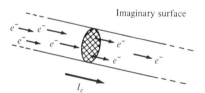

Figure 2.1 Charge flow in a conductor.

EXAMPLE 2.1

Determine the current in amperes through the wire of Fig. 2.1 if 28 C of charge pass through the cross-sectional area in one direction in 4 sec.

Solution:

$$I = \frac{Q}{t} = \frac{28}{4} = 7 \text{ ampere}$$

Before continuing, it must be pointed out that there are two fields of thought regarding the direction to be assigned to the current of a network. The vast majority of educational and industrial institutions employ the *conventional* current approach, which is defined by Fig. 2.2 to be opposite to that of *electron* flow. In an electrical conductor we now know that it is the negatively charged electrons that actually move through the conductor. However, at the time electrical energy was first discovered it was thought the positive carriers were the moving particles and hence the reverse direction. Be assured that even though the conventional flow is used in this text it is not a factor for further consideration since through continued exposure and choosing the proper notation, one system or the other could have been employed with very little added difficulty.

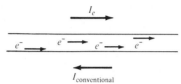

Figure 2.2 Conventional versus electron flow.

An analog often used to develop a clearer understanding of the concept of current is the flow of water through a pipe, such as shown in Fig. 2.3. To *measure* the flow of water the pipe must be separated and a meter inserted as shown in the figure. The same is also absolutely true for the measurement of current, as also indicated in the figure. In other words, if the meter were removed, an open-circuit condition would exist for each. For rather obvious reasons, the current-measuring device is called an *ammeter*. Note that the current direction is such as to enter the positive terminal and leave the negative. If it were connected in the reverse manner, the meter would read downscale below 0 (zero) A. The significance of the positive and negative sign will become apparent in the sections to follow. The *face* of a typical ammeter is shown in Fig. 2.4. The *short* condition simply indicates a direct connection between the meter terminals which totally isolates the meter circuitry from the energized electrical network. With the dial set on 10 A the scale with the 10 to the far right is employed and represents the maximum reading of 10 A. For the dial setting of 1 A the same scale is used except now each reading is divided by 10. That is,

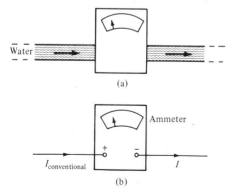

Figure 2.3 Flow measurements: (a) water, (b) currents.

Figure 2.4 Dc ammeter. (Courtesy Simpson Electric Co.)

$10 \rightarrow 1$ A, $9 \rightarrow 0.9$ A, and $4.5 \rightarrow 0.45$ A. In summary, the maximum of a scale has the same numbers as the dial setting, but some reading may have to be adjusted by a scale factor such as $\frac{1}{10}$ in this case.

As with every measurable quantity, there are very large and small levels of current. For high-power applications, hundreds of amperes are typical. The average home has a 100-, 150-, or 200-A service. The service rating indicates the maximum current that can be drawn by that home from the power line. Considering that a single air conditioner may draw 15 A (15% of a 100-A service) makes the choice of installing a larger service in a new home an important consideration. At the other end of the scale of magnitudes is the field of electronics where thousandths and even millionths of an ampere are encountered. For such small values, a system of scientific notation is commonly used which has the labels and abbreviations given in Table 2.1. The notation for very large numbers is also included.

Table 2.1

Scientific Notation

$$1,000,000.0 = 10^6 \quad = \text{mega} = \text{M}$$
$$1000.0 = 10^3 \quad = \text{kilo} = \text{k}$$
$$\frac{1}{1000} = 0.001 = 10^{-3} \quad = \text{milli} = \text{m}$$
$$\frac{1}{1,000,000} = 0.000001 = 10^{-6} \quad = \text{micro} = \mu$$
$$0.000000001 = 10^{-9} \quad = \text{nano} = \text{n}$$
$$0.000000000001 = 10^{-12} = \text{pico} = \text{p}$$

A simple count from the decimal point to the right of the number 1 will result in the proper power of 10.

EXAMPLE 2.2

Determine the proper power of 10 for the following numbers:

Solution:
1. $10,0000 = 10000.0 = 10^{+4}$.
2. $0.00001 = 10^{-5}$.
3. $0.004 = 4(0.001) = 4 \times 10^{-3}$.
4. $520,000 = 52(10,000) = 52 \times 10^{+4}$.

A current of 0.002 A can therefore be labeled as 2×10^{-3} A or 2 mA. This same notation can be applied to the other electrical quantities to be introduced shortly.

Since some applications are limited solely to using very small currents, meters are available that are designed specifically for these levels. Figure 2.5 is a photograph of a milliammeter and microammeter. Consider the difficulty of reading a micro-ampere or milliampere indication on a scale having a maximum level of 1 A. It would, in fact, be impossible. It is therefore quite important that the proper instruments be chosen for the measurements to be performed.

The magnitude of the current will, as one might expect, affect the size of wire

Figure 2.5 Dc milliammeter and microammeter. (Courtesy Simpson Electric Co.)

to be employed for a particular application. The larger the current, the larger the diameter of the wire that must be employed. Obviously, therefore, the current levels encountered in the thin telephone or electric bell wire must be very small, while those found in the heavier wire used with heavy machinery and power distribution will be much higher. Table 2.2 lists the maximum current rating of some commercially available wire. Both solid and stranded types are available for many of the diameters listed. The stranded is made of many strands, rather than being solid, to permit increased flexibility for a wide variety of applications. The No. 12 wire is usually the type used in homes for outlets and such. Note that it can handle a maximum of 25 A. Since a home can have a 200-A service, it is obvious that more than one path will have to be provided to carry the necessary current to all the outlets of the home.

In Table 2.2 you will note the presence of the symbol CM, which is an indication of the cross-sectional area of the wire. It is a shorthand notation for a defined unit of measure called the *circular mil*. By definition, the area of a circular wire having a diameter of 1 mil is 1 circular mil. One mil is simply one thousandth of an inch. That is,

$$\boxed{1 \text{ mil} = 0.001 \text{ in.}} \tag{2.3}$$

Through a short derivation it can be shown that the area in circular mils of a wire can be determined from

$$\boxed{A_{\text{CM}} = (d_{\text{mils}})^2} \qquad (d_{\text{mils}} = \text{diameter in mils}) \tag{2.4}$$

EXAMPLE 2.3

Determine the area in CM of a $\frac{1}{16}$-in.-diameter wire.

Solution:

Before converting inches to mils the diameter should first be put in fractional form by simply performing the indicated division. That is,

$$\frac{1}{16} \text{ in.} = 0.0625 \text{ in.}$$

Table 2.2

American Wire Gage (AWG) Sizes

AWG #	AREA (CM)	$\Omega/1000'$ AT 20°C	MAXIMUM ALLOWABLE CURRENT FOR RHW INSULATION (AMP)*	AWG #	AREA (CM)	$\Omega/1000'$ AT 20°C
0000	211,600	0.0490	360	19	1288.1	8.051
000	167,810	0.0618	310	20	1021.5	10.15
00	133,080	0.0780	265	21	810.10	12.80
0	105,530	0.0983	230	22	642.40	16.14
1	83,694	0.1240	195	23	509.45	20.36
2	66,373	0.1563	170	24	404.01	35.67
3	52,634	0.1970	145	25	320.40	32.37
4	41,742	0.2485	125	26	254.10	40.81
5	33,102	0.3133	—	27	201.50	51.47
6	26,250	0.3951	95	28	159.79	64.90
7	20,816	0.4982	—	29	126.72	81.83
8	16,509	0.6282	65	30	100.50	103.2
9	13,094	0.7921	—	31	79.70	130.1
10	10,381	0.9989	40	32	63.21	164.1
11	8,234.0	1.260	—	33	50.13	206.9
12	6,529.9	1.588	25	34	39.75	260.9
13	5,178.4	2.003	—	35	31.52	329.0
14	4,106.8	2.525	20	36	25.00	414.8
15	3,256.7	3.184		37	19.83	523.1
16	2,582.9	4.016		38	15.72	659.6
17	2,048.2	5.064		39	12.47	831.8
18	1,624.3	6.385		40	9.89	1049.0

From the 1965 National Electric Code published by the National Fire Protection Association.

Then simply move the decimal point three places to the right for the conversion to mils and

$$0.0625 \text{ in.} = 62.5 \text{ mils}$$

The area can then be determined by

$$A_{CM} = (d_{mils})^2 = (62.5)^2 = 3906.25 \text{ CM}$$

which corresponds very closely to a **No. 14 wire.**

EXAMPLE 2.4

Determine the diameter in inches of the No. 12 wire typically used in house wiring.

Solution:

From Table 2.2

$$\text{No. 12 wire} \longrightarrow 6529.9 \text{ CM}$$

and

$$d_{mils} = \sqrt{6529.9} = 80.81 \text{ mils}$$

The diameter in inches is obtained by moving the decimal point three places to the left and

$$80.81 \text{ mils} = \textbf{0.08081 in.}$$

which is approximately 1/12.4 in. and which places it somewhere in the middle between the recognizable dimensions of $\frac{1}{16}$ and $\frac{1}{8}$ in.

The $\Omega/1000$ ft rating appearing in Table 2.2 will be examined in a section to follow in this chapter. The wire gage numbers were chosen so that a drop in three gage numbers corresponds with doubling the area of the wire. For a drop in 10 gage numbers there is a corresponding 10-fold increase in area.

Since each conductor has a maximum current rating, a circuit element had to be developed to limit the current to a safe level. For many years, the fuse was the common mode of protection in the home or industry. Today it has been replaced in all new establishments by a circuit breaker such as those appearing in Fig. 2.6. The fuse contained a metal link of softer metal that would melt open when the current reached too high a level. The fuse itself then had to be replaced. The circuit breaker is a switch designed to open when the current reaches a particular level. Unlike the fuse, however, that has to be replaced the circuit breaker can simply be reset. Although a No. 12 wire can handle a maximum of 25 A, a safety factor is normally introduced in the household circuit-breaker panel by using circuit breakers that will *trip* at 15 or 20 A. Most circuit breakers have this maximum rating clearly visible on the face of the breaker.

Within the home a number of outlets are connected to each circuit breaker. The more fixtures, appliances, etc., attached to these outlets, the higher the current through the circuit breaker. Therefore, the circuit breaker will "pop" the moment a load is applied that brings the current level above the rated value of the breaker.

Figure 2.6 Circuit breaker. (Courtesy Potter and Brumfield.)

Before resetting the breaker it will obviously be necessary to remove some of the load. Even though safety factors are built into every table such as Table 2.2, it would be extremely dangerous to simply replace a 20-A breaker by a 30-A breaker so that the circuit can handle a heavier load. Dangerous side effects such as fire and smoke could result.

2.3 VOLTAGE

The fundamental concept of voltage is one that usually requires a degree of increased concentration and effort to develop a clear understanding of, as compared to the concept of the flow-variable; current. Continuous exposure and a measure of effort, however, will invariably result in a clear working knowledge of this fundamental quantity. For many, the term voltage has only been encountered when referring to a 12-V car battery or reading the label on an appliance where it indicates that it will only work properly on a 120-V outlet. Unlike current which has a flow characteristic, voltage is an *across* variable in the sense that two points in an electrical system have a difference in voltage levels. For the car battery there is a difference in voltage between

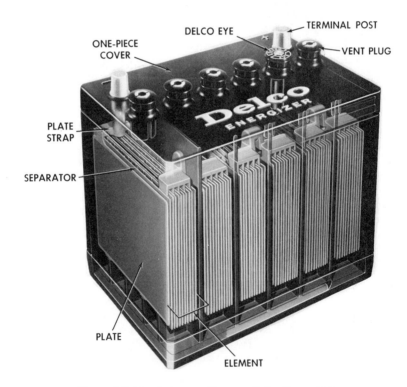

Figure 2.7 12-volt battery (Courtesy Delco-Remy, Division of General Motors.)

the two terminals of 12 V, but the voltage does not flow from one terminal to the other. In fact, voltage is often likened to the pressure that causes the flow of the liquid through the system just as voltage "causes" charge to flow in an electrical system.

The typical car battery has two terminals that are said to have a *potential difference* of 12 V between them or a terminal voltage of 12 V. The term potential is derived from the concept of potential energy, which refers to the energy that a body possesses by virtue of its position. In the battery shown in cross-section in Fig. 2.7 the difference in potential is established between the terminals at the expense of the chemical energy reactions within the battery. The result is an excess of positive charges (ions) on the positive ($\oplus$) terminal and negative charges (electrons) on the negative ($\ominus$) terminal. This *positioning* of the charges at the expense of chemical energy will result in a flow of charge (current) through a conductor placed between the terminals, as shown in Fig. 2.8. The electrons in the copper conductor are relatively free to leave the parent copper atoms and move toward the excessive number of positive charges at the positive terminal (recall from your fundamental physics courses that like charges repel and unlike charges attract). In addition, the negative terminal pressures, through repulsion, the electrons toward the positive terminal. The net result is a flow of charge (current) through the conductor. The chemical action of the battery is designed to absorb the flow of electrons and maintain the distribution of charge at the terminals of the battery. Through all the above, it should be pointed out that the positive ions remaining when an electron leaves the parent atom are able to oscillate only in a fixed mean position and cannot move toward the negative terminal. Of course it is understood that there will be a slight drift in that direction but nothing to compare with the electron flow. In the same figure the directions of electron and conventional flow have been included for further clarification.

The potential difference between, or the voltage across, any two points (*a* and *b*) in an electrical system is determined by

$$V_{ab} = \frac{W_{ab}}{Q_{ab}} \qquad (2.5)$$

where V_{ab} = potential difference in volts
$\quad\;\; W_{ab}$ = energy expended or absorbed in joules (J)
$\quad\;\; Q_{ab}$ = charge measured in coulombs

V_{ab} is the potential difference measured in volts between two points *a* and *b*, where the second subscript *b* is usually recognized as the point of lower potential.

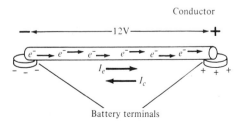

Figure 2.8 Flow of charge established by a 12-V battery.

W_{ab} is the energy expended or absorbed in moving the charge Q_{ab} from point a to b. As an example, if 12 J of energy were required to carry a charge of 1 C from one terminal of the battery to the other, there would be a potential difference of 12 V between the terminals.

EXAMPLE 2.5

Determine the energy required to bring a charge of 20C through a potential difference of 40 V.

Solution:

Per Eq. (2.5),

$$W_{ab} = Q_{ab}V_{ab} = (20 \times 10^{-6})(40) = 800 \times 10^{-6}$$
$$= \mathbf{800\ \mu J}$$

Note the continued use of scientific notation.

The basic difference between current (a flow variable) and voltage (an across variable) has its effect on the measurement of each also. The basic voltmeter as shown in Fig. 2.9 is very similar to the ammeter in its basic appearance, but the measurement techniques are quite different. As shown in Fig. 2.10, the voltmeter does not *break* the circuit but is placed *across* the element for which the potential difference is to be

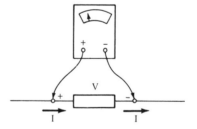

Figure 2.9 Dc voltmeter (Courtesy Simpson Electric Co.)

Figure 2.10 Proper voltmeter connections.

determined. Like the ammeter it is designed to affect the network as little as possible when inserted for measurement purposes. A great deal more will be said about loading effects of meters, their tolerance, and proper use in the last section of this chapter when all the dc instruments have been introduced. The choice of scales is determined in much the same manner as discussed for the ammeter. The dial chooses the maximum voltage to be measured for that setting and determines whether the 100 on the face of the meter will stand for 1, 1000 or 10,000 V. For each reading on the 1-V dial setting the scale indication will have to be divided by 100. That is, 90 is an indication of 0.9 V. The best reading is usually obtained by that dial setting that results in a mid-scale deflection. For the general protection of the meter with unknown voltage levels it is best to start with the highest scale to obtain some idea of the voltage to be measured. Then work down to the best reading possible.

As indicated earlier in the introductory section, our initial discussions will be limited to dc quantities. Voltage levels, therefore, will remain fixed at some predetermined value. Time-varying voltage such as encountered in a home outlet (ac) will be examined in the next chapter. Other sources of dc voltage include dc generators, solar cells, and laboratory supplies, which will all be covered in detail in later chapters. Sources of voltage will employ the capital letter E as their symbol to distinguish it from voltage drops across load elements, which will employ the capital letter V. Both, of course, will still be measured in volts. Voltage sources are often said to have an electromotive force (emf) of so many volts. In abbreviated form, a car battery has an emf of 12 V. Incidentally, a voltmeter cannot be used to check the condition of a 12-V battery. Even though it may have lost a great deal of its charge, the potential difference between the two terminals may still read very close to 12 V. The hydrometer must be used to determine the level of chemical action within the battery. Too low a reading indicates a need for recharging.

As with current levels, voltages can also vary from the microvolt to megavolt range. Therefore, the scientific notation introduced earlier is frequently applied to voltage levels also. In radio and TV receivers, very low voltage (μV and mV) levels are encountered, while at generating stations kilovolt and megavolt readings are encountered. Typically the bare power lines in a residential area carry 22,000 V (ac), while the covered lines into the home from the transformer in the pole carry 220 V (ac). Transformers will be covered in detail in a later chapter.

A photograph of a typical laboratory dc supply appears in Fig. 2.11. The supply voltage can be taken between + and − , or + and ground, or − and ground. On most supplies the output between + and − is said to be *floating* since it is not connected to a common ground or potential level of the network. The term *ground* simply refers to a zero or earth potential level. The chassis or outer shell of most electrical equipment, whether it be a supply or instrument, is grounded through the power cable to the terminal board. The third prong (usually round) on any electrical

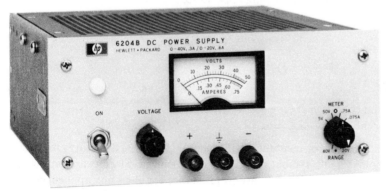

Figure 2.11 Laboratory dc supply. (Courtesy Hewlett Packard.)

equipment or appliance is the ground connection. Due to the above, all connections in the network that are connected directly to the chassis are at ground potential. This is all done primarily for safety reasons. Rather than have a high voltage possibly find ground through the technician, it will have the alternative paths indicated above available. If the + and ground terminals are used and the dial set at 10 V, then the + terminal is 10 V positive with respect to (w.r.t.) the ground (0-V) terminal. However, if the − and ground terminals are used, the − terminal will be 10 V negative w.r.t. to the ground terminal. Perhaps Fig. 2.12 will clarify the above to some degree. In actuality, it simply indicates that voltage levels can be lower than ground potential. The additional connection to the unused terminal is normally required with three-terminal supplies.

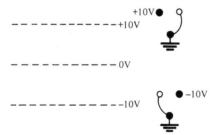

Figure 2.12 Available voltage levels on a dc laboratory supply.

2.4 RESISTANCE AND OHM'S LAW

The two very fundamental quantities of voltage and current can be related through a third quantity of equal importance—resistance. In any electrical system the *pressure* to cause things to happen is the applied difference in potential or emf. The desired result is the flow of charge or current. The degree to which an electrical system is able to accomplish the above is determined by the resistance of the system. The greater the resistance to the flow of charge, the less the resulting current in the system. This effect is immediately obvious when we examine the most fundamental law of electric circuits, *Ohm's law*, which has the following appearance:

$$I = \frac{E}{R}$$

(2.6)

where I = amperes

E = volts

R = resistance in ohms (Ω)

Graphic symbol: ●─────∿∿∿─────●

Note that the symbol for resistance is the capital letter R and that it is measured in ohms and employs as its symbol the Greek letter omega. Since it appears in the denominator of the equation, an increase in resistance for the same applied voltage will certainly cause a reduction in current, as indicated earlier. The reverse is true for a decrease in resistance.

EXAMPLE 2.6

Determine the current drawn by a toaster having an internal resistance of 15 Ω if the applied voltage is 120 V.

Solution:

Eq. (2.6),

$$I = \frac{E}{R} = \frac{120}{15} = \textbf{8A}$$

EXAMPLE 2.7

Determine the internal resistance of an alarm clock that draws 20 mA at 120 V.

Solution:

$$R = \frac{V}{I} = \frac{120}{20 \times 10^{-3}} = 6 \times 10^3 = \textbf{6K}$$

The resistance of a conductor is determined by four quantities: material, length, area, and temperature. The first three are related by the following equation at $T = 20°C$ (room temperature):

$$\boxed{R = \frac{\rho l}{A}} \tag{2.7}$$

where R = resistance in ohms
   ρ = resistivity of the material in CM-ohms/ft
   l = length of the sample in feet
   A = area in CM

Note that the area is measured in CM, which was discussed in Section 2.2. The resistivity ρ (Greek letter rho) is a constant determined by the material used. A few are listed in Table 2.3.

Table 2.3

Resistivity

MATERIAL	ρ
Silver	9.9
Copper	10.37
Gold	14.7
Aluminum	17.0
Tungsten	33.0
Nickel	47.0
Iron	74.0
Nichrome	600.0
Carbon	21,000.0

In words, the equation indicates that an *increase* in *area* or a *decrease* in *length* will reduce the resistance. The column Ω/1000 ft of Table 2.2 will now have some meaning. It is simply the resistance in ohms of a 1000-ft section of the wire. Referring to the table you will note the very low values encountered, although for very long lengths of any one of the wires the resistance can become a consideration. Its magnitude can be totally ignored for applications such as house wiring and general appliances.

EXAMPLE 2.8

Determine the resistance of a $\frac{1}{4}$-mile length of copper wire having a $\frac{1}{8}$-in. diameter.

Solution:

$$l = \tfrac{1}{4} \text{ mile} = \tfrac{1}{4}(5280 \text{ ft}) = 1320 \text{ ft}$$

$$\text{Diameter} = \tfrac{1}{8} \text{ in.} = 0.125 \text{ in.} = 125 \text{ mils}$$

$$A_{\text{CM}} = (d_{\text{mils}})^2 = (125)^2 = 15{,}625$$

$$R = \rho\frac{l}{A} = \frac{(10.37)(1320)}{15{,}625} = \frac{13{,}688.40}{15{,}625.0} = \mathbf{0.876\Omega}$$

EXAMPLE 2.9

Determine the resistance of 2 miles of AWG 00 wire.

Solution:

From Table 2.2,

$$\frac{\Omega}{1000 \text{ ft}} = 0.0780$$

$$2 \text{ miles} = 2(5280) = 10.560 \text{ ft}$$

and

$$\frac{10{,}560}{1000} \times 0.0780 = \mathbf{0.82368\Omega}$$

For most conductors, as the temperature increases, the increased activity of the atoms of the wire will make it more difficult for the charge carriers to pass through and the resistance will increase. The resistance vs. temperature curve for a copper wire is shown in Fig. 2.13. Note that zero resistance is not achieved until absolute zero ($-273°$C) is reached. However, a straight-line approximation to the curve will intersect at $-234.5°$C, the *inferred absolute temperature* of copper. The inferred absolute temperature for various conductors is provided in Table 2.4. Through similar triangles an equation can be developed from the curve of Fig. 2.13 that will permit determining the resistance of a conductor at one temperature if its value is known at some other temperature on the linear (straight-line) approximation. That is,

$$\boxed{\frac{T + t_1}{R_1} = \frac{T + t_2}{R_2}} \tag{2.8}$$

where $T =$ inferred absolute temperature of material in °C (without negative sign)
  $R_1 =$ resistance at temperature t_1
  $R_2 =$ resistance at temperature t_2

If the resistance R_1 is known at a temperature t_1, the resistance R_2 can be determined for a temperature t_2. The minus sign of the inferred absolute temperature is

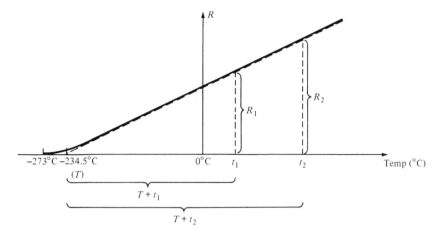

Figure 2.13 The change in resistance of copper as a function of temperature.

Table 2.4

Inferred Absolute Temperatures (T)

MATERIAL	TEMPERATURE, °C
Silver	−243
Copper	−234.5
Gold	−274
Aluminum	−236
Tungsten	−204
Nickel	−147
Iron	−162
Nichrome	−2250

not included when substituting the proper value for the material of interest. Occasionally the following equation is employed rather than Eq. (2.9) to find the resistance at another temperature:

$$R_2 = R_1[1 + \alpha_1(t_2 - t_1)] \tag{2.9}$$

It utilizes a constant α_1, called the *temperature coefficient of resistance*, which has the symbol alpha (α_1) and is an indication of the rate of change of resistance for that material with change in temperature. In other words, the higher its value, the greater the change in resistance per unit change in resistance. A few values of this coefficient are provided in Table 2.5 for different materials. Note that α_1 includes the effect of the inferred absolute temperature T as determined by the material of interest.

EXAMPLE 2.10

The resistance of a copper conductor is 0.3 Ω at room temperature (20°C). Determine the resistance of the conductor at the boiling point of water (100°C).

Solution:

Per Eq. (2.8),

$$\frac{T + t_1}{R_1} = \frac{T + t_2}{R_2} \longrightarrow \frac{234.5° + 20°}{0.3} = \frac{234.5 + 100}{R_2}$$

$$R_2 = \frac{(334.5)(0.3)}{254.5}$$

$$= 0.394\Omega$$

Per Eq. (2.9),

$$R_2 = R_1[1 + \alpha_1(t_2 - t_1)]$$

$$= (0.3)[1 + 0.00393(100 - 20)]$$

$$= (0.3)(1 + 0.3144)$$

$$= 0.394\ \Omega$$

Table 2.5

Temperature Coefficient of Resistance (α_1)

MATERIAL	α_1
Silver	0.0038
Copper	0.00393
Gold	0.0034
Aluminum	0.00391
Tungsten	0.005
Nickel	0.006
Iron	0.0055
Nichrome	0.00044
Carbon	−0.0005

There are numerous applications where resistive values are added to a network to perform a very specific, necessary function. For this purpose, a variety of resistors designed to suit the application has been designed with *tolerance* (on special order) as low as 0.01%. The lower the tolerance, the more care that must go into the manufacture of the resistor. A tolerance of 0.1% (0.001) on a 50-Ω resistor would indicate that the actual value of that resistor will not be off by more than (0.001)(50) = 0.05 Ω or the manufacturer guarantees that it will fall within the range 49.95–50.05 Ω. Tolerances on the order of 5, 10, or 20% are more common. A few of the various types of fixed-type resistors are shown in Fig. 2.14. For the carbon resistors, note the increase in size with wattage rating. We shall find in the next section that the wattage rating of a resistor is an indication of the power it can dissipate in the form of heat before its characteristics would be destroyed—hence the increase in size with the rating to handle the increased heat dissipation. Metal heat conductors called *heat sinks* are often employed in conjunction with electrical and electronic elements to assist in drawing the heat away from the element and prevent unnecessary damage. A few will be shown in a later chapter dealing with electronic devices.

The spacing between the wire wrappings and the large surface area used inside

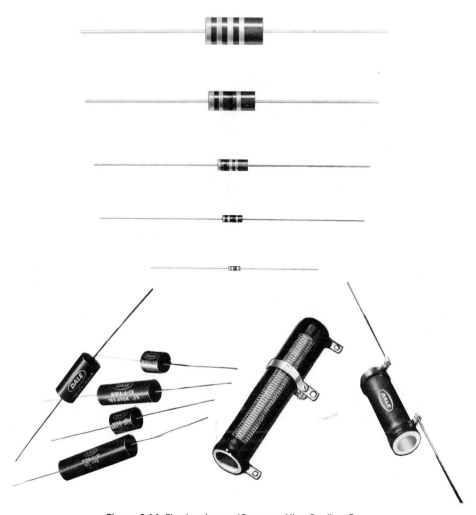

Figure 2.14 Fixed resistors. (Courtesy Allen Bradley Co.,
Ohmite Manufacturing Co.)

and outside of the wire-wound resistors is another technique of increasing the heat
dissipation of the element.

A three-terminal device called the *potentiometer* can be used as a voltage or
potential controlling device (from whence came its name) or as a variable resistor or
rheostat if only two of its terminals are employed. Since the device has three terminals,
as shown in Figure 2.15, it is important that the terminals be used for the variation
in resistance desired. The symbol for the device appearing in Fig. 2.15 indicates quite
clearly that between the two outside terminals the resistance is fixed at the total
value R even if the shaft is turned. Between the center terminal (or wiper arm) and
either outside terminal the resistance will vary between a minimum value of $0 \, \Omega$ when
the contacts touch and a maximum of R when the wiper arm reaches the other

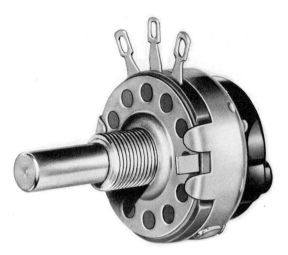

Symbol:

Figure 2.15 Molded Composition Potentiometer. (Courtesy Allen Bradley)

outside terminal that is not part of the overall network. How the two terminals chosen are hooked up will determine whether a right-hand rotation of the shaft will increase or decrease the resistance. When all three terminals are employed in an electrical system its purpose is to control potential levels in the network as determined by the resistance between the respective terminals of the potentiometer. In Fig. 2.16 the voltage V_1 will increase with a right-hand rotation of the shaft, while V_2 will decrease.

Another type of variable resistor is shown in Fig. 2.17. This wire-would variety can also be used, like the three-terminal carbon element of Fig. 2.15, as a potentiometer. The resistance obtained is determined by which two terminals are employed.

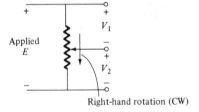

Right-hand rotation (CW)

Figure 2.16 Potentiometer as used to control potential (voltage) levels.

As the knob on one end is tuned the metal contacts on the movable part will slide down the wire-wound resistor. The resistance between the two fixed terminals will again always be fixed, while the resistance between the movable contact and a fixed terminal will very linearly with rotation of the knob. The wire employed in the design of wire-wound resistors has a resistance level many times that of the typical conductor so that sufficient resistance is developed. This type is used for controlling currents and potential levels beyond that of the carbon potentiometer in Fig. 2.15.

Resistors of very small size (but not necessarily small resistance value) can be found in integrated circuits, which have become increasingly common in recent years. The resistor in this case is not a discrete element as above but is manufactured within

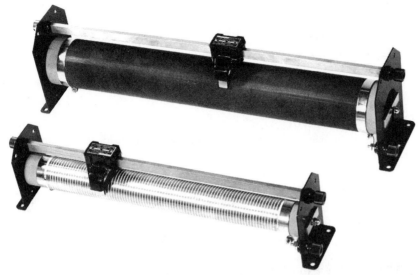

Figure 2.17 Wire-wound variable resistors. (Courtesy James G. Biddle Co.)

a single chip with the other electronic elements of the package that is designed to perform a particular function. A great deal more will be said about ICs in a later chapter.

For some small resistors, it is impossible or impractical to print the numerical value on the casing. Rather, a system of color coding is employed where by certain colors are given the numerical value indicated in Table 2.6. For the most common

Table 2.6

Color Coding

0	Black	7	Violet	
1	Brown	8	Gray	
2	Red	9	White	
3	Orange	0.1	Gold	
4	Yellow	0.01	Silver	
5	Green	5%	Gold	} Tolerance
6	Blue	10%	Silver	

of the fixed resistors, the carbon composition, the color bands will appear as shown in Fig. 2.18. Fortunately, however, the same order is also applied to a number of other types of resistors using color bands. The first and second bands (closest to one end) determine the first and second digits, while the third determines the number of zeros to follow the first two digits or the power of 10 to appear as a multiplying factor. The fourth band is the tolerance, which, as the table indicates, will not appear if the tolerance is $\pm 20\%$.

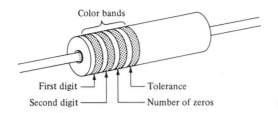

First digit —
Second digit —
Tolerance
Number of zeros

Color bands

Figure 2.18 Molded composition resistor.

EXAMPLE 2.11

Determine the manufacturers-guaranteed range of values for the carbon resistor in Fig. 2.19.

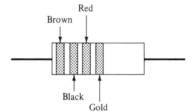

Brown
Red
Black
Gold

Figure 2.19

Solution:

$$\text{Brown} = 1, \quad \text{Black} = 0, \quad \text{Red} = 2, \quad \text{Gold} = \pm 5\%$$

$$\therefore \quad 10 \times 10^2 \pm 5\% = 1000 \pm 50 = \mathbf{950 \ \Omega} \longrightarrow \mathbf{1050 \ \Omega}$$

If the measured value of the resistor were 1040 Ω instead of the 1000-Ω label, it would still meet the manufacturer standards.

As with current and voltage, there is also an instrument called an *ohmmeter* that will measure resistance. The most frequently employed is the wide-range ohmmeter appearing on the typical VOM (volt-ohm-milliameter), such as shown in Fig. 2.20 with its scale. The VOM is simply a single instrument that will measure voltage, resistance, and current (in the milliampere range). Its versatility makes it quite popular with the laboratory technician. Note first, that the ohmmeter scale is non-linear. Therefore, the scale setting should be set to obtain a reading somewhere in the low- or midscale region for the best accuracy. For the $R \times 10$ setting, each reading on the scale is multiplied by a factor of 10. Similarly, the $R \times 100$ and $R \times 1000$ scales have 100 and 1000 unit multipliers. The ohmmeter is an excellent instrument for determining which terminals provide the desired resistance variation on a three-terminal potentiometer. Simply hook up the meter to the terminals of interest and turn the shaft, noting the effect on the scale indication. It must be pointed out, however, that the ohmmeter, unlike the ammeter and the voltmeter, requires an internal battery. If the instrument is left on the resistance scale and the leads should touch, the battery will be drained quite rapidly. Therefore, the VOM should be set on a high voltage scale when not being used to prevent this possibility. It is also very important that *an ohmmeter never be connected to a live network*. The reading will be erroneous since it is calibrated to the internal battery, and chances are the instrument itself will be damaged.

Figure 2.20 VOM (Volt-Ohm-Milliammeter) (Courtesy Simpson Electric Co.)

The ohmmeter can also be used for checking continuity in a network by search-ing for a 0-Ω (or least resistance) indication, which indicates the two leads of the instrument are directly connected by the leg of the network. In addition, it can be used to determine which lead is which when more than one wire appear in a terminal box. By connecting one lead of the ohmmeter to the wire of interest outside the box it can be found inside by touching each lead in the box and finding the 0-Ω indication. A last application of the ohmmeter is searching for a short (the high voltage lead is touching ground) in the house wiring if the circuit breaker keeps *popping* or the system was just installed and you would like to be sure the *hot* lead is not touching ground somewhere in the system. Be sure the power is off, and then simply take one ohmmeter lead and attach it to one of the wires coming out of the electrical outlet and connect the other to the metal box. Only one of the outlet wires should result in a 0-Ω reading when connected in this manner. If not, the circuit and that section of wiring should be rechecked very carefully. A *short* condition is simply one in which an *unwanted* low-resistance path has been established between a voltage level and ground. Through Ohm's law it is quite clear that a very low resistance ($\simeq$ 0Ω) will result in a very high current that can cause dangerous conditions such as fire and smoke.

There are instruments such as shown in Fig. 2.21 that are designed specifically to measure very large resistances such as those in the megohm range. One applica-tion of such an instrument is as in insulation tester.

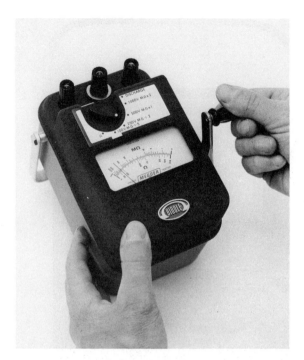

Figure 2.21 The Megger® Tester (Courtesy James G. Biddle Co.)

An *insulator* is any material with a very high resistance characteristic that can be used to cover wires and carry very high voltages in a power line without providing a conduction path to the support element. A few are shown in Fig. 2.22. A lower than usual reading between the yolk or frame of a motor and an internal field conductor would indicate that the insulation between the windings is perhaps breaking down and could offer a very dangerous situation.

The *semiconductor* is a material whose characteristics lie somewhere between those of a good conductor and an insulator. As we shall see when electronic devices are introduced, they are constructed using semiconductor materials. Our interest for the moment, however, lies with semiconductor devices whose terminal resistance will vary with temperature or incident light.

The *thermistor* is a two-terminal device such as shown in Fig. 2.23, with the characteristics and symbol also appearing in the figure. Note, from the characteristics, that as the temperature increases, the resistance decreases significantly with what we refer to as a *negative temperature coefficient*. The temperature can be changed by simply changing the current through the device (increasing currents result in increasing temperatures) or of the surrounding medium.

The photoconductive cell is a two-terminal device whose resistance will decrease almost linearly with increase in incident light on the designed surface. Its appearance, characteristics and symbol are provided in Chapter 6. Both the thermistor and photoconductive devices have numerous applications in control systems where temperature and incident light are determining factors.

Figure 2.22 Insulators. (Courtesy E. F. Johnson Co.)

2.5 POWER, EFFICIENCY, ENERGY

For any electrical, mechanical, or input-output system *power* is a measure of the *rate* of energy conversion of that system. For a simple dc electrical system it is a measure of the rate in which electrical energy is converted into heat by the power-dissipating resistive elements. For a motor its output horsepower rating is a measure of its ability to do mechanical work. For the battery of Fig. 2.24(a) the power delivered is determined by

$$\boxed{P = EI} \qquad \text{(watts)} \qquad\qquad (2.10)$$

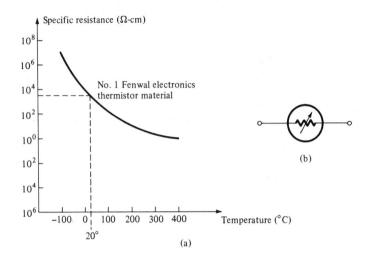

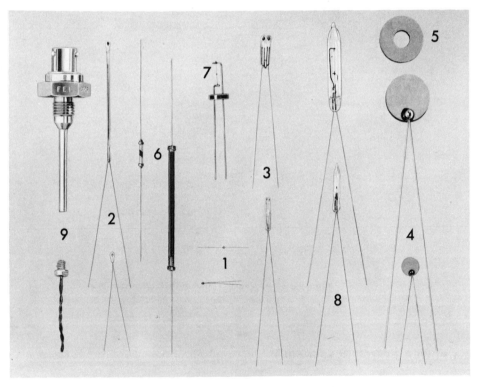

(Courtesy Fenwal Electronics, Inc.)

(c)

Figure 2.23 Thermisters (a) characteristics: (b) symbol:
(c) actual devices; (1) beads, (2) glass probes, (3) iso-curve
interchangeable probes and beads; (4) disks; (5) washers;
(6) rods; (7) specially-mounted beads; (8) vacuum and gas
filled probes; (9) special probe assemblies.

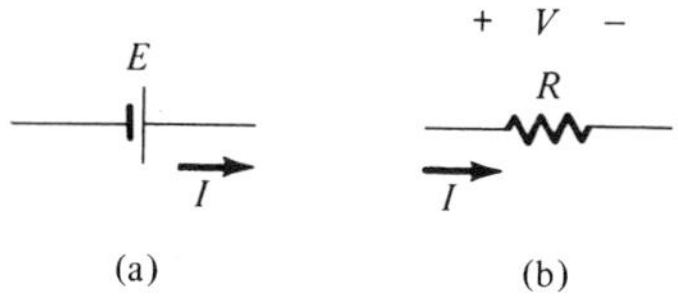

Figure 2.24 Power: (a) supplied by a battery, (b) absorbed by a resistor.

The unit of measurement for power is the *watt* (W), which refers to a rate of energy conversion of 1 J/sec.

For the resistive element of Fig. 2.24(b) the power dissipated by the resistive element is given by

$$P = VI = I^2R = \frac{V^2}{R} \qquad \text{(watts)}$$

where one is derived from the other with a simple application of Ohm's law.

Every electrical appliance in the home from the electric shaver to the motor in the refrigerator have a wattage rating. A few such appliances are listed in Table 2.7 with their average wattage rating.

Table 2.7

Average Wattage Ratings of Some Household Appliances

APPLIANCE	WATTAGE RATING	APPLIANCE	WATTAGE RATING
Air conditioner	2000	Heating equipment	
Clock	2	Furnace fan	320
Clothes dryer		Oil-burner motor	230
Electric, conventional	5000	Hi-fi equipment	200
Electric, high-speed	8000	Iron, dry or steam	1000
Gas	400	Projector	1000
Clothes washer	400	Radio	30
Coffee maker	Up to 1000	Range	
Dishwasher	1500	Free-standing	Up to 16,000
Fan		Wall ovens	Up to 8000
Portable	160	Refrigerator	300
Window	200	Shaver	10
Heater	1650	Tape recorder	60
		Toaster	1200
		TV receiver	
		Black-and-white	200
		Color	400

Courtesy of Con Edison.

EXAMPLE 2.12

Determine the current drawn by a 250-W television when connected to a 120-V outlet.

Solution:

$$P = VI \Rightarrow I = \frac{P}{V} = \frac{250}{120} = \textbf{2.083 A}$$

EXAMPLE 2.13

Determine the resistance of a 1200-W toaster that draws 10 A.

Solution:

$$P = I^2 R \Rightarrow R = \frac{P}{I^2} = \frac{1200}{(10)^2} = \frac{1200}{100} = \textbf{12 }\Omega$$

Power is measured by a device which is called (for obvious reasons) a *wattmeter*, such as shown in Fig. 2.25. It has a minimum of two terminals for the voltage-sensing portion and two terminals for the current reading. For most wattmeters, the current terminals are heavier in appearance than those employed for the voltage sensing. In any case, the potential terminals are connected *across* the load to which the power is to be determined, while the current terminals are connected in *series* (as the ammeter) with the load. The wattmeter of Fig. 2.26 is connected to measure the total power delivered to the resistors R_1, R_2, and R_3. Note the polarities of the terminals of the meter with respect to the direction of assumed current flow and the applied potential.

Figure 2.25 Wattmeter (Courtesy Weston Instruments.)

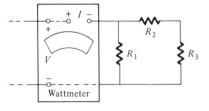

Figure 2.26 Measuring the power to resistors R_1, R_2, and R_3.

There are a number of multiscale wattmeters with more than four terminals. The manual provided should be consulted for the proper hookup, although only four terminals are used for any one measurement, and the connections for the voltage and current sensing are basically the same as described above.

For the homeowner, the electrical bill received each month is an indication of the electrical energy used and *not* the power made available for his use. It is important that the difference between power and energy be clearly understood. Consider a 10-hp motor, for example; unless used over a period of *time* there is *no* energy conversion effected by the machine. Energy and power can therefore be related by the following, which introduces the time-of-use element:

$$\boxed{W = Pt}\qquad\text{(watt-sec)}\qquad\qquad(2.11)$$

where W = energy in joules
$\quad\ P$ = power in watts
$\quad\ t$ = time in seconds

In other words, the longer we use an energy-converting device of a particular power rating, the greater will be the total energy converted. As indicated above, the unit of measure is the watt-second. However, this quantity is usually too small to serve most practical energy consumption measurements so the watt-hour or kilowatt-hour is normally employed. The meter on the side of every residential and industrial building is called a kilowatt-hour meter and has the appearance shown in Fig. 2.27. The four dials combine to indicate the total kilowatt-hours of energy used over a monthly period.

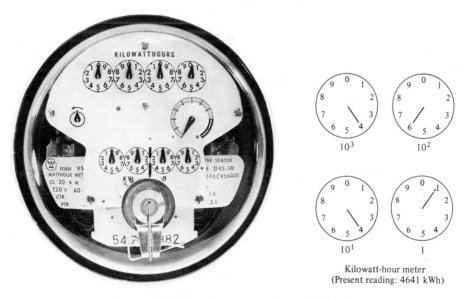

Kilowatt-hour meter
(Present reading: 4641 kWh)

Figure 2.27 Kilowatt-hour meter. (Courtesy Westinghouse Electric Corp.)

The number of kilowatt-hours is determined from the following equation:

$$\text{Kilowatt-hours} = \frac{P \cdot t}{1000} \tag{2.12}$$

where $P = $ watts
$\qquad t = $ hours

The following example should reveal how the monthly bill is determined from the kilowatt-hour reading. However, keep in mind that this example will only use an average cost per kilowatt-hour rather than using the sliding scale of cost per increase in demand. Most utilities are quite willing to forward a rate breakdown to any of its customers.

EXAMPLE 2.14

Determine the cost of using the following appliances for the time indicated if the average cost is 2.8 cents/kWh:

1. 1200-W iron for 2 hr.
2. 400-W color TV for 3 hr and 30 min.
3. Six 60-Watt bulbs for 7 hr.

Solution:

$$\text{Kilowatt-hours} = \frac{(1200)(2) + (400)(3.5) + (6)(60)(7)}{1000}$$

$$= \frac{2400 + 1400 + 2520}{1000} = 6.32 \text{ kW}$$

$$(6.32)(2.8) = \textbf{17.695 cents}$$

For any input-output electrical system, the overall efficiency of operation is a characteristic of primary importance. It is an indication of how much of the energy supplied is being used to perform the desired task. For a motor, for example, the more the output mechanical power for the same electrical power input, the higher the efficiency. In equation form, as a percent,

$$\eta\% = \frac{P_o}{P_i} \times 100\% \tag{2.13}$$

Conservation of energy requires that energy be conserved and that the output energy never be greater than the supplied or input energy. The maximum efficiency is therefore obviously 100% when $P_o = P_i$.

For any system, the following equation is always applicable:

$$P_i = P_o + P_l \tag{2.14}$$

where P_l is the power losses in the system.

For some applications the following conversion must be applied:

$$1 \text{ hp} = 746 \text{ W}$$

EXAMPLE 2.15

Determine the efficiency of operation and power lost in a 5-hp dc motor that draws 18 A at 230 V.

Solution:

$$\eta\% = \frac{P_o}{P_i} \times 100\% = \frac{5(746)}{18(230)} \times 100\% = \frac{3730}{4140} \times 100\% = \textbf{90}\%$$

$$P_l = P_i - P_o = 4140 - 3730 = \textbf{410 W}$$

2.6 SERIES AND PARALLEL dc NETWORKS

Two components in any electrical system are connected in series or parallel. For elements connected in series, there is *only one terminal common to the two components*. A simple series configuration appears in Fig. 2.28. The 12-V battery is in series with each resistor since it has only one terminal (*a* or *c*) in common with a resistor. The resistors R_1 and R_2 are also in series since they have only one terminal (*b*) in common.

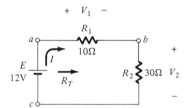

Figure 2.28 Series dc circuit.

Note in the figure that there are no additional elements connected to points *a*, *b*, or *c*. If a third element did appear, the two elements originally connected to that terminal would no longer be in series.

For resistors in series, the total resistance is the sum of the individual resistances. That is,

$$\boxed{R_T = R_1 + R_2 + R_3 + \cdots + R_N} \qquad (2.15)$$

For Fig. 2.28, $R_T = R_1 + R_2 = 10 + 30 = 40\ \Omega$.

The current through a series circuit is the *same* for each element, as indicated in Fig. 2.28. The current I can be determined using Ohm's law:

$$I = \frac{E}{R_T} = \frac{12}{40} = 0.3\ \text{A}$$

For voltage sources in series, those batteries pressuring current in one direction can be added together and the total determined by the algebraic sum of the resultant of each polarity. Consider the provided example of Fig. 2.29.

A law of very basic importance to the analysis of electrical systems is *Kirchhoff's voltage law*, which states, in words, that *the algebraic sum of the voltage rises and drops around a closed path or loop is zero*. The application of the law requires that one must

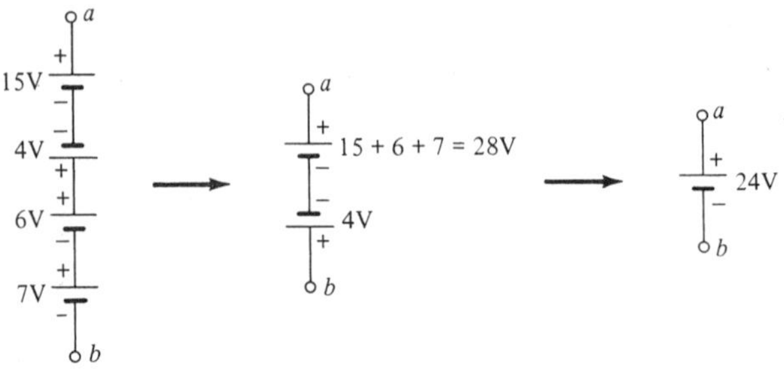

Figure 2.29 Algebraic summation of series voltage sources.

algebraically add the voltages around a closed loop in one direction. For elements in which the minus sign of the voltage polarity is encountered first a plus sign is applied since a — to + indicates a rise in potential. A minus sign is used for the reverse situation or a sequence of + to —. Consider the simple series network of Fig. 2.28 and move clockwise beginning at point *a*,

$$-V_1 - V_2 + E = 0$$

$$\overset{+-}{(ab)} \quad \overset{+-}{(bc)} \quad \overset{-+}{(ca)}$$

or, rearranging,

$$E = V_1 + V_2$$

which states in words that the applied voltage in a series circuit equals the potential drops of the network. Substituting for V_1 and V_2, where

$$V_1 = I_T R_1 = (0.3)(10) = 3 \text{ V}$$
$$V_2 = I_T R_2 = (0.3)(30) = 9 \text{ V}$$

results in

$$E = 12 = V_1 + V_2 = 3 + 9 = 12 \text{ V}$$

verifying the law.

The *voltage-divider rule* allows us to calculate the voltage across one or a combination of series resistors without having to first solve for the current. Its basic format is the following:

$$V_x = \frac{R_x E}{R_x + R} \tag{2.16}$$

where V_x is the voltage across a single resistor R_x or a combination of series resistors having a total resistance R_x. E is the applied voltage across the series circuit, and R is the remaining *total resistance* of the series network.

For Fig. 2.28

$$V_2 = \frac{R_2 E}{R_2 + R_1} = \frac{30(12)}{30 + 10} = \frac{3}{4}(12) = 9 \text{ V}$$

as obtained before. Further examples of some of these basic laws will appear in the sections to follow.

Each law or rule provided above for series circuits has its counterpart in parallel networks. A simple parallel network appears in Fig. 2.30. Two elements are in parallel if they have *two* terminals in common (*a* and *b*). In Fig. 2.30 the supply voltage E is in parallel with R_1 since they have points *a* and *b* in common. However, for the same reason it is also in parallel with R_2 and R_3. The resistors R_1 and R_2 are in parallel with each other and the resistor R_3. Since all the elements are in parallel with each other, we say we have a *parallel network*.

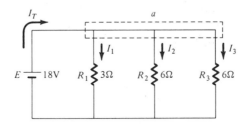

Figure 2.30 Parallel dc circuit.

The total resistance of parallel elements is determined by

$$\frac{1}{R_T} = \frac{1}{R_1} + \frac{1}{R_2} + \frac{1}{R_3} + \cdots + \frac{1}{R_N}$$
(2.17)

which for two parallel elements becomes

$$R_T = \frac{R_1 R_2}{R_1 + R_2}$$
(2.18)

For the network of Fig. 2.30,

$$\frac{1}{R_T} = \frac{1}{R_1} + \frac{1}{R_2} + \frac{1}{R_3} = \frac{1}{3} + \frac{1}{6} + \frac{1}{6}$$
$$= 0.333 + 0.166 + 0.166$$
$$= 0.666$$

and

$$R_T = \frac{1}{0.666} = 1.5\,\Omega$$

or, combining two elements at a time,

$$R_T' = 3\|6 = \frac{3 \cdot 6}{3 + 6} = \frac{18}{9} = 2\,\Omega$$

and

$$R_T = R_T'\|6\,\Omega = 2\|6 = \frac{2 \cdot 6}{2 + 6} = \frac{12}{8} = 1.5\,\Omega$$

The symbol $\|$ simply refers to "in parallel with."

For *equal* parallel resistors, the following equation can be applied:

$$R_T = \frac{R}{N}$$
(2.19)

where R is the value of one of the equal parallel resistors and N is the number of equal parallel resistors.

For the above example, with two parallel 6-Ω resistors,

$$R'_T = 6\|6 = \tfrac{6}{2} = 3\,\Omega$$

and

$$R_T = R'_T\|3 = 3\|3 = \tfrac{3}{2} = 1.5\,\Omega$$

For parallel elements, the voltage is the same across each component. In Fig. 2.30, the voltage across each element is 18 V. This should be somewhat obvious since the ends of each are connected directly to the supply.

Therefore, through Ohm's law,

$$I_1 = \tfrac{18}{3} = 6\text{ A}$$

$$I_2 = \tfrac{18}{6} = 3\text{ A}$$

$$I_3 = \tfrac{18}{6} = 3\text{ A}$$

Note from the above calculations that the *smaller* the parallel resistor, the *greater* the current.

Kirchhoff's current law for parallel elements states that *the current entering a junction must equal the current leaving that junction.* In Fig. 2.30, for example, the current entering terminal a is I_T and the current leaving is I_1, I_2, I_3, and therefore, through application of the law,

$$I_T = I_1 + I_2 + I_3$$

For Fig. 2.30,

$$I_T = \frac{E}{R_T} = \frac{18}{1.5} = 12\text{ A}$$

and

$$I_T = 12 = I_1 + I_2 + I_3 = 6 + 3 + 3 = 12\text{ A}$$

The counterpart of the voltage-divider rule for series circuits is the *current-divider rule* for parallel networks. In works, it states that the current through one of two parallel branches is the other parallel resistor multiplied by the total current entering the parallel branches divided by the sum of the parallel resistors. For the network of Fig. 2.31, for example, the currents I_1 and I_2 are determined by

$$I_1 = \frac{R_2 I}{R_2 + R_1} = \frac{4(6)}{4 + 2} = \frac{4(6)}{6} = 4\text{ A}$$

$$I_2 = \frac{R_1 I}{R_2 + R_1} = \frac{2(6)}{2 + 4} = \frac{2(6)}{6} = 2\text{ A}$$

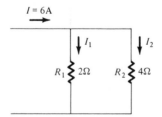

Figure 2.31 Current-divider rule.

Note again that the smaller the parallel resistor, the greater the branch current. Consider also the verification of Kirchhoff's current law:

$$I = I_1 + I_2$$
$$6 = 4 + 2 = 6$$

For the home and industrial applications all outlets, lighting, machinery, etc., are connected in parallel. In the home every outlet has a terminal voltage of 120 V. Although the voltage is what we refer to as an ac voltage and not dc as being considered here, it permits an examination of the effects of parallel elements. As additional appliances are connected to the same circuit, the current drawn through the circuit breaker will increase as determined by Kirchhoff's current law, although each appliance will still receive the necessary 120-V terminal voltage for proper operation. One obvious advantage of the parallel connection of loads is that if one should fail to operate, the others will still operate properly since the terminal voltage is still available. However, in a series connection of elements if one should fail, the remaining appliances will cease to operate also since the current path has been broken. Consider also that with a series configuration of perhaps lighting fixtures the more connected in series, the greater the terminal resistance, the less the current, and the dimmer the lights. The latter condition will not occur for parallel elements since each fixture will continue to receive 120 V and have the same brightness until the attached loads require a total current above that of the circuit breaker. Consider trying to determine which bulb of a series connection of bulbs (such as sometimes encountered with Christmas tree lights) is bad if they are all out. It presents a definite problem not encountered with the parallel connection since only the bad bulb will fail to light.

2.7 SERIES-PARALLEL NETWORKS

If a clear understanding of the definitions of series and parallel elements has been developed, a series-parallel network that has both series and parallel elements should not be overly difficult to analyze. As one might expect after examining the examples provided here there is an infinite variety of series-parallel networks. The analysis of any one of these networks, however, simply requires that the combination of elements be treated in small packages before considering the network as a whole.

For example, consider the network of Fig. 2.32. The battery E and the resistors R_1 and R_5 form a series path since there is only one terminal in common between any of the two elements. The resistors R_2 and R_3 are in parallel because they both have points c and d in common. Examine the network a little more carefully and you will find that R_3 and R_4 are also in parallel since they have both c and d, e (the same point) in common. The network can be redrawn as shown in Fig. 2.33 without losing the unknown quantities of R_T, I_T, I_1, and V_1. Note that the series resistors were combined $(5.6 + 3.2 = 8.8 \ \Omega)$ along with the parallel resistors R_3 and R_4 $(6 \parallel 3 = 2 \ \Omega)$. It is now more obvious that the parallel combination of the 3- and 2-Ω resistors is in series with the 8.8-Ω resistor. For R_T we must first combine the parallel elements,

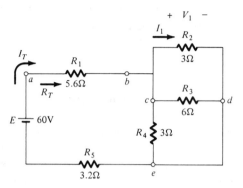

Figure 2.32

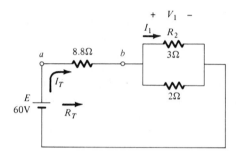

Figure 2.33

$$3 \| 2 = \frac{3 \cdot 2}{3 + 2} = \frac{6}{5} = 1.2 \, \Omega$$

and the series elements,

$$R_T = 8.8 + 1.2 = 10 \, \Omega$$

The source current

$$I_T = \frac{E}{R_T} = \frac{60}{10} = 6 \, \text{A}$$

The current I_1 through the 8.8-Ω resistor (or either of its former series components) is therefore 6 A. The current I_1 can be determined using the current-divider rule:

$$I_1 = \frac{2(I_T)}{2 + 3} = \frac{2(6)}{5} = \frac{12}{5} = 2.4 \, \text{A}$$

The power dissipated in the form of heat by the 3-Ω resistor (R_2) is determined by

$$P = I^2 R = (2.4)^2 \times 3 = 17.28 \, \text{W}$$

and the power delivered by the source voltage by

$$P_S = E \cdot I_T = (60)(6) = 360 \, \text{W}$$

The sum of the powers delivered to each resistive element will equal that provided at the source. That is,

$$P_S = P_{R_1} + P_{R_2} + P_{R_3} + P_{R_4} + P_{R_5}$$

In the network just analyzed, the final configuration after the small parcels were considered is a series circuit.

2.8 CURRENT SOURCES

At one time or another we have all been exposed to a dc voltage source even if only in the form of a car or flashlight battery. There is, however, a second type of source of energy called the *current* source that the average layman, unless he has had some experience in the field, will not have been exposed to. The dc voltage source supplies a fixed voltage to a network, and its current is determined by a resistive load to which it is applied. The *current source supplies a fixed current to a network, and its terminal voltage is determined by the network to which it is applied.*

Although not considered thus far, every source of voltage or current has some internal resistance. The voltage source has a series internal resistor such as shown in Fig. 2.34(a) and the current source a parallel resistance such as shown in Fig. 2.34(b). Ideally, the resistor R_S should be zero (like a short circuit) and the resistor R_P infinite ohms (like an open circuit). The representative network of Fig. 2.35 will demonstrate

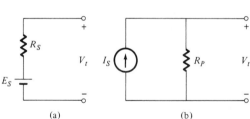

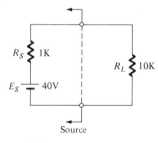

(a)   (b)   Source

Figure 2.34 Sources: (a) voltage, (b) current.   **Figure 2.35**

the validity of the above. Due to the internal resistance, the voltage across the load is not 40 V but, as determined by the voltage-divider rule,

$$V_{10\,\text{K}} = \frac{10\,\text{K}(40)}{10\,\text{K} + 1\,\text{K}} = 36.36\ \text{V}$$

For $R_S = 50\ \Omega$, the terminal voltage more closely approaches the ideal value of 40 V. That is,

$$V_{10\,\text{K}} = \frac{10\,\text{K}(40)}{10\,\text{K} + 0.05\,\text{K}} = 39.8\ \text{V}$$

For $R_S = 0\ \Omega$ (the ideal situation)

$$V_{10\,\text{K}} = 40\ \text{V}$$

Similarly, for the current source of Fig. 2.36 the larger the resistor R_P, the more ideal the source. That is, for $R_P = 1\ \text{K}$ and using the current-divider rule,

$$I_L = \frac{1\,\text{K}(5)}{1\,\text{K} + 1\,\text{K}} = \frac{5}{2} = 2.5\ \text{A}$$

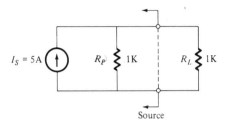

Source

Figure 2.36

and for $R_P = 100$ K,

$$I_L = \frac{(100\text{ K})(5)}{100\text{ K} + 1\text{ K}} = 4.95\text{ A}$$

and for $R_P = \infty \; \Omega$ (the ideal case—open circuit),

$$I_L = 5\text{ A}$$

Understandably, since the voltage source is the more common type of supply, it will take a measure of exposure before you will feel as comfortable with the current source. In a number of sections to follow, when we consider electronic devices and circuits, the current source will appear again, and quite frequently. This additional exposure will serve to clarify the basic characteristics of this source of energy. A commercially available current source is shown in Fig. 2.37. For a specified load variation, the current through the load can be set with its terminal voltage being determined by the load applied.

Figure 2.37 Dc current source. (Courtesy Hewlett Packard)

In circuit analysis, it is often advantageous to convert a voltage source to a current source or vice versa. The conversion equations appear in Fig. 2.38 with their respective sources.

Note that the resistance is the same for each and that the current or voltage is determined by the other using Ohm's law. For the sources of Fig. 2.38, if a conversion were made, a load connected to the terminals would not be able to tell which source was present. An assignment at the end of the chapter will demonstrate this fact.

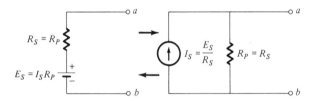

Figure 2.38 Source conversion.

For laboratory supplies, whether they be current or voltage, a current regulation or voltage regulation factor is provided which is determined by the internal resistance of the supply. For the ideal voltage supply the voltage set by the controls should be that value no matter what load current is supplied by the source. This is reflected by the *ideal* characteristic appearing in Fig. 2.39. However, let us define, as shown in Fig. 2.39, the voltage available at the supply terminals with no load attached as E_{NL}. Note that since a path for current flow is nonexistent, there is no voltage drop across R_S and $E_{NL} = E_S$. When a load is applied, however, and adjusted so that the load current increases to its rated value the internal voltage drop across R_S will increase, and the terminal voltage of the supply (or across the load) will decrease. At rated conditions (full-load conditions) the voltage is denoted as E_{FL}. The graph of Fig. 2.39 clearly indicates that the closer the value of E_{NL} is to E_{FL}, the closer the characteristics of the supply to the ideal characteristics. The less the internal resistance of the supply, the closer to ideal the characteristics and the better the supply. The ideal can only be obtained with $R_S = 0\ \Omega$. A measure of the slope of this line is given on data sheets as the *voltage regulation* as defined by the following equation:

$$\text{Voltage regulation} \% = \text{VR} \% = \frac{E_{NL} - E_{FL}}{E_{FL}} \times 100\% \qquad (2.20)$$

As indicated above, the smaller the difference between E_{NL} and E_{FL}, the better the supply as evidenced by a very small numerator in Eq. (2.20) and a smaller voltage regulation. In other words, the smaller the voltage regulation, the more ideal the supply. Supplies with voltage regulations of 0.1% or less are quite common. The current regulation for current sources is determined by an equation having the same format as Eq. (2.20).

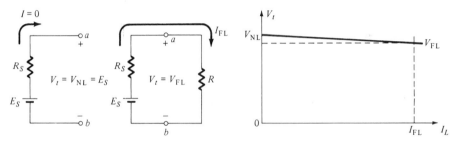

Figure 2.39 Voltage regulation.

2.9 MULTISOURCE NETWORKS

As will be shown in an example to follow, the techniques of circuit analysis introduced thus far cannot be applied to networks with two voltage (or current) sources that are not in series or parallel. Rather, an approach such as *superposition* or *branch-current analysis* must be used. Although the methods will only be demonstrated with two independent sources, each method can be applied to a network with any number of sources, whether they be current or voltage.

The simplest to describe and usually the easiest to apply employs the *superposition theorem.* It states that *the current through any element in a dc network can be found by simply finding the algebraic sum of the currents through that element due to each source independently.* The above can be restated, without change, for the voltage across an element in a dc network. To consider the effects of each source, the remaining sources must be properly removed. This requires that each voltage source be set to zero, reflecting a short-circuit condition, and that each current source be replaced by an open-circuit condition to indicate zero supply current. Any internal resistances associated with either R_S, or R_P, must remain when the effects of the source are removed, as indicated above.

EXAMPLE 2.16

Determine the current through the 6-Ω resistor for the network of Fig. 2.40 using the superposition theorem.

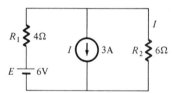

Figure 2.40

Solution:

The effects of the voltage source will first be considered and the current source removed, as indicated in Fig. 2.41. The current

$$I' = \frac{E}{R_1 + R_2} = \frac{6}{4 + 6} = 0.6 \text{ A}$$

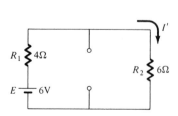

Figure 2.41

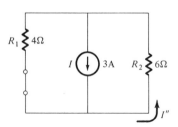

Figure 2.42

The effect of the current source can be examined by removing the voltage source, as shown in Fig. 2.42. Applying the current-divider rule,

$$I'' = \frac{R_1(I)}{R_1 + R_2} = \frac{4(3)}{4 + 6} = \frac{12}{10} = 1.2 \text{ A}$$

Note that I'' is the opposite in direction to that of I' so that $I_{6\,\Omega} = I'' - I' = 1.2 - 0.6 = 0.6$ A in the direction of the larger, I''.

The superposition theorem cannot be applied to power effects since the power and the current are related by a nonlinear (I^2R) relationship. In other words, the net power cannot be determined by simply finding the sum of the powers delivered by each source. It must be determined from the final solution for the voltage or current.

The second method to be applied is called *branch-current analysis*. It requires that all current sources first be converted to voltage sources. Each branch of the network is then assigned a labeled branch current, and Kirchhoff's voltage law is applied around each independent closed loop of the network. Kirchhoff's current law is then applied to provide the remaining necessary information. The term *independent* simply requires that each equation contain information not already present in the other equations. This criteria are usually satisfied by simply applying the law to each *window* of the network. In the following example there are two windows, and therefore two applications of Kirchhoff's voltage law are required. The final result is the currents of each branch and hence the title, branch-current analysis.

EXAMPLE 2.17

Determine the current through each branch of the network of Fig. 2.43.

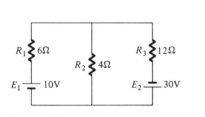

Figure 2.43

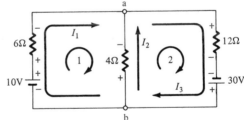

Figure 2.44

Solution:

In Fig. 2.44, the branch current has been assigned, and the polarities across the resistors have been included for the assumed direction of branch current. If the solution indicates that the assumed direction was incorrect, a minus sign will appear.

Applying Kirchhoff's voltage law in the clockwise direction around loop 1,

$$1: \quad 10 - 6I_1 + 4I_2 = 0$$

Note that the voltage source polarity is unaffected by the chosen direction of branch current. In addition, consider that the voltage across one resistor has a negative sign while the other has a positive sign, evidence of the fact that the sign is determined solely by the assumed branch-current direction and not the fact that it is a resistive element.

For loop 2 (clockwise),

$$2: \quad -4I_2 - 12I_3 + 30 = 0$$

Rearranging the two equations,

$$-6I_1 + 4I_2 = -10$$
$$-4I_2 - 12I_3 = -30$$

We now have two equations and three unknowns. A third equation is required that is obtained by an application of Kirchhoff's current law to either terminal a or b of the network of Fig. 2.44.

At node (term for a junction of two or more branches) a an application of Kirchhoff's current law will result in

$$\underbrace{I_1 + I_2}_{\text{entering}} = \underbrace{I_3}_{\text{leaving}}$$

Substituting for I_3 in the second equation (loop 2),

$$-4I_2 - 12(I_1 + I_2) = -30$$
$$-4I_2 - 12I_1 - 12I_2 = -30$$

and

$$-12I_1 - 16I_2 = -30$$

or multiplying through by -1, for loop 1 we have

$$12I_1 - 16I_2 = -30$$
$$-6I_1 + 4I_2 = -10$$

We now have two equations and two unknowns. The solution for I_1 or I_2 can now be obtained using determinants (a mathematical technique for solving simultaneous equations described in a number of available texts but not required to continue with this text) or simply solving for one of the variables and substituting into the other equation, as demonstrated below:

$$1: \quad I_1 = \frac{-30 + 16I_2}{12}$$

$$2: \quad -6\left[\frac{-30 + 16I_2}{12}\right] + 4I_2 = -10$$

$$\frac{180}{12} - \frac{16}{12}I_2 + 4I_2 = -10$$

$$15 - 8I_2 + 4I_2 = -10$$

$$-4I_2 = -10 - 15 = -25$$

$$I_2 = \frac{25}{4} = \textbf{6.25 A}$$

For I_1, we can now use the equation above and the result for I_2:

$$I_1 = \frac{-30 + 16I_2}{12} = \frac{-30 + 16(6.25)}{12}$$

$$= \frac{-30 + 100}{12} = \frac{70}{12}$$

$$= \textbf{5.85 A}$$

The branch current I_3 is obtained from

$$I_1 + I_2 = I_3$$

and

$$I_3 = 6.25 + 5.85 = \mathbf{12\ A}$$

Since each current resulted in a positive sign, the assigned direction for each was correct. The voltage across any resistor can then be determined using Ohm's law.
For the 12-Ω resistor,

$$V_{12\,\Omega} = I_3 R_3 = (12)(12) = \mathbf{144\ V}$$

An advanced technique of circuit analysis called *mesh analysis* has some similarities with the branch-current method described above. However, the application of the technique requires the understanding of certain concepts that must be left to the research of the reader. Once understood, however, it can be applied with less difficulty than branch-current analysis and has application to computer techniques. A technique called nodal analysis will result in the junction voltages from which the currents can be determined. This method will be left, like mesh analysis, to the student's own research. A number of excellent references for each are readily available.

2.10 NETWORK THEOREMS

There are a number of theorems that are extremely useful in the analysis and design of electrical systems. Although the analysis here is for dc networks, the theorems can also be applied to ac networks (Chapter 3) with very little change in their mode of application. The theorems to be included are the *Thevenin* and *maximum power* theorems.

Thevenin's theorem permits the reduction of a two-terminal dc network with any number of resistors and sources to one having only one source and one internal resistor in the series configuration of Fig. 2.45.

The Thevenin resistance R_{Th} is the dc resistance between the output terminals of the network to be reduced, with all sources (current and voltage) set to zero. The Thevenin voltage E_{Th} is the open-circuit voltage between the output terminals with all sources present as in the original network. Consider, as an example, the network of Fig. 2.46 in which the network to the left of points a-a' is to be replaced by a Thevenin equivalent circuit. We shall then see that the current though R_L for various values of R_L can be determined very quickly from the reduced Thevenin equivalent circuit rather that analyzing the entire network again for each source.

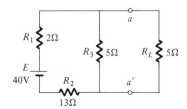

Figure 2.45 Thevenin equivalent circuit.   **Figure 2.46**

The network of Fig. 2.47 will result if we set all sources to zero to determine R_{Th}.

Between terminals a-a' we find that

$$R_{Th} = 5\|(2 + 13) = \frac{(5)(15)}{5 + 15} = \frac{75}{20} = 3.75 \ \Omega$$

For E_{Th}, the sources are replaced and the open-circuit voltage determined as shown in Fig. 2.48:

$$E_{Th} = V_{5\Omega} = \frac{5(40)}{5 + 13 + 2} \qquad \text{(voltage-divider rule)}$$

$$= \frac{200}{20} = 10 \ \text{V}$$

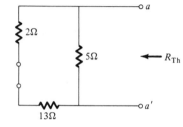
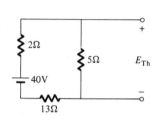

Figure 2.47 Determining R_{Th}.   **Figure 2.48** Determining E_{Th}.

The Thevenin circuit is drawn in Fig. 2.49 with the load resistor R_L replaced between terminals a-a'.

Insofar as the load resistor R_L is concerned, the network to the left of a-a' is the same as before the reduction. The current through R_L is

$$I_L = \frac{E_{Th}}{R_{Th} + R_L} = \frac{10}{3.75 + 5} = 1.143 \ \text{A}$$

An application of series-parallel techniques to the original network would result in the same solution for I_L. However, if R_L should now be changed to 26.5 Ω (for example), the entire series-parallel network would have to be reexamined if it were not for our Thevenin equivalent, which permits the following Ohm's law application for the new load current:

$$I_L = \frac{E_{Th}}{R_{Th} + R_L} = \frac{10}{3.75 + 26.5} = 0.2917 \ \text{A}$$

Consider also the savings in components if the elements of the original network could be replaced by the two required for the Thevenin equivalent. Additional examples in the use of this theorem will appear throughout the text.

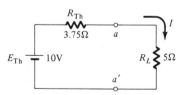

Figure 2.49 Thevenin equivalent circuit.

The *maximum power theorem* is used to ensure that a load receives maximum power from the supply. In words, it states that a load will receive maximum power when its terminal resistance is equal to the Thevenin resistance seen by the load. In the example above, R_L should be 3.75 Ω in order to receive maximum power. For greater or lesser values the power delivered to the load will decrease. In general, for maximum power to a load,

$$R_L = R_{\text{Th}} \qquad (2.21)$$

In Fig. 2.50, where $R_L = R_{\text{Th}}$, we find that

$$I_L = \frac{E_{\text{Th}}}{R_L + R_{\text{Th}}} = \frac{E_{\text{Th}}}{2R_{\text{Th}}}$$

and

$$P_L = I_L^2 R = \left(\frac{E_{\text{Th}}}{2R_{\text{Th}}}\right)^2 R_{\text{Th}}$$

or

$$P_{L_{\max}} = \frac{E_{\text{Th}}^2}{4R_{\text{Th}}} \qquad (2.22)$$

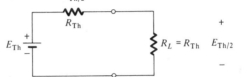

Figure 2.50 Maximum power transfer conditions.

For the example provided above the maximum power that can be delivered to R_L is

$$P_{L_{\max}} = \frac{E_{\text{Th}}^2}{4R_{\text{Th}}} = \frac{(10)^2}{4(3.75)} = \frac{100}{15} = 6.67 \text{ W}$$

Always keep in mind that the power that does not reach a load connected directly to a supply is lost in the internal resistance of the source and serves no useful purpose (reducing the overall efficiency of the system). The accepted level of efficiency must be carefully weighed, particularly if large amounts of power are involved.

2.11 Δ-Y CONVERSIONS

Occasionally, a network configuration is encountered where the elements do not appear to be in series or parallel. One such configuration is shown in Fig. 2.51. Note, in particular, that R_1 and R_3 are not in parallel since they do not have two points in common and yet they are not in series since a third element is connected to the common point between the two resistors. In fact, in the entire network no two elements are in series or parallel. A solution can be obtained by converting the delta (Δ) configuration in the upper part of the figure to a wye (Y) configuration, as shown in the

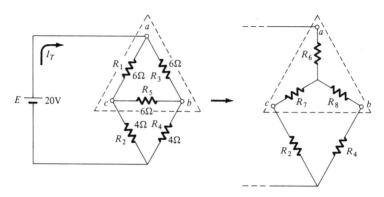

Figure 2.51 Δ-Y conversion.

adjoining figure. The above description accounts for the section title, Δ-Y conversions. It is only necessary to have the converting equations as listed below, which refer to the resistors as labeled in Fig. 2.51:

$$\Delta \longrightarrow Y \text{ conversion}$$

$$R_6 = \frac{R_1 R_3}{R_1 + R_2 + R_3}, \qquad R_7 = \frac{R_1 R_5}{R_1 + R_2 + R_3},$$
$$R_8 = \frac{R_3 R_5}{R_1 + R_2 + R_3} \tag{2.23}$$

It is sometimes necessary to convert from a Y to Δ configuration. The necessary converting equations are listed below:

$$Y \longrightarrow \Delta \text{ conversion}$$

$$R_1 = \frac{R_6 R_8 + R_7 R_8 + R_6 R_7}{R_8}, \qquad R_3 = \frac{R_6 R_8 + R_7 R_8 + R_6 R_7}{R_7},$$
$$R_5 = \frac{R_6 R_8 + R_7 R_8 + R_6 R_7}{R_6} \tag{2.24}$$

The relative positions of the resistors with regard to terminals a, b, and c must be carefully noted to properly use the conversion equations. Once the resistor values of the Y configuration are known the delta (R_1, R_3, and R_5) configuration is completely removed and *replaced* between the same three terminals by the Y (R_6, R_7, and R_8) configuration. You will the find that R_4 and R_8 are in series and as a unit in parallel with the series combination of R_2 and R_7. R_T and I_T can then be obtained using series-parallel techniques, discussed earlier. A solution could have been obtained using a method of analysis such as branch-current or nodal analysis, but either of these techniques would have resulted in a set of at least three simultaneous equations, which would prove very time-consuming to those without a background in determinants.

If the configuration is such that $R_1 = R_3 = R_5$, or, in words, each element of the delta has the same value, then the equations reduce to the following simple form:

$$\boxed{R_Y = \frac{R_\Delta}{3}} \quad \text{or} \quad \boxed{R_\Delta = 3R_Y} \tag{2.25}$$

For the network of Fig. 2.51, where this condition is satisfied,

$$R_Y = \frac{R_\Delta}{3} = \frac{6}{3} = 2\,\Omega$$

and the configuration of Fig. 2.52 will result, where $R_T' = R_4 + R_8 = R_2 + R_7 = 6\,\Omega$ and $R_T = R_6 + (R_T'/2) = 2 + \frac{6}{2} = 5\,\Omega$ with $I_T = E/R_T = \frac{20}{5} = 4\,\text{A}$.

The unique configuration of Fig. 2.51 is referred to as a *bridge* network. A very interesting and useful condition results when the following relationship is satisfied:

$$\boxed{\frac{R_1}{R_2} = \frac{R_3}{R_4}} \tag{2.26}$$

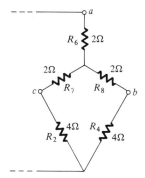

Figure 2.52

When the ratio condition of Eq. (2.26) is satisfied, the current through (and consequently the voltage across) the connected arm R_5 is zero. The resistor R_5 can then be replaced by either an open circuit ($i = 0$) or a short circuit ($v = 0$), and the analysis leading to R_T or I_T will be unaffected. The bridge network is said to be *balanced* when Eq. (2.26) is satisfied. Note in Fig. 2.51 that the ratios result in

$$\frac{R_1}{R_2} = \frac{6}{4} = \frac{R_3}{R_4}$$

and if we substitute an open circuit for R_5, we have

$$R_T = (6 + 4) \| (6 + 4) = 10 \| 10 = 5\,\Omega$$

obtained earlier, or using the short-circuit equivalent of R_5 we have

$$R_T = 6 \| 6 + 4 \| 4 = 3 + 2 = 5\,\Omega \qquad \text{as above}$$

This condition can be put to good use in a resistance-measuring device called the *Wheatstone bridge*. A galvanometer (sensitive ammeter) is inserted in the network in place of R_5 and the unknown resistance substituted for R_4 (or any of the other three), as shown in Fig. 2.53 with a photograph of a typical commercial bridge. The resistors, R_1, R_2, and R_3 are then adjusted until the current through the movement is zero. The equation can then be applied and the unknown resistor determined by $R_4 = R_{UNK} = R_2R_3/R_1$. As indicated in the figure the actual instrument is not made as basically simple as indicated by the network of Fig. 2.53, but the principle of operation is exactly the same. The bridge configuration will see further application in this chapter and in those to follow. For now, take note of its unique characteristics and appearance.

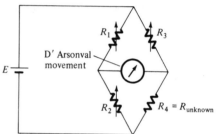

Figure 2.53   Wheatstone bridge. (Courtesy James G. Biddle Co.)

2.12 CAPACITORS

Thus far our interest has been solely in the basic dc sources and resistors. There are two other very important elements called the capacitor and inductor that we must now consider since they play a very important role in the behavior of electrical systems. This section will introduce the capacitor, while Section 2.13 will be devoted to the inductor.

The construction of each is basically quite simple. The capacitor in its most fundamental form has simply two conducting surfaces separated by a dielectric (a type of insulator) as shown in Fig. 2.54. The label capacitor comes from the *capacity* of the element to store a charge on its plates. The larger its capacitance, the more charge will be deposited on its plates for the same voltage across the plates. In equation form, capacitance is defined by

$$C = \frac{Q}{V} \qquad \text{farads (F)} \qquad (2.27)$$

Note that the unit of measurement is the *farad*, although microfarads (μF) and picofarads (pF) are more common values of capacitor values.

For the system of Fig. 2.54, we shall assume that the plates are initially uncharged

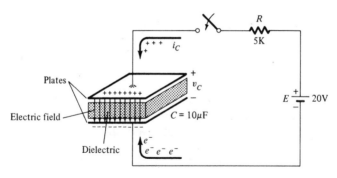

Figure 2.54 Charging capacitor.

and the switch open. When the switch is closed positive charges from the battery will be deposited on the top plate of the capacitor and negative charges on the lower plate. The result of this charge flow from source to capacitor is a current i_C that initially jumps to a value limited by the resistor R and then decreases toward zero. In other words, the charging rate of the plates is initially very heavy and then decreases quite rapidly to zero. The result of the deposited charge is an *electric field*, shown between the plates of the capacitor in Fig. 2.54, that extends from the positive to the negative charges. Since the voltage is directly related to the charge on the plates through Eq. (2.27) and the charging of the plates cannot occur instantaneously, the *voltage of a capacitor cannot change instantaneously*. It takes a definite amount of time determined by the circuit elements, which will be examined later in this section.

A plot of the current i_C vs. time would result in the exponential curve of Fig. 2.55. Note that the current jumped to a peak value of $E/R = 20/5 \text{ K} = 4 \text{ mA}$ and then decreased to zero exponentially. For most practical applications the time required to decay to zero is approximately 5τ, where τ (tau), called the *time constant* of the circuit, is determined by

$$\boxed{\tau = RC} \qquad \text{(seconds)} \qquad\qquad (2.28)$$

For the network of Fig. 2.54,

$$\tau = RC = (5 \text{ K})(10 \; \mu\text{F}) = (5 \times 10^3)(10 \times 10^{-6}) = 50 \times 10^{-3} = 50 \text{ ms}$$

and the discharge time is

$$5\tau = 5(50 \text{ ms}) = 250 \text{ ms} = 0.25 \text{ sec} = \tfrac{1}{4} \text{ s}$$

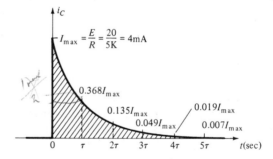

Figure 2.55 Charging current i_c of Fig. 2.58.

In equation form, the current i_C is the following:

$$i_C = \frac{E}{R}e^{-t/\tau} \tag{2.29}$$

Since the exponential function in the form e^{-x}, or $1 - e^{-x}$ appears quite frequently in the analysis of electrical systems, a plot of the function and its value at each time constant appears in Fig. 2.56. Note that after one time constant, the curve, and therefore i_C above, drops to 36.8% of its peak value, while at five time constants it is down to 0.7% of its peak value (less than 1%).

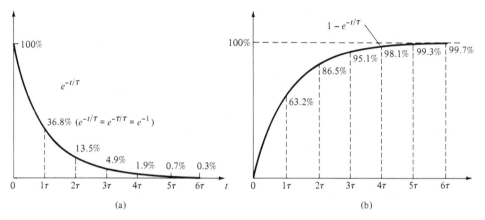

Figure 2.56 Functions: (a) $e^{-t/\tau}$, (b) $1 - e^{-t/\tau}$.

The voltage v_C will build up to a maximum value of $E = 20$ V at a rate directly related to the rate in which charge is deposited on the plates. Note from the plot of v_C in Fig. 2.57 that it reaches 63.2% of its final value in only one time constant and 99.3% in five time constants. The curve of Fig. 2.57 can be described by the following equation:

$$v_C = E(1 - e^{-t/\tau}) \tag{2.30}$$

Note the presence of the same experimental function which will dictate that v_C reach its final value of E volts at the same instant i_C decays to zero. Consider the importance of the following statement, which will be used to full advantage in the chapters to follow: When $v_C = E = 20$ V and $i_C = 0$ A (after approximately five time constants) the capacitor has the characteristics of an *open circuit*. In electronic systems this characteristic can be used to *block* dc from reaching certain points in the network.

Since $v_R = i_R R = i_C R$ and R is a constant, the voltage of v_R will have the same shape as the i_C curve and a peak value of 20 V.

The general behavior of the capacitor in a simple dc switching situation is thus described. The factors that affect the capacitance of a capacitor appear in the following equation:

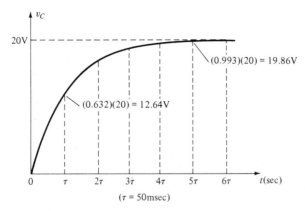

Figure 2.57 Increasing voltage v_c of Fig. 2.58.

$$C = \epsilon_0 \epsilon_r \frac{A}{d}$$ (2.31)

where $\epsilon_0 =$ permittivity of air (8.85×10^{-12})
$\epsilon_r =$ relative permittivity
$A =$ area in square meters
$d =$ distance between plates in meters

The equation indicates quite clearly that the larger the area of the plates or the smaller the distance between the plates, the larger the capacitance. The factor ϵ, called the permittivity of the material, is a measure of how well the dielectric between the plates will "permit" the establishment of field lines between the plates. The insertion of these insulators between the plates will increase the capacitance above that with simply air between the plates by a factor of ϵ_r, the *relative permittivity*, given by $\epsilon_r = \epsilon/\epsilon_0$. Table 2.8 lists a few dielectrics with their ϵ_r values.

Table 2.8

Relative Permittivity

DIELECTRIC	ϵ_r
Vacuum	1.0
Air	1.0006
Teflon	2.0
Paper, paraffined	2.5
Rubber	3.0
Transformer oil	4.0
Mica	5.0
Porcelain	6.0
Bakelite	7.0
Glass	7.5
Water	80.0

The most common types of capacitors are the mica, paper electrolytic, ceramic, and air, all of which appear in Fig. 2.58(b) with the symbol for the fixed and variable device in Fig. 2.58(a). The basic construction of every commercial capacitor leans toward making the capacitance its largest value while still limiting its size. For the ceramic type, two conducting surfaces are "rolled" together (with a ceramic dielectric between them) to make the area factor a maximum. For the mica capacitor, maximum area is obtained by "stacking" the plates (with the mica dielectric between them) and connecting the proper plates together. For the air capacitor the common area between

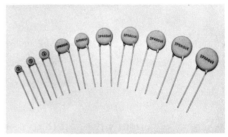

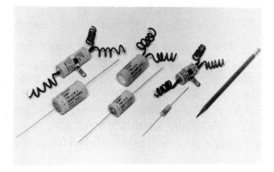

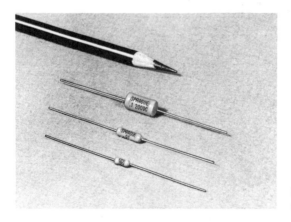

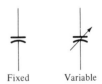

Fixed   Variable

Figure 2.58 Commercially available capacitors with their graphic symbols.

the stacked plates is made adjustable (for various values of capacitance) by a control knob that can move one set of plates.

If possible, the capacitance value is stamped right on the capacitor. However, some very small mica and ceramic capacitors require the use of a color-coding system, described in Fig. 2.59. Note that the numerical values associated with each color are exactly the same as applied to the color coding of resistors.

Capacitors connected in parallel are treated like resistors in series when it comes to determining the total capacitance. That is, for the parallel capacitors of Fig. 2.60, the total capacitance is determined by

$$C_T = C_1 + C_2 + C_3 + \cdots + C_N$$ (2.32)

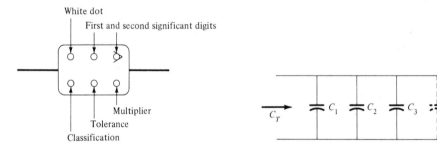

Figure 2.59 Capacitor color coding.   **Figure 2.60** Parallel capacitors.

For capacitors in series, the equation relates directly to that obtained for parallel resistors. That is, for the capacitors of Fig. 2.61, the total capacitance is determined by

$$\frac{1}{C_T} = \frac{1}{C_1} + \frac{1}{C_2} + \frac{1}{C_3} + \cdots \frac{1}{C_N}$$ (2.33)

For two series capacitors,

$$C_T = \frac{C_1 C_2}{C_1 + C_2}$$ (2.34)

The above equations can understandably be quite useful if a particular capacitance value is not available and must be obtained through a combination of available values.

Although an insulator (the dielectric) is placed between the plates, there is a

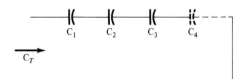

Figure 2.61 Series capacitors.

very small flow of charge between the plates called *leakage current.* If a capacitor is fully charged and then removed from the network and set aside, it will only be able to hold onto its charge for a limited amount of time due to the resulting leakage current through the dielectric that will deplete the plates of their charge. Keep in mind, however, that this leakage current is usually very very small and can be ignored for most practical applications.

In addition to the capacitance obtained through commercial design, there is a measure of capacitance between any two conducting or charged surfaces in an electronic system. This usually undesirable capacitance is called *stray capacitance* and can have, if not carefully considered, a pronounced effect on the performance of some electronic systems. In both the transistor and tube the stray capacitance between the elements of each can severely limit the range of operation of each in an electronic system unless properly considered.

For networks in which the value of R and C to be used in determining τ are not so evident an application of Thevenin's theorem can often be quite helpful, as demonstrated in Example 2.18.

EXAMPLE 2.18

Determine the mathematical expression for v_C if the switch is closed at $t = 0$ in Fig. 2.62.

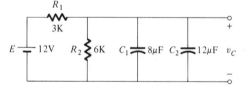

Figure 2.62

Solution:

The total capacitance of the parallel capacitors is given by

$$C = C_1 + C_2 = 8 + 12 = 20 \ \mu F$$

The value of R to be used is not evident. Applying Thevenin's theorem to the network of E_1, R_1, and R_2 we find that $R_{Th} = R_1 \parallel R_2 = 3 \parallel 6 = 2$ K (remember that the source E is set to zero when R_{Th} is determined).

$$E_{Th} = \frac{R_2(E)}{R_2 + R_1} = \frac{6(12)}{6 + 3} = \frac{2}{3}(12) = 8 \text{ V}$$

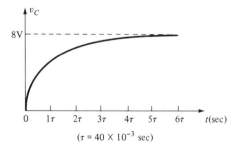

$(\tau = 40 \times 10^{-3} \text{ sec})$

Figure 2.63 Voltage v_c of Fig. 2.62.

and

$$\tau = R_{\text{Th}}C = (2\ K)(20 \times 10^{-6}) = 40 \times 10^{-3} = 40\ \text{ms}$$

with

$$v_C = E(1 - e^{-t/\tau}) = E_{\text{Th}}(1 - e^{-t/\tau})$$

or

$$v_C = 8(1 - e^{-t/40 \times 10^{-3}})$$

A plot of v_C appears in Fig. 2.63.

One very interesting characteristic of the capacitor is that the current i_C is not related directly to the magnitude of the voltage across the capacitor but to its *rate of change* of voltage across the capacitor. That is, the faster the rate of change of voltage across the capacitor, the greater the resulting i_C. An applied dc voltage of 100 V would (after 5τ of switching the network on) result in effectively zero current since the voltage does not change with time. In equation form, i_C and v_C are related by

$$i_C = C\frac{dv_C}{dt} \qquad (2.35)$$

where dv_C/dt is a measure of the instantaneous rate of change of v_C with time. The notation d is employed in *differential calculus*, which most college-level students have had some exposure to but which will not be a prerequisite for the material to follow. The equation clearly indicates that if v_C fails to change, $dv_C/dt = 0$ and hence i_C will be zero.

The capacitor and the inductor (to be described in the next section) are quite different from the resistor when we consider the transfer of energy to these elements. We have learned in the previous sections that the resistor dissipates the energy delivered to it in the form of heat. The ideal capacitor, however (if we ignore the leakage current), *does not dissipate* the energy delivered to it but simply stores it in the form of the electric field that through design can be returned to the system as electrical energy. In other words, the capacitor, unlike the resistor, is an energy-storing element and not a dissipative one like the resistor.

2.13 INDUCTORS

The inductor, or coil, as it is often called, has terminal characteristics that are in many ways similar to those of the capacitor, although the roles of the voltage and current are interchanged. Fundamentally, the inductor is a coil of wire such as shown in Fig. 2.64 with or without a core in the center. When the switch is closed in the circuit of Fig. 2.64 the current i_L will pass through the wire of the coil and establish a magnetic field *linking* (passing through) the coil as shown in the figure. Magnetic effects will be discussed in detail in chapter 4. Suffice it to say here that magnetic flux lines are continuous lines that indicate through their density the strength of a magnetic field in a particular region. For the inductive network there is an interchange between the electrical energy of the network and the magnetic field of the inductive elements. Like the capacitor, the inductor (ideally speaking) *does not dissipate electrical*

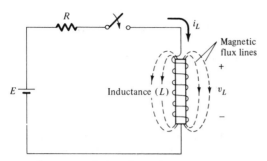

Figure 2.64 Series inductive network.

energy but stores it in a form (the magnetic field) that can be returned to the electrical system as required. Practically speaking, there is some dissipation in the dc resistance of the many turns of wire, but this loss can often be ignored.

For the capacitor we found that the voltage could not change instantaneously. For the inductor, *the current i_L cannot change instantaneously.* For some applications, where surge currents cannot be tolerated, inductors can be used to full advantage. In Fig. 2.64, therefore, a period of time will be required before the current can reach its maximum continuous value. In fact, for most practical applications a period of time equal to five constants (5τ) is sufficient, where

$$\tau = \frac{L}{R} \qquad \text{(seconds)} \tag{2.36}$$

The equation for i_L is similar to that for v_C for the capacitor, as indicated below:

$$i_L = \frac{E}{R}(1 - e^{-t/\tau}) \tag{2.37}$$

Note that the final or maximum value is E/R, which is independent of the inductor. In fact, when the current reaches its final value the voltage across the inductor will have decayed to zero. The inductor then has the characteristics of a *short circuit* and can be replaced as such in the network of Fig. 2.64. The current is then quite obviously simply determined by the source voltage E and the resistor R. A plot of Eq. (2.37) appears in Fig. 2.65(a). Since the equation is exactly the same in its exponential format as that obtained for v_C of the capacitor, the current will reach the same percent of maximum value after each time constant.

The inductance L of the coil, measured in *henries* (H), is a measure of the rate of change of flux established by a change in current through the inductor. In other words, the greater the inductance, the greater the change in flux linking a coil due to a change in current through the coil.

At the closing of the switch, the voltage v_L will equal the impressed voltage since the only remaining voltage drop $v_R = i_R R = i_L R = 0R = 0$ (recall that i_L cannot change instantaneously from its zero value). However, as the current i_L increases, the voltage $v_R = i_L R$ increases, causing a corresponding decrease in v_L since the source voltage E is fixed. The curve for v_L appears in Fig. 2.65(b). The voltage v_R will plot a curve the same as that for i_L since they are related by the constant R.

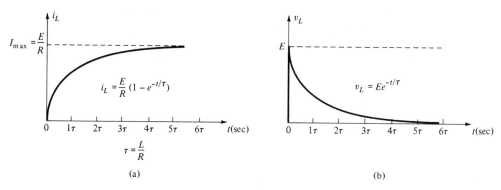

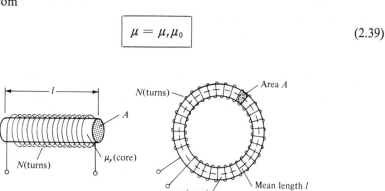

Figure 2.65 Functions: (a) i_L, (b) v_L for Fig. 2.64.

Note that each curve reaches its final steady-state value simultaneously. That is, after approximately five time constants, the voltage v_L is zero, i_L is E/R, and v_R is E V. In addition, we can conclude that after 5τ the inductor in a dc network can be replaced by a short-circuit equivalent.

The basic constructions of the typically available commercial inductors appear in Fig. 2.66. The equation for the inductance of either is determined by

$$L = \frac{N^2 \mu A}{10^8 l}$$ (2.38)

where N = number of turns
 μ = permeability of core
 A = area of core in square inches
 l = length of core in inches

and where all the parameters are self-descriptive except the parameter μ, called the *permeability* of the material, which is a measure of its magnetic properties. It is determined from

$$\mu = \mu_r \mu_0$$ (2.39)

Figure 2.66 Construction of some commercially available coils.

where μ_0 is the permeability of air [320 for the English system of Eq. (2.38)] and μ_r is the *relative permeability* of the material as compared to air. The better the magnetic properties of the material, the higher the relative permeability factor. Some materials such as steel and iron have permeabilities hundreds and even thousands of times that of air. That is, $\mu_r \geq 100$. For coils without a core, $\mu_r = 1$.

EXAMPLE 2.19

Determine the inductance L of the coil of Fig. 2.67.

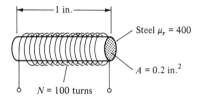

Figure 2.67 Coil for Example 2.19.

Solution:

$$L = \frac{N^2 \mu A}{10^8 l} = \frac{N^2 \mu_r \mu_0 A}{10^8 l}$$

$$= \frac{(100)^2 (400)(3.20)(0.2)}{10^8 \times 1} = 0.0256$$

$$= 25.6 \text{ mH}$$

The circuit symbol for the inductor (fixed and variable) appears in Fig. 2.68 with photographs of a number of types of commercially available inductors. In most cases, the inductance can be stamped right on the element, although in some cases a color-coding system very similar to that described for capacitance is employed. One technique of obtaining a variable inductance is to have a movable core within the coil. The less core within the coil, the less flux linking the coil and the less will be the resulting inductance.

Inductors in *series* (Fig. 2.69) have a total inductance value determined in the same manner as series resistors. That is,

$$L_T = L_1 + L_2 + L_3 + \cdots + L_N \tag{2.40}$$

For parallel inductors, the total inductance is found in the same manner as parallel resistors. That is, for the coils of Fig. 2.70 the total inductance is given by

$$\frac{1}{L_T} = \frac{1}{L_1} + \frac{1}{L_2} + \frac{1}{L_3} + \cdots + \frac{1}{L_N} \tag{2.41}$$

while for two parallel inductors,

$$L_T = \frac{L_1 L_2}{L_1 + L_2} \tag{2.42}$$

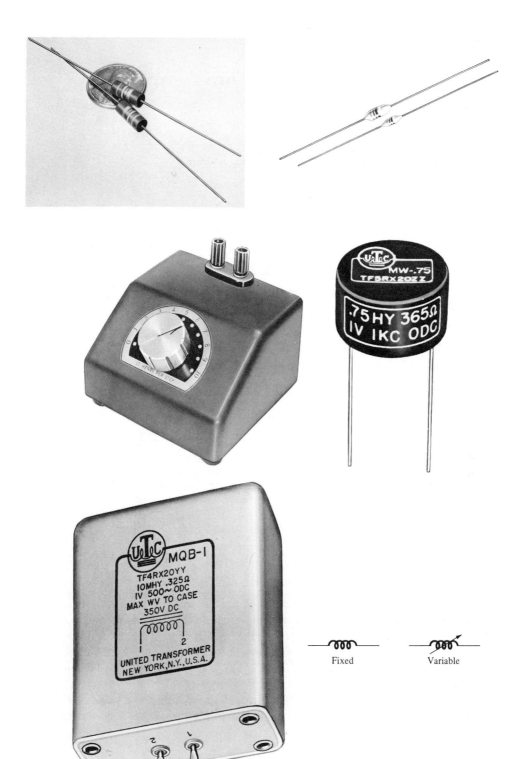

Figure 2.68 Commercially available inductors with their graphic symbols.

Fixed   Variable

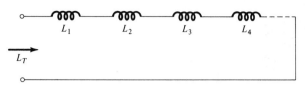

Figure 2.69 Series inductors.

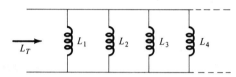

Figure 2.70 Parallel inductors.

EXAMPLE 2.20

Determine the total inductance of the series-parallel connection of elements in Fig. 2.71.

Solution:

For the parallel inductors L_2 and L_3

$$L'_T = \frac{L_1 L_2}{L_1 + L_2} = \frac{(40 \times 10^{-3})(20 \times 10^{-3})}{40 \times 10^{-3} + 20 \times 10^{-3}}$$

$$= \frac{800 \times 10^{-6}}{60 \times 10^{-3}} = 13.3 \text{ mH}$$

and

$$L_T = L_1 + L'_T = 10 \text{ mH} + 13.3 \text{ mH} = \textbf{23.3 mH}$$

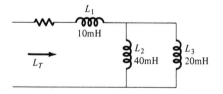

Figure 2.71 Example 2.20.

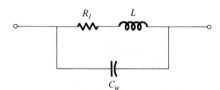

Figure 2.72 Total inductor equivalent circuit.

As with the capacitor, the inductor is not an *ideal* device. In addition to its inductance there is a measure of dc resistance due to the many turns of fine wire. In addition, stray capacitance exists between the parallel conducting turns of wire of the coil. The addition of these two elements would result in the equivalent circuit of Fig. 2.72. However, for many practical applications the inductor can still be considered ideal.

For networks in which it is difficult to determine the proper value of L and R for the time constant an application of Thevenin's theorem can be the solution. Consider the following exercise, for example.

EXAMPLE 2.21

Sketch the curve for i_L in Fig. 2.73 if the switch is closed at $t = 0$.

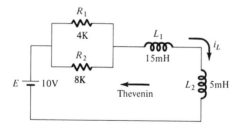

Figure 2.73 Example 2.21.

Solution:

The total inductance is

$$L_T = L_1 + L_2 = 15 + 5 = 20 \text{ mH}$$

Applying Thevenin's theorem,

$$R_{\text{Th}} = R_1 \| R_2 = \frac{4 \times 8}{4 + 8} = \frac{32}{12} = 2.66 \text{ K}$$

$$E_{\text{Th}} = E = 10 \text{ V}$$

Therefore,

$$\tau = \frac{L}{R} = \frac{20 \times 10^{-3}}{2.66 \text{ K}} = 7.5 \ \mu\text{sec}$$

and

$$i_L = \frac{E}{R}(1 - e^{-t/\tau}) = \frac{10}{2.66 \text{ K}}(1 - e^{-t/7.5 \times 10^{-6}})$$

$$= 3.76 \times 10^{-3}(1 - e^{-t/7.5 \times 10^{-6}})$$

The plot of i_L appears in Fig. 2.74.

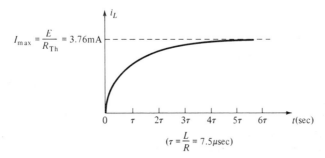

Figure 2.74 Current i_L for the network of Fig. 2.73.

For the inductor, the instantaneous voltage and current are related by

$$\boxed{v_L = L \frac{di_L}{dt}} \qquad (2.43)$$

Note again the appearance of the derivative. It clearly indicates that the magnitude of the voltage v_L is directly proportional to the rate of change of current through the coil and not simply the magnitude of the current through the coil. In other words, for an applied dc voltage the current will fail to change after steady-state conditions are reached, and the voltage v_L will decay to zero.

In an effort to emphasize the effect of a coil and capacitor in a dc network after the changing (transient) behavior has passed, consider the network in Fig. 2.75. As indicated in the adjoining figure the capacitor has been replaced by an open circuit and the inductor by a short circuit. The resulting currents and voltages are as follows:

$$I_1 = \frac{E}{R_1 + R_2} = \frac{10}{2 + 3} = \frac{10}{5} = 2 \text{ A} = I_L$$

$$V_C = V_{R_2} = \frac{R_2(E)}{R_2 + R_1} = \frac{3(10)}{3 + 2} = \frac{30}{5} = 6 \text{ V}$$

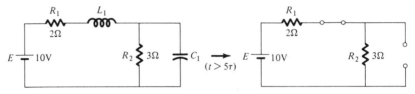

Figure 2.75 Steady-state behavior of a dc L-C network $(t > 5\tau)$

We shall find when we analyze electronic networks that the capacitor can play an important role in isolating dc and ac signals in the same system by acting as an open circuit to dc and a short circuit (approximately) for ac. For final emphasis, keep in mind that both the ideal inductor and capacitor *do not* dissipate energy like a resistor but simply store it in a form that can be returned to the system as required by the network design.

2.14 METER CONSIDERATIONS

In the pertinent section of this chapter the ammeter, voltmeter, and ohmmeter were introduced. We must now examine those considerations that will determine whether the proper meter is being used for the quantity to be measured.

For ammeters, it is usually sufficient to choose a meter that can handle the expected current level. Certainly a mA or μA meter should never be used to measure currents in the ampere range. In addition, the chosen meter should provide a reading in the midrange rather than the very low or high region of the scale. If in doubt as to the magnitude of the current, simply start with the highest scale and carefully work down to prevent damage through the *pinning* of the meter. An ammeter's internal resistance is usually so small that it can be ignored in comparison to the other series elements. *Remember:* The ammeter is *always* connected in series with the branch in which the current is to be determined.

The voltmeter presents a totally different situation, however, since its internal resistance can affect the network to which it is applied. On the face of every voltmeter (or VOM) is an ohm-per-volt rating. To determine the internal resistance of the meter the ohm-per-volt rating can simply be multiplied by the largest possible voltage reading of a particular scale. For example, if the 50-V scale of a meter with an ohm-per-volt rating of 1000 Ω is to be used, the internal resistance is simply

$$50(1000) = 50 \text{ K}$$

No matter what reading is obtained on that scale (whether it be 2 or 48 V) the internal resistance will be 50 K. It is determined solely by the ohm-per-volt rating and the maximum reading of that scale. Why this concern? Consider the network of Fig. 2.76 in which the voltage V_1 is to be measured using the meter described above. The internal resistance of the voltmeter will combine with the 50 K of the network, resulting in an equivalent resistance of 25 K. (Recall that voltmeters are always placed in parallel with the element across which the voltage is to be determined.)

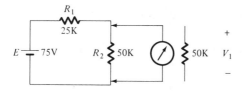

Figure 2.76 Meter loading.

Therefore

$$V_1 = \frac{25 \text{ K}(75)}{25 \text{ K} + 25 \text{ K}} = 37.5 \text{ V}$$

rather than the

$$V_1 = \frac{50 \text{ K}(75)}{50 \text{ K} + 50 \text{ K}} = 50 \text{ V}$$

as it should be.

In cases such as this it is absolutely necessary to ensure that the internal resistance of the voltmeter is *much greater* than the resistance across which the voltage is to be determined. A voltmeter with an ohm-per-volt rating of 50,000 would result in an internal resistance of $(50)(50 \text{ K}) = 2500 \text{ K}$, which when placed in parallel with the 50 K would result in $50 \text{ K} \parallel 2500 \text{ K} \simeq 50 \text{ K}$, and the reading would be 50 V, as it should be. The ohm-per-volt rating of 1000 would be fine for networks with resistive values in the neighborhood of 25 and 50 Ω rather than 25 and 50 K.

Ohmmeters are never used to measure resistance in an energized network. The meter may be damaged, or at the very least the reading will be meaningless. In addition if the resistance of a single element is to be determined, it is best to remove that element from the network to prevent the other elements from affecting the reading.

There are numerous instruments available today, such as shown in Fig. 2.77, that are quite multifunctional. The most frequantly employed is the VOM (volt-ohm-milliammeter), which typically has an ohm-per-volt rating of 50 K. Before using the ohmmeter section the zero ohms adjust knob should be used to set the pointer on the scale. By touching the two probes together, simulating 0 Ω, the 0-Ω knob can be turned to set this indication. Remember that the ohmmeter section employs an internal battery for its measurements. If the meter should be set aside in this setting, it is possible for the probes to touch and discharge the internal emf. Generally, the meter

Figure 2.77 Multifunctional meters. (Courtesy Simpson Electric Co. and Triplett Electric Co.)

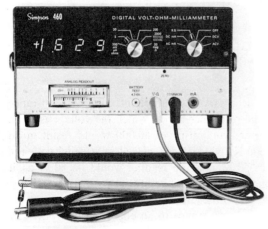

should be left in the high voltage scale when not in use. If the meter fails to properly respond to the 0-Ω scale, the battery will probably have to be replaced. This replacement is usually not a difficult task.

A digital multimeter that can measure dc and ac voltages and resistances is also shown in Fig. 2.77.

The EVM (electronic voltmeter) must be plugged into a 120-V outlet to work properly (also the case for the digital VOM mentioned above). This meter has an internal resistance of about 11 MΩ, which makes it an excellent instrument in high-resistance networks.

On most meters, the numbers appearing opposite the dial setting are the same numbers appearing at the high end of the scale to be used, although a particular scale may be used for more that one dial setting. Consider a meter with a 5-, 50-, and 500-V dial setting. In each case, the scale to be used may have a maximum indication of 50 V. In one case 40 V is read as 40 V if the 50-V dial setting is used, while the 40 may represent 4 V if the 5-V dial setting is used and 400 V for the 500-V setting. In addition, some numbers on the face of the meter are applicable to the scale above and below even though they may indicate different quantities. Finally, no matter how many divisions appear between two numbers on any scale they *equally* divide the difference between the numbers. For example, if five divisions appear between 40 and 50 V, then each division represents 2 V, while if the scale is so tight that only one division appears, then they are each indicating a difference of 5 V. The proper use of any of these instruments can be developed only through continual use and exposure.

PROBLEMS

§ 2.2

1. If 4×10^{20} electrons pass through a conductor in 40 sec,
   (a) Determine the charge in coulombs.
   (b) Calculate the current.

2. Find the time required (in seconds) for 32 C of charge to pass through a copper conductor if the current is 50 mA.

3. (a) Determine the diameter in mils of a No. 8 solid copper wire (refer to Table 2.2).
   (b) Repeat part (a) for the diameter in inches.

§ 2.3

4. Determine the energy expended (in joules) to bring a charge of 12 C through a potential difference of 40 mV.

5. What is the potential difference between two points in an electric circuit if 170×10^{-3} J of energy are required to bring a charge of 200 μJ from one point to the other?

§ 2.4

6. Determine the internal resistance of a battery-operated clock if a current of 0.4 mA results from an applied voltage of 1.2 V.

7. Determine the current through a soldering iron if 120 V are applied. The iron has a resistance of 80 Ω.

8. Find the voltage drop across a 5-K resistor with a current of 30 μA passing through it.

9. What resistance would be required to limit the current to 0.4 A if the applied voltage is 64 V?

10. Determine the resistance of 200 ft of $\frac{1}{16}$-in.-diameter copper wire.

11. Calculate the resistance of 600 ft of No. 14 wire using Table 2.2.

12. If the resistance of a copper conductor is 0.25 Ω at room temperature ($T = 20°C$), what is its resistance at 100°C?

13. At what temperature will the resistance of a 400-ft length of No. 8 copper wire double if its resistance at $T = 20°C$ is 0.5 Ω?

14. (a) Determine the resistance of a molded composition resistor with the following color bands: first, red; second, orange; third, brown; fourth, gold.

(b) Indicate its possible range of values.

15. (a) Determine the resistance of the thermistor appearing in Fig. 2.23 at 200°C.

(b) At what temperature will the resistance increase by a factor of 10 from the 200°C value?

§ 2.5

16. Determine the power delivered by a 12-V battery at a current drain of 6 A.

17. Calculate the power dissipated by a 4-K resistor having a current of 4 mA passing through it.

18. A 300-W television in connected to a 120-V outlet. Determine the current drawn by the set.

19. Determine the total energy dissipated by a 1600-W toaster used for 10 min.

20. Calculate the cost of using the following appliances for the indicated time period if the unit cost is 2.8 cents/kWh.

(a) Six 60-Watt bulbs for 5 hr.

(b) 8-W clock for 24 hr.

(c) 400-W television for 6 hr.

(d) 5000-W clothes dryer for 1 hr.

21. (a) A 3-hp motor has an input power demand of 2400 W. Determine its efficiency.

(b) If the applied voltage is 120 V, determine the input current.

(c) What is the power lost in the energy transfer (in watts)?

§ 2.6

22. (a) Determine the total resistance of the series circuit in Fig. 2.78.

(b) Calculate the current I.

(c) Find the voltage V_1 using Ohm's law.

(d) Calculate V_1 using the voltage-divider rule.

(e) Find the voltage across each resistor and verify Kirchhoff's voltage law around the closed path.

(f) Calculate the power supplied by the battery and to the resistor R_1.

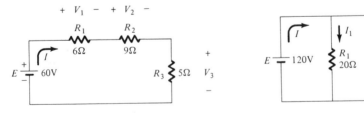

Figure 2.78 Problem 22.   Figure 2.79 Problem 23.

23. (a) Determine the total resistance of the parallel network of Fig. 2.79.

(b) Calculate the current I.

(c) Find the current I_1 using Ohm's law.

(d) Find the current I_1 using the current-divider rule.

(e) Determine the current through each resistor and verify Kirchhoff's current law at either node.

(f) Calculate the power supplied by the battery and note whether it equals the sum of that delivered to each resistor.

24. Find the total resistance of the networks of Fig. 2.80.

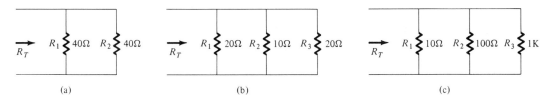

(a)   (b)   (c)

Figure 2.80 Problem 24.

§ 2.7

25. For the series parallel network of Fig. 2.81,
   (a) Determine R_T.
   (b) Calculate I.
   (c) Find I_1, I_2, and I_3.
   (d) Calculate the power to the 3-Ω resistor.

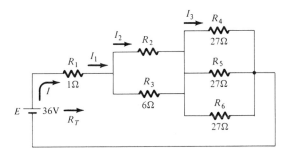

Figure 2.81 Problem 25.

§ 2.8

26. For the network of Fig. 2.82, determine
   (a) I_1.
   (b) I_2.
   (c) I_3.
   (d) V_S.

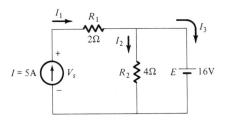

Figure 2.82 Problem 26.

27. Convert a voltage source of series elements $E_S = 36$ V, $R_S = 9$ Ω to a current source.

28. Determine the voltage regulation of a supply whose terminal voltage drops 2.4 V from open circuit to full-load conditions if the full-load voltage is 220 V.

§ 2.9

29. Using the superposition theorem, determine the current I in the network of Fig. 2.83.

30. Determine the current I in the network of Fig. 2.84 using branch-current analysis.

31. Determine the voltage at node a in Fig. 2.84.

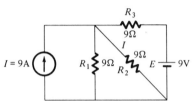

Figure 2.83 Problem 29.

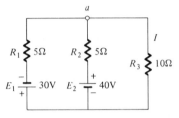

Figure 2.84 Problem 30.

§ 2.10

32. (a) Find the Thevenin network for the portion of the network in Fig. 2.85 to the left of resistor R_1.
    (b) Using the Thevenin network, determine I_1.
    (c) Use series-parallel techniques on the original network to determine I_1 and compare the results with part (b).

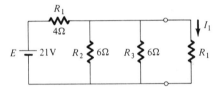

Figure 2.85 Problem 32.

33. (a) Determine the value of R_1 in Fig. 2.85 that would result in maximum power to R_1.
    (b) Calculate the maximum power that could be delivered to R_1 if it were changed to that value.
    (c) Find the power delivered to R_1 with the resistance value appearing in Fig. 2.85 and verify that it is less than the maximum value of part (b).

§ 2.11

34. Determine the total resistance of the network of Fig. 2.86.

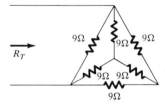

Figure 2.86 Problem 34.

35. Calculate the total resistance of the network of Fig. 2.87.

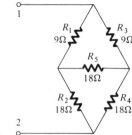

Figure 2.87 Problem 35.

36. Is the bridge of Problem 35 balanced?

37. (a) If $R_1 = 10\ \Omega$, $R_2 = 40\ \Omega$, $R_3 = 4\ \Omega$, and $R_4 = 16\ \Omega$ in Fig. 2.87, is the bridge balanced?

    (b) If 20 V were applied to points 1, 2, determine the current through R_4 [for the values of part (a)].

    (c) With the 20 V applied, determine the current through R_1 [for the values of part (a)].

§ 2.12

38. If 2400 μC of charge are deposited on the plates of a capacitor having a potential drop of 120 V across the plates, determine the capacitance of the capacitor.

39. For the network of Fig. 2.88,

    (a) Determine the time constant of the network.

    (b) For the changing phase, write the mathematical expression for the current i_C and voltages v_C and v_R.

    (c) Sketch the waveform of each quantity in part (b).

    (d) What is i_C in magnitude after one time constant?

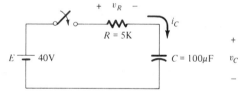

Figure 2.88 Problem 39

40. Find the capacitance of a capacitor having 4-cm² plates, a dielectric of mica, and a d (distance between plates) of 0.04 in.

41. (a) Find the total capacitance of four 6-μF capacitors in series.

    (b) Repeat part (a) for capacitors in parallel.

42. Find the expression for v_C in the network of Fig. 2.89 following the closing of the switch.

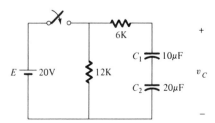

Figure 2.89 Problem 42.

§ 2.13

43. For the network of Fig. 2.90,
    (a) Determine the time constant.
    (b) Determine the mathematical expression for i_L, v_L, and v_R during the charging phase.
    (c) Sketch the waveforms of part (b).
    (d) What is the magnitude of v_R after one time constant?

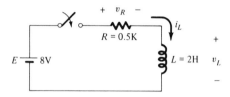

Figure 2.90 Problem 43.

44. Determine the inductance of an inductor with 500 turns, $\mu_r = 100$, area $= 0.25$ in². , and $l = 1.5$ in.

45. (a) Determine the total inductance of two series coils with values of 0.6 and 0.3 H.
    (b) Repeat part (a) for the coils in parallel.

46. Determine the expression for v_L when the switch is closed in Fig. 2.91.

47. Determine I_1 and V_1 for the network of Fig. 2.92 when the steady-state condition is reached.

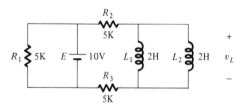

Figure 2.91 Problem 46.   **Figure 2.92** Problem 47.

§ 2.14

48. (a) Sketch the location and connections of ammeters and voltmeters used to measure the currents I_1 and I_2 and voltages V_1 and V_2 in Fig. 2.93.
    (b) Using a voltmeter with an ohm-per-volt rating of 1000, determine the indication of the meter if placed across the 4-K resistor if the 50-V scale is used.
    (c) Repeat part (b) for a meter employing an ohm-per-volt rating of 20,000.

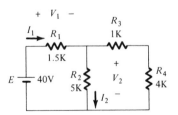

Figure 2.93 Problem 48.

3

ac Networks

3.1 INTRODUCTION

Our analysis thus far has been limited to networks with fixed, nonvarying, time-independent currents and voltages. We shall now begin to consider the effects of alternating voltages and currents such as shown in Fig. 3.1.

Each waveform in Fig. 3.1 is called an alternating waveform since it varies above and below the horizontal zero axis. The first [Fig. 3.1(a)] is called a *sinusoidal ac voltage* and is the type universally available at home outlets, industry, etc. In this chapter we shall deal specifically with this type of waveform. The second [Fig. 3.1(b)] is called, for obvious reasons, a *square wave*, which has its application in a number of areas to be introduced in later chapters. The last of the three [Fig. 3.1(c)] and perhaps the least encountered is called a *triangular* waveform. Compare each with the dc voltage of Fig. 3.2, which is fixed in magnitude and does not vary with time.

Although a multitude of questions of how to analyze networks with an applied signal that varies with time may now begin to surface, be assured that after a careful

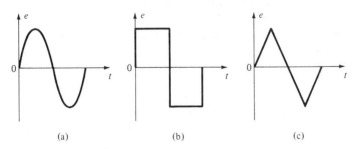

Figure 3.1 Alternating waveforms: (a) sinusoidal, (b) square wave, (c) triangular.

examination of this chapter the analysis of ac networks will ride on a level of difficulty only slightly greater than that encountered for dc networks. In fact, the only added difficulty will be mathematical and not in the application of the theorems already described for dc networks.

Although not described in detail as we progress through the chapter, meters are also available for measuring ac voltages and currents. They are so specified on the face of the meter with their ranges of measurement. They are hooked up in exactly the same manner and their effect insofar as meter loading is concerned corresponds directly with the discussion for dc networks. An ohmmeter is not specified to be either ac or dc since resistance values do not change with the type of signal applied. A commercially available ac voltmeter and ammeter appear in Fig. 3.3. Note the similarities in basic appearance with the corresponding dc instruments. Other instruments needed to measure ac quantities will be introduced and considered in limited detail in this chapter.

For the remainder of this chapter the phrase *ac* voltage or current will refer specifically to the sinusoidal alternating waveform of Fig. 3.1(a).

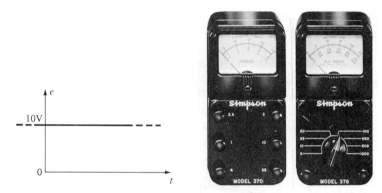

Figure 3.2 Dc voltage.   **Figure 3.3**

3.2 THE SINUSOIDAL (ac) WAVEFORM

The sinusoidal ac voltage is available at the output terminals of a laboratory supply, which employs an oscillator network to be described in Chapter 10, or an ac generator much as shown in Fig. 3.4, which will be described in detail in Chapter 5. The shaft of the generator is turned by some external means [steam turbine, hydro (water), etc.]

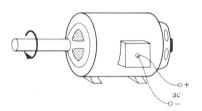

Figure 3.4 Ac generator (alternator).

resulting in the output ac voltage shown. In this case there is a conversion of mechanical energy at the shaft to electrical energy at the output terminals.

Note in the expanded view of a sinusoidal voltage in Fig. 3.5 that the waveform repeats itself in the negative sense from π to 2π and has the same peak value of 120 V above and below the axis. In its simplest mathematical form, an equation for the sinusoidal voltage is given by

$$v = V_{peak} \sin \theta \qquad\qquad (3.1)$$

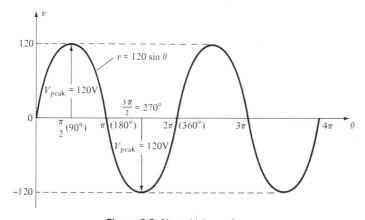

Figure 3.5 Sinusoidal waveform.

Note at 90° that $\sin \theta = \sin 90° = 1$ and that $v = V_{peak}(1) = V_{peak} = 120$ V. The same occurs at 270° but $\sin 270° = -\sin 90° = -1$. Certainly at 0°, 180°, 360° $\sin \theta = 0$, and therefore $v = 0$ V. The axis of Fig. 3.5 is more frequently labeled in radians than degrees. The equivalent radian values appear for 90°, 180°, 270°, and 360°. The following equations will permit a direct conversion from one to the other:

$$\text{Radians} = \left(\frac{\pi}{180°}\right)(\text{degrees}) \qquad\qquad (3.2)$$

or

$$\text{Degrees} = \left(\frac{180°}{\pi}\right)(\text{radians}) \qquad\qquad (3.3)$$

Figure 3.6 pictorially compares the degree and radian measure.

EXAMPLE 3.1

1. Convert the following from degrees to radians:

$$45°: \quad \text{Radians} = \left(\frac{\pi}{180°}\right)(45°) = \frac{\pi}{4}$$

$$270°: \quad \text{Radians} = \left(\frac{\pi}{180°}\right)(270°) = \frac{3}{2}\pi$$

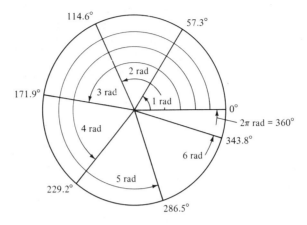

Figure 3.6 Radian vs degree measure.

2. Convert the following from radians to degrees:

$$\frac{\pi}{3}: \quad \text{Degrees} = \left(\frac{180°}{\pi}\right)\left(\frac{\pi}{3}\right) = \mathbf{60°}$$

$$\frac{4}{3}\pi: \quad \text{Degrees} = \left(\frac{180°}{\pi}\right)\left(\frac{4}{3}\pi\right) = \mathbf{240°}$$

The sinusoidal waveform can also be plotted vs. time, as shown in Fig. 3.7(a).

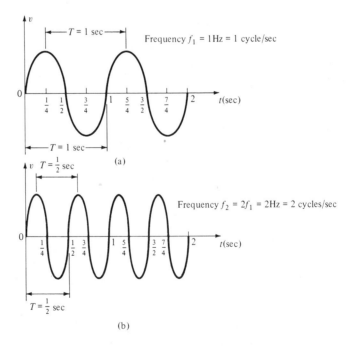

Figure 3.7 Sinusoidal voltages of different frequencies plotted against time.

By definition, the *period* (T) of a sinusoidal waveform is the time required for one complete appearance of the waveform, which is called a *cycle*. For the waveform of Fig. 3.7(a) the period is 1 sec. For the adjoining waveform [Fig. 3.7(b)] the period is $\frac{1}{2}$ sec. The *frequency* (f) of an alternating waveform is the number of cycles that appear in a time span of 1 sec. For Fig. 3.7(a) the frequency is 1 cycle per second, and for Fig. 3.7(b) it is 2 cycles per second. The units *cycle per second* has almost universally been replaced by the label *Hertz (Hz)*; that is,

$$1 \text{ Hertz } (Hz) = 1 \text{ cycle per second} \qquad (3.4)$$

A moment of reexamination of the above paragraph will reveal that as the time per cycle decreased, the frequency increased by a corresponding amount. In equation form, the two are related by

$$T = \frac{1}{f} \qquad (3.5)$$

where T = period in seconds
      f = frequency in hertz

EXAMPLE 3.2

Determine the period and the frequency of the waveform of Fig. 3.8.

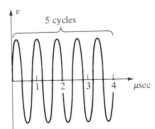

Figure 3.8 Figure for Example 3.2.

Solution:

In Fig. 3.8 five cycles encompass the indicated time of 4 μsec. Therefore

$$T = \tfrac{1}{5}(4 \ \mu s) = \textbf{0.8 } \boldsymbol{\mu}\textbf{s}$$

and

$$f = \frac{1}{T} = \frac{1}{0.8 \times 10^{-6}} = \frac{10^6}{0.8} = 1.2 \times 10^6 \text{ Hz} = \textbf{1.2 MHz}$$

For each cycle of a sinusoidal waveform we pass through 360° or 2π rad. If the frequency (cycles per second) increases, the number of degrees or radians traversed will increase per unit time. Or we can say that we shall circulate at an increased angular velocity around the 360° or 2π-rad circle of Fig. 3.6. For the latter reason a term called the *angular velocity* is defined by

$$\omega = 2\pi f \qquad \text{(radians per second)} \qquad (3.6)$$

The angle θ traversed at an angular velocity ω for a particular period of time t can therefore be determined by

$$\boxed{\theta = \omega t}$$ (3.7)

Equation (3.1) for the sinusoidal waveform can then be written as

$$\boxed{v = V_p \sin \omega t}$$ (3.8)

or

$$\boxed{v = V_p \sin 2\pi f t}$$ (3.9)

For home and industrial use in North America 60 Hz is the available frequency. From the above we now know that the sinusoidal voltage available at our home outlet completes 60 cycles in one second, and has a period

$$T = \frac{1}{f} = \frac{1}{60} = \underline{\textbf{16.67 ms}}$$

and an angular velocity

$$\omega = 2\pi f = (6.28)(60) = \underline{\textbf{377 rad/s}}$$

It also has a peak voltage of about 170 V, so the entire sinusoidal function can be represented by

$$v = 170 \sin 377t$$

The waveform of Fig. 3.5 cuts through the horizontal axis with increasing magnitude at time t or θ equal to zero. If the waveform should cut through the axis before or after 0°, a phase shift term must be added to the general mathematical expression for a sine wave. For intersections before 0°, such as shown in Fig. 3.9(a), the following format is used,

$$\boxed{v = V_p \sin(\omega t + \theta)}$$ (3.10)

while for intersections to the right of the zero axis [Fig. 3.9(b)] the following format is employed,

$$\boxed{v = V_p \sin(\omega t - \theta)}$$ (3.11)

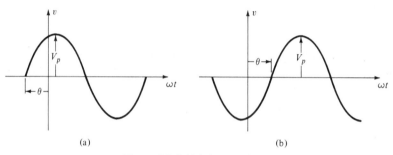

(a)   (b)

Figure 3.9 Initial phase angles.

For the two sinusoidal waveforms plotted on the same horizontal axis, a *phase relationship* between the two can be determined. For the waveforms of Fig. 3.10(a) the sinusoidal voltage crosses the horizontal axis 90° before the sinusoidal current. The voltage v is said to *lead* the current i by 90°, or the current i *lags* the voltage v by 90°. Note that positive (increasing values with time) slope crossings must be compared. In Fig. 3.10(b), since positive slopes are 210° apart, v is said to lead i by 210° or i lag v by the same angle. Note that if one positive slope intersection and one negative slope intersection were compared, an incorrect answer of 30° could be obtained. It is also correct to say that i leads v by 360° − 210° = 150°, as shown in the figure.

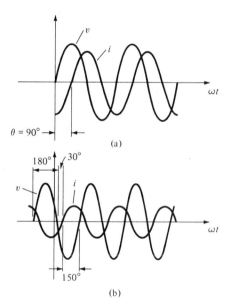

Figure 3.10 Phase relationships.   (b)

3.3 EFFECTIVE (rms) VALUE

Since a sinusoidal voltage or current has the same shape above and below the axis, the question of how power can be delivered may be a bothersome one since it appears that net flow to a load over one full cycle will be zero. However, this question is quickly answered if we simply keep in mind that *at each instant* of the positive portion of the waveform power is being delivered and dissipated by the load. Although the current may reverse direction during the negative portion, power will still be delivered and absorbed by the load at each instant of time. In other words, the power delivered at each instant of time is additive even though the current may reverse in direction.

In an effort to determine a single numerical value to associate with the time-varying sinusoidal voltage and current a relationship was developed experimentally between a dc and ac quantity that would result in each delivering the *same* power to a load. In other words, what sinusoidal voltage would have to be applied to deliver the same power as a particular dc voltage? The results indicated that if a 10-V dc source were applied to a load, the same power could be delivered to that load with a sinusoidal voltage having a peak value of 14.14 V. The waveforms of the two voltages just described appear in Fig. 3.11. In equation form, if we found that the *equivalent dc* or *effective* value of a sinusoidal voltage is equal to 0.707 times the peak value, for the example above,

$$0.707(V_p) = 0.707(14.14) = \underline{\mathbf{10\ V}}$$

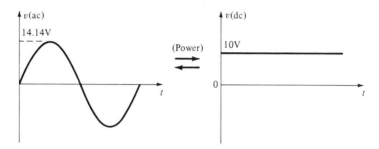

Figure 3.11 Relating effective values.

In equation form

$$\boxed{V_{\text{dc equivalent}} = V_{\text{eff}} = 0.707(V_{\text{peak}}) = \frac{1}{\sqrt{2}}(V_{\text{peak}})} \qquad (3.12)$$

or for currents

$$\boxed{I_{\text{dc equivalent}} = I_{\text{eff}} = 0.707(I_{\text{peak}}) = \frac{1}{\sqrt{2}}(I_{\text{peak}})} \qquad (3.13)$$

For an equivalent dc voltage an ac voltage must satisfy the following relationship:

$$\boxed{V_{\text{peak}} = 1.414V_{\text{eff}} = \sqrt{2}\,V_{\text{eff}}} \qquad (3.14)$$

or

$$\boxed{I_{\text{peak}} = 1.414I_{\text{eff}} = \sqrt{2}\,I_{\text{eff}}} \qquad (3.15)$$

For the function introduced earlier,

$$v = 170 \sin 377t$$

$$V_{\text{eff}} = 0.707(170) = \underline{\mathbf{120\ V}}$$

Note that the frequency does not play a part in determining the equivalent dc value.

Another label for the effective value is the *rms* value, which is derived from the mathematical procedure for determining the effective value of any waveform. First, the *square* of the waveform is found and then the *mean* value by finding the area under the squared curve and dividing by the period T. The square *root* of the resulting value is then obtained, which equals the effective value. The abbreviation rms is derived from the first letters of the italicized words above in the reverse order that they appear (root-mean-square).

The manufacturer's label on all electrical equipment, machinery, instruments, appliances, etc., lists the effective values of the voltage or current. The home outlet is rated 120 V, 60 Hz indicating an effective value of 120 V and a peak value of $\sqrt{2}(120) = 170$ V. Some appliances such as an air conditioner are rated at 220 V. For such units a separate line must be run from the meter panel to provide the increased voltage. All ac voltmeters and ammeters indicate effective values or peak-to-peak (p-p) values unless otherwise specified.

ac Meters which employ D'Arsonval movements (such as used to measure dc quantities) must have an additional electronic circuitry to give the ac waveform an average (or dc) level so that calibration can be introduced and a reading of some meaning obtained. It should be somewhat obvious that the average value of the sinusoidal waveform over one full cycle is zero. More will be said about the D'Arsonval movement and average values in the sections to follow. There are two other types of movements used with ac instruments that do not require this additional circuitry. They are the *iron-vane* and *dynomometer* types. They will read the effective value or dc equivalent value of any waveform including sinusoidal, dc, square wave, etc., even though a particular instrument may be labeled ac voltmeter for those who don't fully understand the versatility of the movement.

Since we are into meters, the *wattmeter*, and *frequency meter*, should be introduced. The wattmeter, with its dynomometer-type movement, can read the watts dissipated by a dc or ac network with the same set of connections. Recall that the voltage terminals are always placed in parallel with the load to which the power is to be determined, while the current terminals are connected in series to that load. The frequency meter is an instrument much as shown in Fig. 3.12 used to determine the frequency of the sinusoidal signal. It is hooked up directly across the supply and the reed in the meter that vibrates the most indicates the frequency of the signal.

Incidentally, you will note as we progress through the text that very little will

Figure 3.12 Frequency meter (Courtesy Simpson Electric Co.).

be said about the internal construction of meters. Priorities require that this type of detail be left to the reader using one of the many references available in most libraries.

3.4 AVERAGE VALUES

The average value or dc component of a waveform will be of some importance in the more advanced analysis to follow. For the individual waveform of Fig. 3.13(a) the average value (as indicated earlier) is zero over one cycle. For the waveform of Fig. 3.13(b), the average value is 5 V. The waveform of Fig. 3.13(b) can be obtained by simply placing a dc voltage of 5 V in series with a peak value of 4 V as shown in Fig. 3.13(c). For electronic systems where both dc and ac signals are present waveforms such as Fig. 3.13(b) will appear quite frequently.

For waveforms such as shown in Fig. 3.14(a), the average value is not so obvious, but it can be determined using the following equation:

$$G \text{ (average value)} = \frac{\text{Area (algebraic sum)}}{T \text{ (period)}} \qquad (3.16)$$

The algebraic sum is the sum of those areas above the axis less the sum of the areas below the axis. The time T is the time interval for which the average value is to be determined.

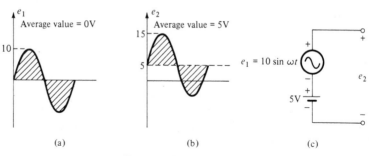

(a)   (b)   (c)

Figure 3.13 Average value.

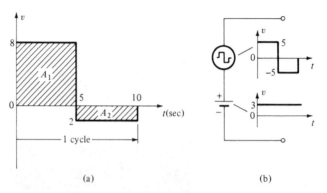

(a)   (b)

Figure 3.14 Average value.

For the complete cycle of Fig. 3.14(a),

$$G = \frac{A_1 - A_2}{T} = \frac{(8)(5) - 2(5)}{10} = \frac{40 - 10}{10} = \frac{30}{10} = \underline{3 \text{ V}}$$

indicating that the square wave of Fig. 3.14(a) could be generated using the two components of Fig. 3.14(b).

For waveforms where the area cannot be obtained as directly as in the above example, try to approximate the odd shape using figures such as rectangles and triangles for which the area can easily be obtained. This technique will at least result in a numerical value that will permit further examination of the network at least on an approximate basis. Some mathematics texts list the areas under some of the more complicated mathematical functions, or if you are so versed, integral calculus can be employed.

3.5 THE *R, L,* AND *C* ELEMENTS

The effect of an ac sinusoidal signal on the basic *R, L,* and *C* elements can now be examined. Although certain statements of fact will have to be made without full derivation or explanation, be assured that the material will be more than complete to permit a full understanding of the more advanced material to follow.

In Fig. 3.15, a sinusoidal voltage has been applied across the resistor. As shown by the resulting current in the same figure, the peak values are related directly by Ohm's law. In addition, there is no phase shift introduced by the resistor, so v_R and i_R are said to be *in phase.* Note that the frequency of both v_R and i_R is also the same.

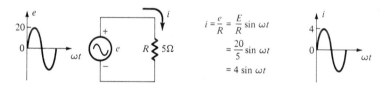

Figure 3.15 Simple ac resistive network.

The power to the resistor can be determined from the *effective* values of the voltage and current in the following manner:

$$\boxed{P_R = I_R^2 R = \frac{V_R^2}{R} = V_R I_R} \qquad \text{(watts)} \qquad (3.17)$$

For the case of Fig. 3.15,

$$P_R = V_R I_R = \left[\frac{20}{\sqrt{2}} \right]\left[\frac{4}{\sqrt{2}} \right] = \frac{80}{2} = \underline{40 \text{ W}}$$

Note the similarities between the forms of Eq. (3.17) and those used for dc networks. The *only* difference is that here we have effective values to determine from the sinusoidal functions.

The reaction of a coil or capacitor to an ac signal is quite different from the reaction of a resistor. Both elements react to the sinusoidal signal in a manner somewhat similar to the resistor in the sense that they can limit the current but neither (ideally speaking) will dissipate any of the electrical energy delivered to it. It will simply store it in the form of a magnetic field for the inductor, or electric field for the capacitor, with the ability to return it to the electrical system as required by design.

For the inductor the *reactance* to an ac signal is determined by

$$X_L = \omega L = 2\pi f L \qquad \text{(ohms)} \qquad\qquad (3.18)$$

Although reactance is similar to resistance in its ability to limit the current, always keep in mind that it is not an energy-dissipating form of opposition as noted in the previous paragraph.

Note that inductive reactance is directly proportional to the frequency of the applied signal. Increasing frequencies result in an increasing reactance magnitude. Recall that the ideal inductor had a short-circuit equivalent in dc networks. Certainly for dc, $f = 0$, and we find that the reactance of the inductor is $X_L = 0L = 0\,\Omega$, corresponding to that conclusion.

A sinusoidal voltage has been applied across a 0.5 H inductor in Fig. 3.16. First, the reactance of the inductor at the applied frequency is 188.5 Ω. Ohm's law can be applied to determine the peak value of the current. That is,

$$I_{\text{peak}} = \frac{V_{\text{peak}}}{X_L} = \frac{20}{188.5} = 106 \times 10^{-3} = \mathbf{106\ mA}$$

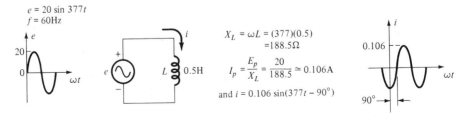

Figure 3.16 Simple ac inductive network.

as shown in Fig. 3.16. Note that the voltage applied across the inductor *leads* the resulting current by 90°. The inductor, therefore, introduces a 90° phase shift between the two quantities not present for the pure resistor.

For an ac system the basic equation for power becomes the following due to the effect of the reactive elements such as the inductor and capacitor:

$$P = \frac{V_p I_p}{2} \cos\theta = V_{\text{eff}} I_{\text{eff}} \cos\theta \qquad \text{(watts)} \qquad\qquad (3.19)$$

As indicated by Fig. 3.17, V_{eff} is the voltage across the elements or network to which the power is to be determined, while I_{eff} is the current drawn by the network.

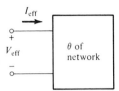

Figure 3.17 Power in an ac network.

The angle θ is the phase angle between V_{eff} and I_{eff}. For a pure resistor in the container, we found that v and i were *in phase* and that $\theta = 0°$. Substituting into the equation will result in $P = VI \cos \theta = VI(1) = VI$ (from this point on the symbols V and I will refer to V_{eff} and I_{eff}, respectively), as indicated earlier. For the pure inductor $\theta = 90°$ and $P = VI \cos 90° = VI(0) = 0$ W, indicating, as stated earlier, that the pure inductor does not dissipate electrical energy but simply stores it in the form of a magnetic field.

For the pure capacitor, the reactance is determined by

$$X_C = \frac{1}{\omega C} = \frac{1}{2\pi f C} \quad \text{(ohms)} \tag{3.20}$$

indicating that as the frequency increases the reactance of a capacitor decreases (opposite to that of the inductor). In addition, for $f = 0$, corresponding with the dc condition, $X_C = 1/0C = \infty \, \Omega =$ very large value corresponding with an open-circuit equivalent as described for dc networks.

A sinusoidal voltage has been applied across a 10-μF capacitor in Fig. 3.18.

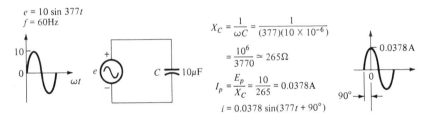

Figure 3.18 Simple capacitive ac network.

The reactance X_C as shown in the figure is 265 Ω, and the peak value of the current can be determined through a simple application of Ohm's law. That is,

$$I_p = \frac{V_p}{X_C} = \frac{10}{265} = 0.0378 = \underline{\textbf{37.8 mA}}$$

as shown in Fig. 3.18. Note in this case that a 90° phase shift has been introduced but that i_C *leads* v_C by 90°, which is the reverse of that encountered for the inductor. Substituting into the general power equation,

$$P_C = VI \cos \theta = VI \cos 90° = VI(0) = 0 \text{ W}$$

substantiating the previous comments for the capacitor also.

The factor, $\cos \theta$, in the general power equation is called the *power factor* of the network and has the symbol appearing in Eq. (3.21):

$$\text{Power factor} = \cos \theta = F_p \qquad (3.21)$$

Its largest value is 1, which results when the network appears to be purely resistive at its input terminals and $\theta = 0°$. The smallest value is zero as obtained from a network whose terminal characteristics are purely reactive (capacitive or inductive). For a network composed of resistors and reactive elements the power factor can vary from 0 to 1. For more reactive loads the power factor will drift toward zero, while for more resistive loads it will drift toward 1.

The example to follow will demonstrate the variation in solution resulting from an initial phase shift in the applied voltage or current.

EXAMPLE 3.3

The current i_L through a 10-mH inductor is $5 \sin(200t + 30°)$. Find the voltage v_L across the inductor.

Solution:

The peak value of the voltage is determined by Ohm's law:

$$X_L = \omega L = (200)(10 \times 10^{-3}) = 2 \ \Omega$$

and

$$V_p = I_p X_L = (5)(2) = 10 \text{ V}$$

Since v_L always leads i_L by 90°, an additional 90° must be added to the 30° already present in the sinusoidal expression for the current. Therefore,

$$\mathbf{v}_L = \mathbf{10 \sin(200t + 120°)}$$

The curves for v_L and i_L appear in Fig. 3.19, clearly indicating the 90° phase shift between the voltage and current.

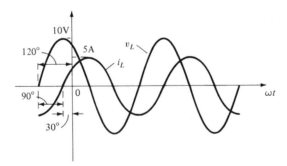

Figure 3.19 v_L and i_L for Example 3.3.

3.6 PHASORS AND COMPLEX NUMBERS

For single-element networks the proper phase angle can be determined with a limited amount of effort. For complex networks, however, this could become a serious difficulty if it were not for a technique to be introduced in this section. It incorporates the use of vectors, which can be used to represent ac voltages or currents and element reactances. Recall that a *vector* is a quantity that has both *magnitude* and *direction*. In Fig. 3.20 we have the vector representation of resistance, inductive reactance, and capacitive reactance. The angle shown with each is determined by the phase shift introduced between the voltage and current for each element. For the resistor, you will recall that the voltage and current are in phase; hence there is no phase shift,

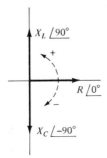

Figure 3.20 Impedance diagram.

and the associated angle is 0°. Since the angle is measured from the right horizontal axis, the resistance vector appears right on the axis. Its length is determined by the magnitude of the resistance R. For X_L and X_C the angle included is the angle by which the voltage across the element will lead the current through the component. For X_L it is +90°, and for X_C it is −90°, since the current i_C leads v_C for a capacitor by 90°. The magnitudes (or lengths) of the vectors is determined by the reactance of each element. Note that the angle is always measured from the same axis. Any one or combination of the resistive or reactive elements of Fig. 3.20 is called an *impedance* and given the symbol $\mathbf{Z}$. It is a measure of the ability of the ac network to *impede* the flow of charge or current through the network. The diagram of Fig. 3.20 with one or any number of the elements is called an *impedance diagram*. Only resistances and reactances appear on an impedance diagram. Voltages and currents appear on a *phasor* diagram such as shown in Fig. 3.21 for each element. The angle associated with each quantity is the phase angle appearing in the sinusoidal time domain. The magnitude is the effective value of the sinusoidal quantity. Each magnitude and associated angle is called a *phasor*. The phasor diagram for the pure resistor indicates

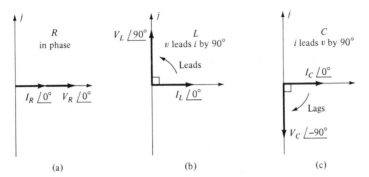

Figure 3.21 Phasor diagrams (a) pure resistor (b) pure coil (c) pure capacitor.

quite clearly that v_R and i_R in phase since they have the same angle and therefore direction. For the inductor, the voltage phasor appears 90° ahead of the current phasor, indicating that v_L leads i_L by 90°. The leading and lagging directions are indicated on the same figure. For the capacitor the reverse is true: v_C lags i_C by 90°, as shown in the figure. Each quantity on a phasor diagram relates directly to a sinusoidal function. Note the absence of resistive and reactive elements, which are reserved for the impedance diagram.

If the phasor or vector notation introduced is to be of any assistance in our future analysis, we must be able to perform certain basic mathematical operations with the phasors or vectors.

A vector as shown in Fig. 3.22 can be determined (or represented) by either providing its magnitude and angle measured from the positive (+) (right) horizontal axis or providing the horizontal and vertical component of the vector. The former is called the *polar* form, while the latter is called the *rectangular* form. The equations necessary to convert from one form to the other are provided below:

$$\boxed{\begin{aligned} a &= c \cos \theta \\ b &= c \sin \theta \end{aligned}} \qquad \boxed{\begin{aligned} c &= \sqrt{a^2 + b^2} \\ \theta &= \tan^{-1} \frac{a}{b} \end{aligned}} \tag{3.22}$$

polar $\longrightarrow$ rectangular $\qquad$ rectangular $\longrightarrow$ polar

The letter j is included in the rectangular form to distinguish between the real (horizontal) and imaginary (vertical) components of the rectangular form. The terms real and imaginary are related solely to mathematical definition and do not require further discussion here to continue with the material. To perform the mathematical operations to be described the letter j defined mathematically by $\sqrt{-1}$ must be examined in its various forms. That is,

$$\begin{aligned} j &= \sqrt{-1} \\ \text{so that} \quad j^2 &= (\sqrt{-1})^2 = -1 \\ \text{and} \quad j^3 &= j^2 j^1 = (-1)(\sqrt{-1}) = -\sqrt{-1} = -j \\ \text{with} \quad j^4 &= j^2 j^2 = (-1)(-1) = +1 \\ &\qquad \vdots \end{aligned} \tag{3.23}$$

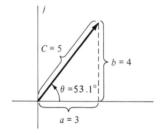

$C = 5$

$b = 4$

$\theta = 53.1°$

$a = 3$

Figure 3.22 Vector conversion.

Although all four operations of addition, subtraction, multiplication, and division can be performed using the rectangular form, only the operations of addition and subtraction will be described since they are accomplished most directly. The multiplication and division operations will be described in the polar form for the same reason. This will avoid a measure of mathematical confusion caused by considering all possibilities. If both quantities to be added appear in the polar form, they will simply have to be converted to the rectangular form before the operation can be performed.

First, let us examine a few descriptive examples of conversion.

EXAMPLE 3.4

Convert the following from polar to rectangular form:

1. $10 \underline{/53.1°}$:
$$a = 10 \cos 53.1° = 10(0.6) = 6$$
$$b = 10 \sin 53.1° = 10(0.8) = 8$$
and
$$10\underline{/53.1°} = 6 + j8$$

2. $16 \underline{/-30°}$:
$$a = 16 \cos 30° = 16(0.866) = 13.8$$
$$b = 16 \sin 30° = 16(0.5) = 8$$
and
$$16 \underline{/-30°} = 13.8 - j8$$

Convert the following from rectangular to polar form:

1. $30 + j40$:
$$c = \sqrt{(30)^2 + (40)^2} = 50$$
$$\theta = \tan^{-1} \tfrac{40}{30} = \tan^{-1} 1.33 = 53.1°$$
and
$$30 + j40 = 50 \underline{/53.1°}$$

2. $4 - j20$:
$$c = \sqrt{(4)^2 + (20)^2} = 20.4$$
$$\theta = \tan^{-1} \tfrac{20}{4} = 78.7°$$
and
$$4 - j20 = 20.4 \underline{/-78.7°}$$

Addition (Rectangular Form)

In rectangular form, addition is carried out by simply algebraically (paying heed to the signs of the quantities to be added) adding the real and imaginary components independently. That is,

$$(a_1 + jb_1) + (a_2 + jb_2) = (a_1 + a_2) + j(b_1 + b_2) \qquad (3.24)$$

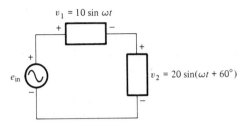

Figure 3.23 Series ac circuit.

Consider the following addition of sinusoidal voltages (Fig. 3.23) necessary to determine the applied sinusoidal voltage e_{in}. Applying Kirchhoff's voltage law,

$$e_{in} = v_1 + v_2$$

In phasor form,

$$V_1 = 0.707(10) \underline{/0°} = 7.07 \underline{/0°}$$
$$V_2 = 0.707(20) \underline{/60°} = 14.14 \underline{/60°}$$

Converting to rectangular form for addition,

$$V_1 = 7.07 + j0$$
$$V_2 = 14.14 \cos 60° + j14.14 \sin 60°$$
$$= 14.14(0.5) + j14.14(0.866)$$
$$= 7.07 + j12.25$$

and

$$E_{in} = V_1 + V_2 = [7.07 + j0] + [7.07 + j12.25]$$
$$= [7.07 + 7.07] + j[0 + 12.25]$$
$$= 14.14 + j12.25$$

For polar form,

$$|E_{in}| = \sqrt{(14.14)^2 + (12.25)^2} \simeq 18.7$$
$$\theta = \tan^{-1}\frac{12.25}{14.14} = \tan^{-1}0.866 = 40.9°$$

and

$$E_{in} = \mathbf{18.7} \underline{\mathbf{/40.9°}}$$

In the sinusoidal time domain,

$$e_{in} = \sqrt{2}(18.7) \sin(\omega t + 40.9°)$$
$$= \mathbf{26.44 \sin(\omega t + 40.9°)}$$

Now you might ask at this point if vector algebra is of any use whatsoever if such a lengthy series of calculations were required to simply add to sinusoidal functions. However, what are the alternatives? Certainly each function could be carefully graphed on the same axis in the proper phase relationship and the two waveforms added on a point-by-point basis. This would obviously prove to be a very lengthy procedure, and the accuracy would be totally dependent on the care with which the network were handled. Admittedly, the introduction to vector algebra and its application to electrical systems has been brief and is still somewhat vague. However, in the analysis to follow there will be continual application of this technique, and in time the confusion should vanish.

If both phasors were drawn on the same set of axes as shown in Fig. 3.24, the resultant could also be obtained by vector addition. However, this still requires the use of the phasor domain and a measure of artwork.

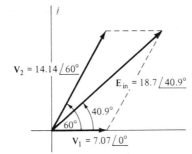

Figure 3.24 Phasor algebra solution for e_{in}.

Subtraction (Rectangular Form)

As with addition, subtraction requires that the real part be treated separately from the imaginary component. That is, in equation form,

$$(a_1 + jb_1) - (a_2 + jb_2) = (a_1 - a_2) + j(b_1 - b_2) \qquad (3.25)$$

EXAMPLE 3.5

Determine the current I_1, for the network of Fig. 3.25 in rectangular form.

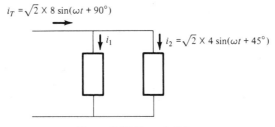

Figure 3.25 Example 3.5.

Solution:

Applying Kirchhoff's current law,

$$I_T = I_1 + I_2$$

and

$$I_1 = I_T - I_2$$
$$= 8 \underline{/90°} - 4 \underline{/45°}$$
$$= [0 + j8] - [4 \cos 45° + j4 \sin 45°]$$

$$= [0 + j8] - [2.828 + j2.828]$$
$$= (0 - 2.828) + j(8 - 2.828)$$
$$I_1 = -2.828 + j5.172$$

Note in Fig. 3.26 the location of the current vector. The real part is negative and therefore to the left of the vertical axis, while the imaginary part is positive and above the horizontal axis. In other words, it is in the third quadrant. In Fig. 3.26 the result is obtained through a vector addition.

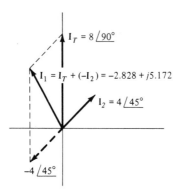

Figure 3.26 Vector solution for Example 3.5.

Multiplication and Division (Polar Form)

The following equations are applied for the indicated operations in the polar form:

$$\boxed{(c_1 \underline{/\theta_1})(c_2 \underline{/\theta_2}) = c_1 c_2 \underline{/\theta_1 + \theta_2}} \qquad (3.26)$$

$$\boxed{\frac{c_1 \underline{/\theta_1}}{c_2 \underline{/\theta_2}} = \frac{c_1}{c_2} \underline{/\theta_1 - \theta_2}} \qquad (3.27)$$

If rectangular forms are encountered, it will simply be necessary to first convert to the polar form before performing the operation. Of course, if a phasor is simply multiplied by or divided by a constant, the operation is quite direct. For example,

$$\frac{5 + j6}{2} = 2.5 + j3$$

and

$$6(2 - j4) = 12 - j24$$

EXAMPLE 3.6

Determine the result of the following operations:

1. $(10 \underline{/60°})(6 \underline{/-20°}) = (10)(6) \underline{/60° + (-20°)} = \mathbf{60 \underline{/40°}}$.
2. $(0.4 \underline{/-30°})(600 \underline{/60°}) = (0.4)(600) \underline{/-30° + 60°} = \mathbf{240 \underline{/+30°}}$.
3. $(40 \underline{/20°})/(5 \underline{/-30°}) = \frac{40}{5} \underline{/20° - (-30°)} = \mathbf{8 \underline{/50°}}$.

Let us now apply the phasor algebra to the basic *RLC* elements and note the effect of associating an angle with the resistive and reactive elements. Recall the following from Fig. 3.21:

$$\boxed{\begin{aligned} R &= R\underline{/0^\circ} \\ X_L &= X_L\underline{/90^\circ} \\ X_C &= X_C\underline{/-90^\circ} \end{aligned}} \tag{3.28}$$

By associating an angle with each impedance vector, it will no longer be necessary to remember the phase relationships between the currents and voltages of each element.

Consider the resistor of Fig. 3.27 for which the voltage is provided and the current is to be determined:

$$I = \frac{V}{R} = \frac{V\underline{/\theta}}{R\underline{/0^\circ}} = \frac{V}{R}\underline{/\theta - 0^\circ} = \frac{V}{R}\underline{/\theta}$$

Note that V and I are in phase since they have the same angle θ. Associating 0° with the resistance permitted the use of phasor algebra and maintained the proper phase relationships between the current and voltage.

For the inductor of Fig. 3.28,

$$I = \frac{V}{X_L} = \frac{V\underline{/\theta}}{X_L\underline{/90^\circ}} = \frac{V}{X_L}\underline{/\theta - 90^\circ}$$

The solution indicates that I lags the voltage by 90° and its magnitude (effective value) is V/X_L. Note the proper phase shift in the waveforms of Fig. 3.28.

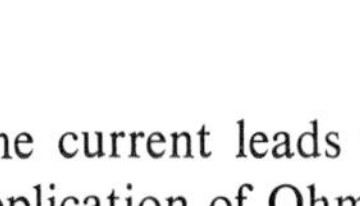

Figure 3.27 Pure resistor. 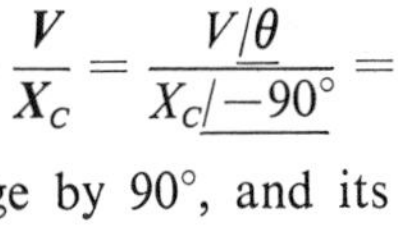   Figure 3.28 Pure inductor. 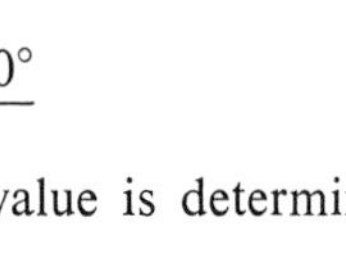   Figure 3.29 Pure capacitor.

And finally for the capacitor of Fig. 3.29,

$$I = \frac{V}{X_C} = \frac{V\underline{/\theta}}{X_C\underline{/-90^\circ}} = \frac{V}{X_C}\underline{/\theta + 90^\circ}$$

The current leads the voltage by 90°, and its effective value is determined by an application of Ohm's law.

A few examples will clarify the use of the phasor notation.

EXAMPLE 3.7

Determine the current through a 20-Ω resistor if the voltage across the resistor is $40\sin(200t + 20^\circ)$.

Solution:

In phasor notation,

$$V = (0.707)(40)\underline{/20^\circ} = 28.28\underline{/20^\circ}$$

Ohm's law:

$$I = \frac{V}{R} = \frac{28.28\ /20°}{20\ /0°} = 1.414\ /20°$$

In the time domain,

$$i = (\sqrt{2})(1.414)\sin(200t + 20°)$$
$$= 2\sin(200t + 20°)$$

EXAMPLE 3.8

Determine the voltage across a 20-mH coil if the current through the coil is $10 \times 10^{-3}\sin(500\ t + 60°)$.

Solution:

$$X_L = \omega L = (500)(20 \times 10^{-3}) = 10,000 \times 10^{-3} = 10\ \Omega$$

In phasor notation,

$$I = (0.707)(10 \times 10^{-3})\ /60° = 7.07 \times 10^{-3}\ /60°$$

Ohm's law:

$$V = IX_L = (7.07 \times 10^{-3}\ /60°)(10\ /90°) = 70.7 \times 10^{-3}\ /150°$$

In the time domain,

$$v = (\sqrt{2})(70.7 \times 10^{-3})\sin(500\ t + 150°)$$
$$= 100 \times 10^{-3}\sin(500\ t + 150°)$$
$$= \textbf{0.1 sin (500}\textit{t} \textbf{+ 150°)}$$

The waveforms appear in Fig. 3.30.

Note how nicely the impedance notation has carried along the proper phase shift.

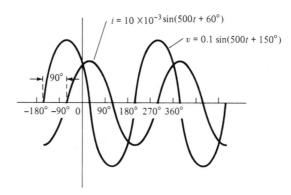

Figure 3.30 Solution for Example 3.8.

EXAMPLE 3.9

Determine the current through a 5-μF capacitor if the voltage across the capacitor is 20 sin 377t.

Solution:

$$X_C = \frac{1}{\omega C} = \frac{1}{(377)(5 \times 10^{-6})} = \frac{10^6}{1885} = 5.305 \times 10^2 = 530.5\ \Omega$$

Phasor notation:

$$V = (0.707)(20) \underline{/0°} = 14.14 \underline{/0°}$$

Ohm's law:

$$I = \frac{V}{X_C} = \frac{14.14 \underline{/0°}}{530.5 \underline{/-90°}} = 0.0267 \underline{/90°} = 26.7 \times 10^{-3} \underline{/90°}$$

In the time domain,

$$i_C = (\sqrt{2})(26.7 \times 10^{-3}) \sin (377\,t + 90°)$$
$$= \mathbf{37.75 \times 10^{-3} \sin (377t + 90°)}$$

Many texts written for this area of interest totally avoid the introduction to vector algebra that we have covered here. Rather they simply state the solution for very particular situations and assume that other possibilities will not be encountered. However, it is a continually applied technique in the field, and one should be somewhat familiar with the notation and technique. For those who fully understand the examples above there is no limitation with regard to the type of single-element networks he may encounter. Consider also, how would one perform the necessary task of simply adding or subtracting two sinusoidal waveforms? For the reasons above, the authors felt that a few pages of introduction to phasor algebra would be in the best interests of the reader. Incidentally, now that the technique has been introduced the material to be considered now will follow a more logical sequence, and many unanswered questions resulting from a "given solution" technique will not result.

3.7 SERIES ac NETWORKS

In this section we shall demonstrate that the analysis of series ac networks is very similar to that for series dc networks. It is simply necessary to use the vector notation for each quantity rather than just the magnitude as employed for dc networks.

For a series ac network, the current is the same through each element, and the total impedance is the vector sum of the impedances of the series elements. That is,

$$\boxed{Z_T = Z_1 + Z_2 + Z_3 + \cdots + Z_N} \tag{3.29}$$

Consider the series RL network of Fig. 3.31. The reactance of the inductor is

$$X_L = \omega L = (2\pi f)(L) = (377)(10.61 \times 10^{-3})$$
$$= 4\,\Omega$$

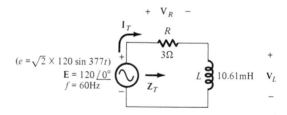

Figure 3.31 Series R–L network.

For more complex networks, it is usually quite helpful to find the solutions for a network in terms of block impedances as shown in Fig. 3.32 before substituting magnitudes and angles. Less errors are usually the result of this type of approach. There is also a more direct relationship with the analysis of dc networks.

If Z_1 and Z_2 in Fig. 3.32 were resistors R_1 and R_2, respectively, in a dc network, the total resistance would simply be the sum of the two: $R_T = R_1 + R_2$. Treating Z_1 and Z_2 "as if" they were pure resistors would result in

$$Z_T = Z_1 + Z_2$$

Substituting,

$$Z_T = (3 + j0) + (0 + j4)$$
$$= 3 + j4 = 5\,\underline{/53.1°}$$

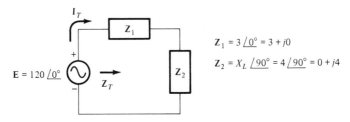

$$Z_1 = 3\,\underline{/0°} = 3 + j0$$
$$Z_2 = X_L\,\underline{/90°} = 4\,\underline{/90°} = 0 + j4$$

$E = 120\,\underline{/0°}$

Figure 3.32

The impedance diagram of Fig. 3.33 clearly reveals that the total impedance Z_T can also be determined graphically through a simple vector addition. Ohm's law:

$$I_T = \frac{E}{Z_T} = \frac{120/0°}{5/53.1°} = 24\underline{/-53.1°}$$

which in the time domain is the following:

$$i_T = \sqrt{2}(24)\sin(\omega t - 53.1°)$$
$$= 33.94\sin(\omega t - 53.1°)$$
$$V_R = I_T Z_1 = (24\,\underline{/-53.1°})(3\,\underline{/0°}) = 72\,\underline{/-53.1°}$$

which in the time domain would be

$$v_R = \sqrt{2}(72)\sin(\omega t - 53.1°)$$
$$= 101.81\sin(\omega t - 53.1°)$$

Note that v_R and i_R are *in phase*,

$$V_L = I_T Z_2 = (24\,\underline{/-53.1°})(4\,\underline{/90°}) = 96\,\underline{/36.9°}$$

which in the time domain is the following:

$$v_L = \sqrt{2}(96)\sin(\omega t + 36.9°)$$
$$= 135.74\sin(\omega t + 36.9°)$$

A phasor diagram of the voltages and current appears in Fig. 3.34. Note that

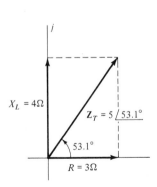

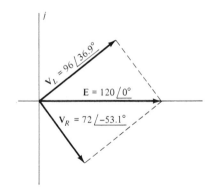

Figure 3.33 Impedance diagram for a series R–L network.

Figure 3.34 Phasor diagram for series R–L network.

the applied voltage E is the *vector* sum of V_R and V_L as determined by Kirchhoff's voltage law:

$$E = V_R + V_L$$

The *voltage-divider rule* is applied in exactly the same manner as described for dc networks. In this example, if we were interested in V_R, the following would result using the block impedances:

$$V_R = \frac{Z_1(E)}{Z_1 + Z_2} = \frac{(3\underline{/0°})(120\underline{/0°})}{3 + j4} = \frac{360\underline{/0°}}{5\underline{/53.1°}}$$

$$= 72\underline{/-53.1°}$$

as obtained before. Compare the above to an application of the rule to a dc network with two series resistors.

Before continuing with the analysis of this network, take a moment to carefully examine the waveforms of the voltages and current plotted on the same axis in Fig. 3.35. Note especially that v_R and i_R are in phase and that v_L leads i_L by 90°. Consider also that since the network is *inductive*, the input current also lags the applied voltage by an angle of 53.1°. The more inductive the network (or the less resistive), the larger the angle by which i_T would lag the applied voltage e. At any point on the axis, the instantaneous values of e, v_R, and v_L will satisfy Kirchhoff's voltage law. For instance, at $t = 0$ or $\theta = 0°$, $e = 0$, and

$$e = v_R + v_L$$

becomes

$$0 = v_R + v_L$$

and

$$v_R = v_L$$

which is reflected at this point in Fig. 3.35.

The power to the network can be determined using any one of a number of equations such as listed on page 95:

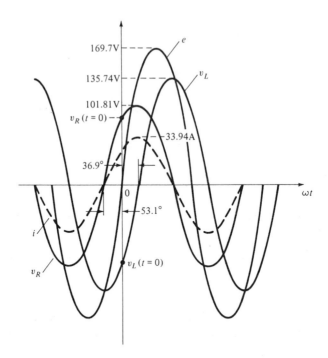

Figure 3.35 Waveforms for series R–L network.

$$P = EI_T \cos \theta_T = I_T^2 R = V_R I_R = \frac{V_R^2}{R} \qquad \text{(watts)} \qquad (3.30)$$

In each case, the same result will be obtained, as shown below for two of the equations. Keep in mind that power is only delivered to (or dissipated by) the resistive elements.

$$P = EI_T \cos \theta_T = (120)(24) \cos 53.1°$$

where θ_T is the phase angle between the applied voltage and current drawn by the network

$$= (120)(24)(0.6)$$

$$= \textbf{1728 W}$$

and

$$P = I_T^2 R = (24)^2 \times 3 = \textbf{1728 W}$$

The power factor of the network is

$$F_p = \cos \theta_T = \textbf{0.6}$$

indicating that it is far from purely resistive ($F_p = 1$) yet it is certainly not purely reactive ($F_p = 0$). For networks (such as this one) where the applied voltage leads the source current the networks are said to have a *lagging* F_p to indicate the inductive characteristics. For capacitive networks the label *leading* F_p is applied. For the network just investigated a lagging power factor of 0.6 is said to exist.

For series circuits the network power factor can also be determined by the following ratio:

$$F_p = \frac{R}{Z_T}$$

(3.31)

which for this example is

$$F_p = \tfrac{3}{5} = 0.6$$

Occasionally, in practical applications, a network is designed so that for a particular frequency range the inductive reactance is much greater than the impedance of any elements in series with it. In this example, for instance, if $X_L = 400\ \Omega$ and $R = 3\ \Omega$, the total impedance is

$$Z_T = \sqrt{R^2 + X_L^2} = \sqrt{(3)^2 + (400)^2} \simeq 400$$

and the network from a practical viewpoint is purely inductive. In addition,

$$F_p = \frac{R}{Z} \simeq \frac{3}{400} = 0.0075 \simeq 0$$

If time and space allocation would permit, a simple RC series network should now be examined. However, this exercise has been left to the Problems at the end of the chapter. The only major differences will appear in the impedance diagram when the capacitive reactance vector appears below the axis and in the phasor diagram where the series current will lead the capacitive voltage by 90° and the applied voltage by a number of degrees determined by the network.

Let us now go a step further and examine the series RLC network of Fig. 3.36. Replacing each element by a block impedance will result in the configuration of Fig. 3.37. The total impedance is the vector sum, and

$$\begin{aligned} Z_T = Z_1 + Z_2 + Z_3 &= (5\ K + j0) + (0 + j4\ K) + (0 - j16\ K) \\ &= 5\ K + j4\ K - j16\ K \\ &= 5\ K - j12\ K \\ &= 13\ K\ \underline{/-67.4°} \end{aligned}$$

The impedance diagram appears in Fig. 3.38. Note that the capacitive and inductive reactances are directly opposing and the difference is the net reactance of the network. The current I_T is

$$I_T = \frac{E}{Z_T} = \frac{60\ \underline{/0°}}{13\ K\ \underline{/-67.4°}} = 4.615 \times 10^{-3}\ \underline{/67.4°}$$

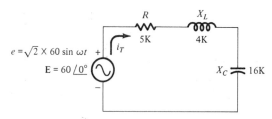

Figure 3.36 Series R–L–C network.

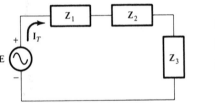

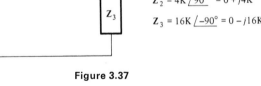

$Z_1 = 5K \underline{/0°} = 5K + j0$

$Z_2 = 4K \underline{/90°} = 0 + j4K$

$Z_3 = 16K \underline{/-90°} = 0 - j16K$

Figure 3.37

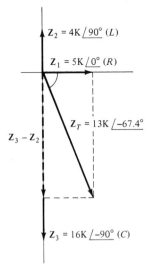

Figure 3.38 Impedance diagram for a series R–L–C network.

which in the time domain is the following:

$$i_T = \sqrt{2}(4.615 \times 10^{-3}) \sin(\omega t + 67.4°)$$
$$= 6.52 \times 10^{-3} \sin(\omega t + 67.4°)$$

The voltage across each element can then be obtained directly using Ohm's law:

$$V_R = I_T R = (4.615 \times 10^{-3} \underline{/67.4°})(5 \text{ K } \underline{/0°})$$
$$= 23.075 \underline{/67.4°}$$
$$V_L = I_T X_L = (4.615 \times 10^{-3} \underline{/67.4°})(4 \text{ K } \underline{/90°})$$
$$= 18.46 \underline{/157.4°}$$
$$V_C = I_T X_C = (4.615 \times 10^{-3} \underline{/67.4°})(16 \text{ K } \underline{/-90°})$$
$$= 73.84 \underline{/-22.6°}$$

The phasor diagram of the voltages and current appears in Fig. 3.39. Note that, like the reactances, the voltages V_L and V_C are in vector opposition and that the series current I_T lags the voltage V_L by 90°, leads the voltage V_C by 90°, and is in phase with V_R.

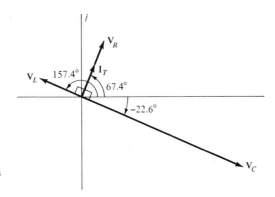

Figure 3.39 Phasor diagram for a series
R–L–C network.

Any of the voltages found above could also be obtained using the voltage divider
rule, which would remove the necessity to find I_T first. For V_L,

$$V_L = \frac{Z_2(E)}{Z_1 + Z_2 + Z_3} = \frac{(4 \text{ K }/90°)(60 /0°)}{13 /-67.4°} = \frac{240 \text{ K }/90°}{13 /-67.4°}$$

$$= 18.46 /157.4°$$

In the time domain,

$$v_L = \sqrt{2}\,(18.46)\sin(\omega t + 157.4°)$$

$$= 26.2 \sin(\omega t + 157.4°)$$

For the power to the network,

$$P = EI_T \cos\theta$$

$$= (60)(4.615 \times 10^{-3})\cos 67.4°$$

$$= (276.9 \times 10^{-3})(0.3843)$$

$$= 106.5 \text{ mW}$$

or

$$P = (I_T)^2\, R$$

$$= (4.615 \times 10^{-3})^2(5 \times 10^3)$$

$$= (21.3 \times 10^{-6})(5 \times 10^3)$$

$$= 106.5 \text{ mW}$$

The network power factor:

$$F_p = \cos\theta = \cos(67.4°) = 0.3843 \text{ leading}$$

or

$$F_p = \frac{R}{Z} = \frac{5 \text{ K}}{13 \text{ K}} = 0.3843 \text{ leading}$$

The low power factor indicates that the terminal characteristics reflect a highly
reactive (leading) network.

3.8 PARALLEL ac NETWORKS

The analysis of parallel ac networks will also be very similar to that introduced for dc networks. The reciprocal of impedance, called *admittance*, is defined by the following equation and, as indicated, is measured in *mhos*:

$$Y = \frac{1}{Z} \qquad \text{(mhos)} \qquad (3.32)$$

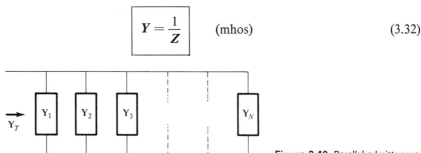

Figure 3.40 Parallel admittances.

For parallel ac networks such as appearing in Fig. 3.40 the total admittance is determined by

$$Y_T = Y_1 + Y_2 + Y_3 + \cdots + Y_N \qquad (3.33)$$

or

$$\frac{1}{Z_T} = \frac{1}{Z_1} + \frac{1}{Z_2} + \frac{1}{Z_3} + \cdots + \frac{1}{Z_N} \qquad (3.34)$$

which for two parallel impedances becomes

$$Z_T = \frac{Z_1 Z_2}{Z_1 + Z_2} \qquad (3.35)$$

The *voltage is the same across parallel ac branches*, and the total input current can be determined using Kirchhoff's current law or first finding the total input impedance (or admittance) and then applying Ohm's law.

The reciprocal of resistance in an ac network is called conductance (as for dc networks) and has the angle 0° associated with it as shown:

$$G = G \underline{/0°} = \frac{1}{R \underline{/0°}} \qquad \text{(mhos)} \qquad (3.36)$$

The reciprocal of reactance is called *susceptance* and is measured in mhos. For each element it has the notation and angle appearing in Eqs. (3.37) and (3.38). The term susceptance comes from the word susceptible since one over the reactance is a measure of how susceptible the branch or element is to the flow of charge through it. An increase in its value is parallel with a rise in the branch current.

$$\boxed{B_L = B_L \underline{/-90°} = \frac{1}{X_L \underline{/90°}}} \quad \text{(mhos)} \qquad (3.37)$$

$$\boxed{B_C = B_C \underline{/90°} = \frac{1}{X_C \underline{/-90°}}} \quad \text{(mhos)} \qquad (3.38)$$

An *admittance diagram* is defined as shown in Fig. 3.41 for a parallel *RLC* network.

Consider the parallel *RL* network of Fig. 3.42. Inserting block impedances will result in the configuration of Fig. 3.43. The total admittance and impedance can be determined as follows:

$$Y_1 = G_1 = \frac{1}{R_1} = \frac{1}{R\underline{/0°}} = \frac{1}{3\,\text{K}\underline{/0°}} = 0.333 \times 10^{-3} \underline{/0°}$$

$$Y_2 = B_L = \frac{1}{X_L} = \frac{1}{X_L\underline{/90°}} = \frac{1}{4\,\text{K}\underline{/90°}} = 0.25 \times 10^{-3} \underline{/-90°}$$

and

$$Y_T = Y_1 + Y_2$$
$$= 0.333 \times 10^{-3}\underline{/0°} + 0.25 \times 10^{-3}\underline{/-90°}$$
$$= 0.333 \times 10^{-3} - j0.25 \times 10^{-3}$$
$$= \mathbf{0.416 \times 10^{-3}\underline{/-36.9°}}$$

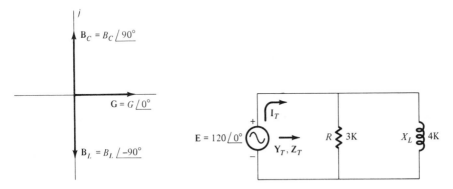

Figure 3.41 Admittance diagram.   **Figure 3.42** Parallel R–L network.

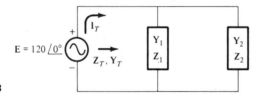

Figure 3.43

and

$$Z_T = \frac{1}{Y_T} = \frac{1}{0.416 \times 10^{-3} \underline{/-36.9°}} = 2.4\ \text{K}\ \underline{/36.9°}$$

or

$$Z_T = \frac{(Z_1)(Z_2)}{Z_1 + Z_2} = \frac{(3\ \text{K}\ \underline{/0°})(4\ \text{K}\ \underline{/90°})}{3\ \text{K} + j4\ \text{K}} = \frac{12\ \text{K}^2\ \underline{/90°}}{5\ \text{K}\ \underline{/53.1°}} = 2.4\ \text{K}\ \underline{/36.9°}$$

The admittance diagram is provided in Fig. 3.44.

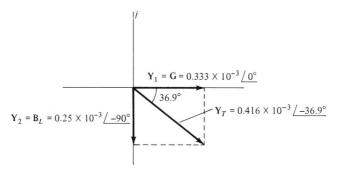

Figure 3.44 Admittance diagram for a parallel R–L network.

Note again that the total admittance can be determined using relatively simple vector algebra. The current I_T is

$$I_T = \frac{E}{Z_T} = \frac{120\ \underline{/0°}}{2.4\ \text{K}\ \underline{/36.9°}} = 50 \times 10^{-3}\ \underline{/-36.9°}$$

or

$$I_T = \frac{E}{Z_T} = E(Y_T) = (120\ \underline{/0°})(0.416 \times 10^{-3}\ \underline{/-36.9°})$$

$$= 50 \times 10^{-3}\ \underline{/-36.9°}$$

The current through either element can then be determined by Ohm's law:

$$I_R = EG = \frac{E}{R} = \frac{120\ \underline{/0°}}{3\ \text{K}\ \underline{/0°}} = 40 \times 10^{-3}\ \underline{/0°}$$

and

$$I_L = EB_L = \frac{E}{X_L} = \frac{120\ \underline{/0°}}{4\ \text{K}\ \underline{/90°}} = 30 \times 10^{-3}\ \underline{/-90°}$$

A phasor diagram of the currents and voltage can then be drawn as shown in Fig. 3.45. Note that I_R is in phase with E and I_L lags E by 90°. The diagram itself demonstrates the validity of Kirchhoff's current law as a method for determining I_T from

$$I_T = I_1 + I_2$$

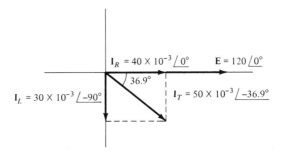

Figure 3.45 Phasor diagram for parallel
R–L network.

The power to the network can be determined using the same equation applied
to series ac networks. That is,

$$P = EI_T \cos \theta_T = I_R^2 R = \frac{V_R^2}{R} = V_R I_R \qquad (3.39)$$

where all voltages and currents are effective values.

For this example,

$$P = EI_T \cos \theta_T = (120)(50 \times 10^{-3}) \cos 36.9° = (6)(0.7997)$$
$$= \mathbf{4.8\ W}$$

or

$$P = \frac{V_R^2}{R} = \frac{E^2}{R} = \frac{(120)^2}{3\ K} = \mathbf{4.8\ W}$$

The power factor for parallel ac networks can be determined using the following
equation:

$$F_p = \cos \theta_T = \frac{G}{Y_T} \qquad (3.40)$$

which for this example results in

$$F_p = \cos \theta_T = \cos 36.9° = \mathbf{0.7997\ lagging}$$

or

$$F_p = \frac{G}{Y_T} = \frac{0.333 \times 10^{-3}}{0.416 \times 10^{-3}} = \mathbf{0.7997\ lagging}$$

The term lagging reflects the fact that the input voltage leads the input current
by a number of degrees—in other words, an *inductive* network.

It should, by now, be somewhat obvious that the analysis of dc and ac networks
is quite similar once the vector notation is introduced. This will continue to be true
as we progress through the text.

A final example in parallel ac networks will be the detailed examination of the
RLC network of Fig. 3.46.

Applying the block admittance labels will result in the configuration of Fig.
3.47, where

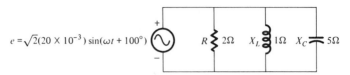

$$e = \sqrt{2}(20 \times 10^{-3}) \sin(\omega t + 100°)$$ $R \lessgtr 2\Omega$ $X_L \lessgtr 1\Omega$ $X_C \overset{\perp}{\top} 5\Omega$

Figure 3.46 Parallel R–L–C network.

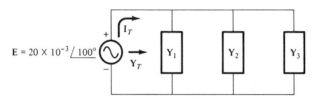

$E = 20 \times 10^{-3} \underline{/100°}$ Y_1 Y_2 Y_3

Figure 3.47

$$Y_1 = \frac{1}{Z_1} = \frac{1}{2\underline{/0°}} = 0.5\underline{/0°} = G$$

$$Y_2 = \frac{1}{Z_2} = \frac{1}{1\underline{/90°}} = 1\underline{/-90°} = B_L$$

$$Y_3 = \frac{1}{Z_3} = \frac{1}{5\underline{/-90°}} = 0.2\underline{/90°} = B_C$$

The total admittance is

$$Y_T = Y_1 + Y_2 + Y_3$$
$$= (0.5 + j0) + (0 - j1) + (0 + j0.2)$$
$$= 0.5 + j(-1 + 0.2)$$
$$= 0.5 - j0.8$$
$$= \mathbf{0.943\underline{/-58°}}$$

and

$$Z_T = \frac{1}{Y_T} = \frac{1}{0.943\underline{/-58°}} = \mathbf{1.06\underline{/58°}}$$

or combining two parallel elements at a time,

$$Z'_T = Z_1 \| Z_2 = \frac{Z_1 Z_2}{Z_1 + Z_2}$$

and

$$Z_T = Z'_T \| Z_3 = \frac{Z'_T Z_3}{Z'_T + Z_3} = \mathbf{1.06\underline{/58°}}$$

The admittance diagram appears in Fig. 3.48.

Note that the susceptances are in direct opposition. Since the inductive susceptance is the larger, the network will have lagging characteristics. That is, the applied
voltage E will lead the input current I_T by a number of degrees determined by the
driving point impedance of the network, Z_T.

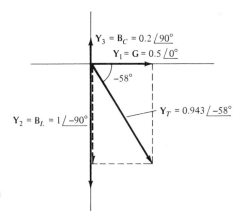

Figure 3.48 Admittance diagram for a parallel R–L–C network.

The input current can again be determined by Ohm's law:

$$I_T = \frac{E}{Z_T} = E(Y_T) = (20 \times 10^{-3} \underline{/100°})(0.943 \underline{/-58°})$$
$$= \mathbf{18.86 \times 10^{-3} \underline{/42°}}$$

which in the time domain is

$$i_T = \sqrt{2}(18.86 \times 10^{-3})\sin(\omega t + 42°)$$
$$= \mathbf{26.7 \times 10^{-3} \sin(\omega t + 42°)}$$

The current through each element can again be determined by Ohm's law:

$$I_R = \frac{E}{R} = \frac{20 \times 10^{-3} \underline{/100°}}{2 \underline{/0°}} = 10 \times 10^{-3} \underline{/100°}$$

$$I_L = \frac{E}{X_L} = \frac{20 \times 10^{-3} \underline{/100°}}{1 \underline{/90°}} = \mathbf{20 \times 10^{-3} \underline{/10°}}$$

$$I_C = \frac{E}{X_C} = \frac{20 \times 10^{-3} \underline{/100°}}{5 \underline{/-90°}} = \mathbf{4 \times 10^{-3} \underline{/190°}}$$

The current I_T could then be determined using Kirchhoff's current law,

$$I_T = I_R + I_L + I_C$$

or as determined above.

The phasor diagram of the currents and voltage appears in Fig. 3.49. Note that the phase angle of E has simply rotated (in the counterclockwise direction) the entire phasor diagram by 100°. The current I_R is still in phase with $V_R = E$, I_L lags the voltage V_L (= E) by 90°, and I_C leads the voltage V_C (= E) by 90°. The vector addition required to determine I_T is shown in the figure. Compare its vector solution with that obtained above.

The power to the network can be determined using any one of the equations

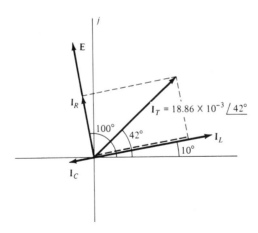

Figure 3.49 Phasor diagram for parallel R–L–C network.

introduced earlier. That is,

$$P = EI_T \cos \theta_T$$
$$= (20 \times 10^{-3})(18.86 \times 10^{-3}) \cos(100° - 42°)$$
$$= (377.2 \times 10^{-6})(\cos 58°) = (377.2 \times 10^{-6})(0.5299)$$
$$= 200 \times 10^{-6} = \mathbf{200\ \mu W}$$

or

$$P = \frac{E^2}{R} = \frac{(20 \times 10^{-3})^2}{2} = \frac{400 \times 10^{-6}}{2} = \mathbf{200\ \mu W}$$

The power factor of the network is

$$F_p = \cos \theta_T = \cos 58° = \mathbf{0.5299\ lagging}$$

For any series or parallel network a change in frequency will change the reactance of the capacitive and inductive elements and therefore the response of the network. In fact, at another frequency, a network may have a leading rather than lagging characteristic, indicating an increase in the terminal effect of the capacitive elements over those of the inductive components. For the range of frequencies being considered here, the effect of frequency, on resistive elements, can be ignored.

3.9 SERIES-PARALLEL ac NETWORKS (LADDER, Δ-Y, BRIDGE)

As pointed out for dc networks there exists an infinite variety of series-parallel configurations. We must therefore limit our attention in this section to those configurations encountered most frequently, such as the *ladder*, *Δ-Y*, and *bridge*.

The network of Fig. 3.50(a) includes the first two loops of the *ladder* network shown in Fig. 3.50(b). The choice of names should be quite obvious from Fig. 3.50(b).

The total (or input) current and the network terminal impedance can be determined by working from the termination (X_C) back to the source (E). The network is redrawn in Fig. 3.51 with the block impedances, where

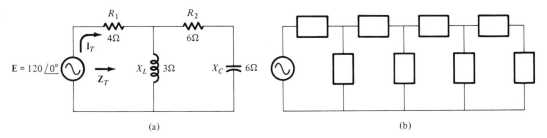

(a)   (b)

Figure 3.50 Series-parallel ac networks (a) two-loop configuration
(c) ladder network.

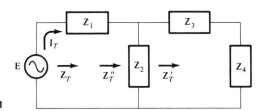

Figure 3.51

$$\mathbf{Z}_1 = R_1 \underline{/0°} = 4 \underline{/0°}$$
$$\mathbf{Z}_2 = X_L \underline{/90°} = 3 \underline{/90°}$$
$$\mathbf{Z}_3 = R_2 \underline{/0°} = 6 \underline{/0°}$$
$$\mathbf{Z}_4 = X_C \underline{/-90°} = 6 \underline{/-90°}$$

The impedance approach will be employed. First, the total impedance of the
series combination of $\mathbf{Z}_3$ and $\mathbf{Z}_4$ is determined:

$$\mathbf{Z}_T' = \mathbf{Z}_3 + \mathbf{Z}_4 = (6 + j0) + (0 - j6) = 6 - j6 = 8.484 \underline{/-45°}$$

This total impedance is in parallel with $\mathbf{Z}_2$, and

$$\mathbf{Z}_T'' = \mathbf{Z}_2 \| \mathbf{Z}_T' = \frac{\mathbf{Z}_2 \mathbf{Z}_T'}{\mathbf{Z}_2 + \mathbf{Z}_T'} = \frac{(3 \underline{/90°})(8.484 \underline{/-45°})}{(0 + j3) + (6 - j6)}$$

$$= \frac{25.452 \underline{/45°}}{6 - j3} = \frac{25.452 \underline{/45°}}{6.71 \underline{/-26.6°}}$$

$$= 3.793 \underline{/18.4°} \simeq 3.6 + j1.2$$

$$\mathbf{Z}_T = \mathbf{Z}_1 + \mathbf{Z}_T'' = (4 + j0) + (3.6 + j1.2) = 7.6 + j1.2 = \mathbf{7.7 \underline{/9°}}$$

and

$$\mathbf{I}_T = \frac{\mathbf{E}}{\mathbf{Z}_T} = \frac{120 \underline{/0°}}{7.7 \underline{/9°}} = \mathbf{15.58 \underline{/-9°}}$$

which in the time domain is

$$i_T = \sqrt{2}\,(15.58) \sin(\omega t - 9°)$$

$$= \mathbf{22.03 \sin(\omega t - 9°)}$$

If the voltage across Z_4 were desired, we would have to work back to that load element. That is,

$$V_{Z_2} = I_T Z_T'' = (15.58\ \underline{/-9°})(3.793\ \underline{/18.4°}) = 59.1\ \underline{/9.4°}$$

and using the voltage-divider rule,

$$V_4 = \frac{Z_4(V_{Z_2})}{Z_4 + Z_3} = \frac{(6\ \underline{/-90°})(59.1\ \underline{/9.4°})}{-j6 + (58.3 + j9.65)}$$

$$= \frac{354.6\ \underline{/-80.6°}}{58.3 + j3.65} \simeq \frac{354.6\ \underline{/-80.6°}}{58.3}$$

$$= 6.08\ \underline{/-80.6°}$$

Perhaps you will recall from the treatment of Δ and Y configurations in dc networks that neither configuration had series or parallel combinations of elements as their building blocks. In fact, without the conversions from one form to another it appeared as if a solution for the total impedance or driving current could not be obtained.

Consider the configuration of Fig. 3.52, for example. A close examination will reveal that within the right-hand structure there are no series or parallel elements. However, a conversion from one form of Fig. 3.53 to the other would permit a

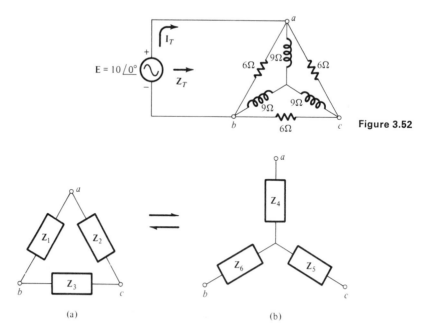

Figure 3.52

Figure 3.53 (a) Δ configuration (b) Y configuration.

reduction in the network (as described in detail for dc networks).

The conversion equations for the Δ and Y configurations of Fig. 3.53 are the following:

$$Z_4 = \frac{Z_1 Z_2}{Z_1 + Z_2 + Z_3} \qquad Z_1 = \frac{Z_4 Z_5 + Z_4 Z_6 + Z_5 Z_6}{Z_5}$$

$$Z_5 = \frac{Z_2 Z_3}{Z_1 + Z_2 + Z_3} \qquad Z_2 = \frac{Z_4 Z_5 + Z_4 Z_6 + Z_5 Z_6}{Z_6} \qquad (3.41)$$

$$Z_6 = \frac{Z_1 Z_3}{Z_1 + Z_2 + Z_3} \qquad Z_3 = \frac{Z_4 Z_5 + Z_4 Z_6 + Z_5 Z_6}{Z_4}$$

$$\Delta \to Y \qquad\qquad Y \to \Delta$$

If all three impedances of a Δ or Y are equal, the equation for conversion reduces to the following:

$$\boxed{Z_Y = \frac{Z_\Delta}{3}} \quad \text{or} \quad \boxed{Z_\Delta = 3Z_Y} \qquad (3.42)$$

For the example provided, the Y will be converted to a Δ between the same three terminals, resulting in the configuration of Fig. 3.54.

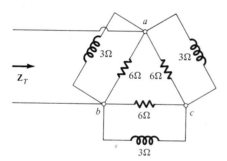

Figure 3.54

The parallel combination of each 6-Ω resistor and 3-Ω inductor will result in the following total impedance:

$$Z = \frac{(6\ \underline{/0°})(3\ \underline{/90°})}{6 + j3} = \frac{18\ \underline{/90°}}{6.7\ \underline{/26.6°}} = 2.686\ \underline{/63.4°}$$

and

$$Z_T = Z \| (Z + Z)$$

$$= \frac{Z(2Z)}{Z + 2Z} = \frac{2Z^2}{3Z} = \frac{2}{3} Z = \frac{2}{3}(2.686\ \underline{/63.4°})$$

$$= 1.7906\ \underline{/63.4°}$$

and

$$I_T = \frac{E}{Z_T} = \frac{10\ \underline{/0°}}{1.7906\ \underline{/63.4°}}$$

$$= 5.585\ \underline{/-63.4°}$$

The Δ could also be converted to a Y. The two center points (of each Y) could then be connected since the loads are balanced (same impedance in each branch).

For unbalanced situations the $Y \longrightarrow \Delta$ conversion should be used since the center-points cannot be connected using a $\Delta \longrightarrow Y$ conversion.

The Δ-Y conversion can also be applied to the *bridge* configuration of Fig. 3.55, which was also introduced for dc networks. As with the network just examined, there are no two elements in series or parallel. Since all three elements of the upper delta are equal, we shall use the $\Delta \longrightarrow Y$ conversion to obtain the network of Fig. 3.56.

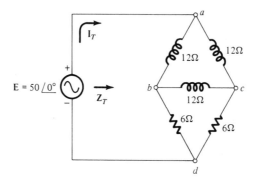

Figure 3.55 Bridge network.

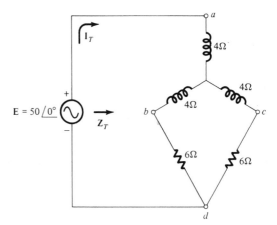

Figure 3.56

The impedance of each branch of the new Y were obtained from Eq. (3.42):

$$Z_Y = \frac{Z_\Delta}{3} = \frac{12\,/90°}{3} = 4\,/90°$$

We can now use series-parallel techniques to find the total impedance and current. That is, for the series 4-Ω inductance and 6-Ω resistance, we have

$$Z_1 = 6 + j4$$

The parallel branches of Z_1 will result in a total impedance determined by

$$Z_2 = \frac{Z_1}{2} = \frac{6 + j4}{2} = 3 + j2$$

The total impedance is then

$$Z_T = 4\,\underline{/90°} + Z_2$$
$$= +j4 + (3 + j2)$$
$$= 3 + j6$$
$$= 6.7\,\underline{/63.4°}$$

with the result that

$$I_T = \frac{E}{Z_T} = \frac{50\,\underline{/0°}}{6.7\,\underline{/63.4°}} = 7.463\,\underline{/-63.4°}$$

In our analysis of dc networks we found that there was a condition set by the branch impedances that would result in zero voltage drop across the bridge arm (*a-b*) or zero current through that branch impedance. For ac networks, the general form of the equation is the same, except now we are dealing with impedances rather than fixed resistive values as encountered for dc networks. For the general network of Fig. 3.57, the balance condition is determined by

$$\boxed{\frac{Z_1}{Z_2} = \frac{Z_3}{Z_4}} \qquad\qquad (3.43)$$

For the network of Fig. 3.52 the following ratios will result:

$$\frac{12\,\underline{/90°}}{6\,\underline{/0°}} = \frac{12\,\underline{/90°}}{6\,\underline{/0°}}$$

and

$$2\,\underline{/90°} = 2\,\underline{/90°} \qquad \text{(checks)}$$

Since Eq. (3.43) is satisfied, there is zero voltage drop across branch *a-b* or zero current through that branch impedance. The latter condition permits substituting an open circuit for that branch, as shown in Fig. 3.58.

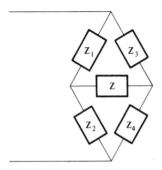

Figure 3.57

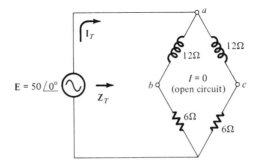

Figure 3.58 Network of Figure 3.55 with open-circuit substituted for "balance" arm.

Now the 12-Ω inductance and 6-Ω resistor are in series, and the series combination of each results in

$$\mathbf{Z}_1 = 6 + j12 = 13.4 \, \underline{/63.4°}$$

and

$$\mathbf{Z}_T = \frac{\mathbf{Z}_1}{2} = \frac{13.4 \, \underline{/63.4°}}{2} = 6.7 \, \underline{/63.4°}$$

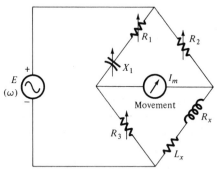

$$R_x = \frac{\omega^2 C^2 R_1 R_2 R_3}{1 + \omega^2 C^2 R_1^2} \quad \begin{array}{l}\text{(at balance)}\\ I_m = 0\end{array}$$

$$L_x = \frac{C R_2 R_3}{1 + \omega^2 C^2 R_1^2}$$

(a) Hay

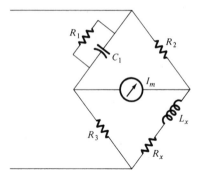

$$L_x = C R_2 R_3 \quad (I_m = 0)$$

$$R_x = \frac{R_2 R_3}{R_1}$$

(b) Maxwell

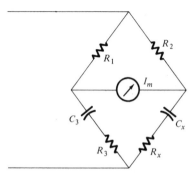

$$C_x = C_3 \frac{R_1}{R_2} \quad (I_m = 0)$$

$$R_4 = \frac{R_2 R_3}{R_1}$$

(c) Capacitance comparison bridge

Table 3.1

as obtained before. Placing a short between terminals a and b (for the condition $v = 0$) will also result in the same solution, but this will be left as an exercise to appear in the Problems of this chapter.

For a situation where the ratios of Eq. (3.43) result in

$$2 \,\underline{/90°} = 2 \,\underline{/-90°}$$

the equality is *not* equal. Both the magnitude and the angle must be equal for the condition $v = 0$ or $i = 0$ to be applicable. In rectangular form both the real and imaginary components must be equal, as demonstrated by the following:

$$3 + j4 = 3 + j4$$

When employed in a resistance-measuring capacity for dc networks the configuration was referred to as a Wheatstone bridge. You will recall that it was only necessary to ensure that the bridge current from a-b was zero before the equation could be applied to determine the unknown resistance from the other three known values. For ac networks, this bridge can be used to measure resistance, inductance, and capacitance.

Three commercially employed bridge configurations appear in Table 3.1. When the bridge arms are adjusted until $I_{Gal} = 0$ the unknown quantity denoted by the x subscripts can be determined using the equations provided.

A photograph of a commercially available RLC bridge appears in Fig. 3.59. The user's manual provided by the manufacturer will make the determination quite straightforward for each of the three elements.

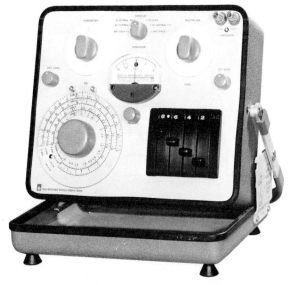

Figure 3.59 Impedance bridge (Courtesy General Radio Co.).

3.10 MULTISOURCE ac NETWORKS

Once the various reactive and resistive elements of an ac network are replaced by their block impedances, the application of the branch-current method is the same as for dc networks. You will recall that a procedure such as one of the above was necessary to solve for the currents and voltages of a network with two sources that were not in series or parallel.

Branch-Current Analysis

Consider the network of Fig. 3.60 with two ac voltage sources. Substituting the block impedances will result in the network of Fig. 3.61 A review of the parallel dc section may be necessary before continuing with this approach since it will be assumed that the general procedure is understood. Assigning the labeled branch currents with an assumed direction and including the polarities will result in the configuration of Fig. 3.62.

Writing Kirchhoff's voltage law around each closed loop in the indicated direction,

$$E_1 - I_1Z_1 - I_1Z_2 - I_3Z_3 = 0$$
$$E_2 - I_2Z_4 - I_3Z_3 = 0$$

and Kirchhoff's current law to one node is

$$I_1 + I_2 = I_3$$

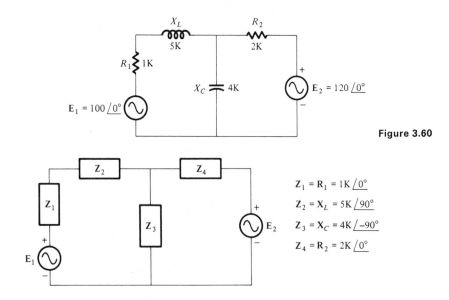

Figure 3.60

$$Z_1 = R_1 = 1K\ \underline{/0°}$$
$$Z_2 = X_L = 5K\ \underline{/90°}$$
$$Z_3 = X_C = 4K\ \underline{/-90°}$$
$$Z_4 = R_2 = 2K\ \underline{/0°}$$

Figure 3.61

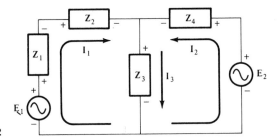

Figure 3.62

The first two equations can then be written as

$$I_1(Z_1 + Z_2) + I_3Z_3 = E_1$$
$$I_2Z_4 + I_3Z_3 = E_2$$

Substituting for I_3,

$$I_1(Z_1 + Z_2) + (I_1 + I_2)Z_3 = E_1$$
$$I_2(Z_4) + (I_1 + I_2)Z_3 = E_2$$

and rearranging terms,

$$I_1(Z_1 + Z_2 + Z_3) + I_2Z_3 = E_1$$
$$I_1Z_3 + I_2(Z_3 + Z_4) = E_2$$

which is two equations and two unknowns and can be solved using either technique discussed for dc networks. Once the solution is determined in terms of the Z parameters their numerical values can be substituted and the solution obtained in phasor notation and converted to the time domain if required.

Source Conversion

Converting a voltage source to a current source or vice versa can also be accomplished in the same manner as for dc networks. The converting equations are provided next to each configuration of Fig. 3.63.

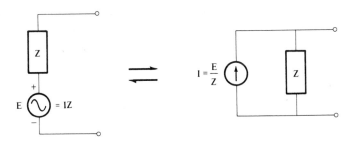

Figure 3.63 Source conversion.

EXAMPLE 3.10

Convert the voltage source of Fig. 3.64 to a current source.

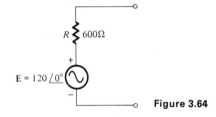

Figure 3.64

Solution:

The solution appears in Fig. 3.65.

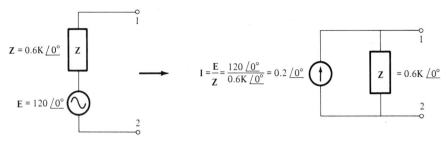

Figure 3.65

3.11 NETWORK THEOREMS

The superposition, Thevenin, and maximum power theorems will all be examined in this section to parallel the coverage for dc networks.

Superposition

As you will recall, the superposition theorem is another technique that can be applied to solve multisource ac networks or to determine the effect of each source on the behavior of a network. As an example consider the network of Fig. 3.66 in which the current I_1 is to be determined. The block impedances have been assigned in Fig. 3.67. Considering the effects of the voltage source will result in the configuration of Fig. 3.68 in which the current source has been replaced by its open-circuit equivalent.

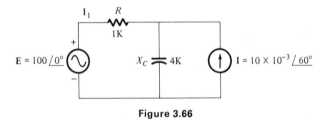

Figure 3.66

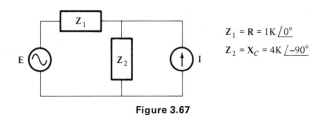

Figure 3.67

The current I_1' is

$$I_1' = \frac{E}{Z_1 + Z_2}$$

For the current source the configuration of Fig. 3.69 will result in which the volt-age source has been replaced by a short circuit. Applying the current-divider rule,

$$I_1'' = \frac{Z_2(I)}{Z_2 + Z_1}$$

The desired current is then determined by

$$I_1 = I_1' - I_1''$$

since the currents have opposite directions. Substitution of numerical values is all that remains to determine I_1.

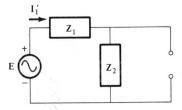

Figure 3.68   **Figure 3.69**

Thevenin's Theorem

Thevenin's theorem permits the reduction of a linear ac network to the series configuration of Fig. 3.70. Any network connected to this reduced network will be totally unaware of the change. In addition to providing a network composed of fewer

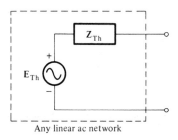

Figure 3.70 Thevenin equivalent circuit.   Any linear ac network

elements that has the same terminal characteristics, it provides a technique for quickly determining the effect of changing the load on the load current or voltage.

The Thevenin voltage or impedance can be determined in the same manner as for dc networks. Consider the networks of Fig. 3.71, where the Thevenin network is to be determined for the network external to the load R_L. The block impedances have been assigned in Fig. 3.72.

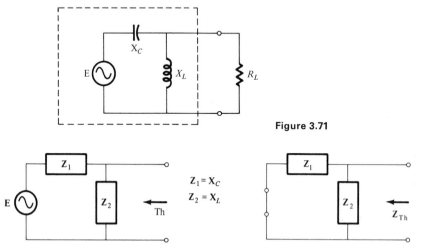

Figure 3.71

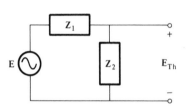

Figure 3.72

Figure **3.73** Determining the Thevenin impedance.

$$Z_1 = X_C$$
$$Z_2 = X_L$$

The Thevenin impedance is determined by setting E to zero as shown in Fig. 3.73 and finding the terminal impedance. That is,

$$Z_{\text{Th}} = Z_1 \| Z_2 = \frac{Z_1 Z_2}{Z_1 + Z_2}$$

The Thevenin voltage is the open-circuit voltage between the network terminals as shown in Fig. 3.74:

$$E_{\text{Th}} = \frac{Z_2 E}{Z_2 + Z_1} \qquad \text{(voltage-divider rule)}$$

Substitution of the numerical values will result in the configuration of Fig. 3.75.

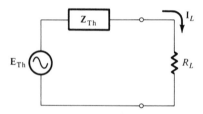

Figure 3.74 Determining the Thevenin voltage.

Figure 3.75 Construction of the Thevenin network.

The current through R_L, $I_L = E_{Th}/(Z_{Th} + R_L)$ can now be determined for different values of R_L without having to analyze the entire network for each value of R_L.

Maximum Power Theorem

For the network of Fig. 3.76 maximum power will be transferred to Z_L when the magnitude of Z_L equals that of Z_{Th} and their associated angles are such that $\theta_L = -\theta_{Th}$. That is, for maximum power transfer

$$\boxed{\begin{array}{c} |Z_L| = |Z_{Th}| \\ \theta_L = -\theta_{Th} \end{array}} \tag{3.44}$$

Since the Thevenin network can be found for any two-terminal linear network, it is only necessary to connect a load that satisfies the above conditions in order to transfer maximum power.

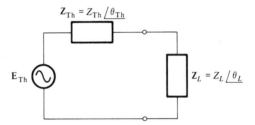

Figure 3.76 Establishing maximum power transfer criteria.

In rectangular form the conditions are

$$R_L = R_{Th}$$

and

$$X_{L_{(L)}} = X_{C_{Th}} \tag{3.45}$$

or

$$X_{C_{(L)}} = X_{L_{Th}}$$

In Fig. 3.77 these conditions have been satisfied for the Thevenin impedance resulting from an inductive network.

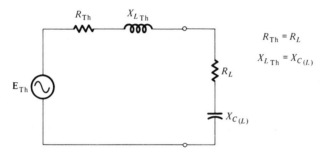

$$R_{Th} = R_L$$
$$X_{L\,Th} = X_{C\,(L)}$$

Figure 3.77 Maximum power transfer conditions.

The resulting impedance is

$$Z = (R_{Th} + jX_{L_{Th}}) + (R_L - jX_{C_{(L)}})$$
$$= (R_{Th} + R_L) + \underbrace{j(X_{L_{Th}} - X_{C_{(L)}})}$$

0 for maximum power conditions

$$\boxed{Z = 2R_{Th}}\qquad \text{(for } R_{Th} = R_L) \qquad\qquad (3.46)$$

and

$$I_L = \frac{E_{Th}}{Z} = \frac{E_{Th}}{2R_{Th}}$$

and the power to the load (the maximum value)

$$P_L = I_L^2 R_L = I_L^2 R_{Th}$$
$$= \left(\frac{E_{Th}}{2R_{Th}}\right)^2 R_{Th}$$

$$\boxed{P_{L_{max}} = \frac{E_{Th}^2}{4R_{Th}}}\qquad\qquad (3.47)$$

EXAMPLE 3.11

For the network in Fig. 3.78, determine the load for maximum power transfer and the maximum power delivered.

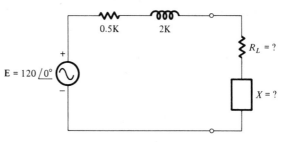

Figure 3.78

Solution:

$$R_L = 0.5 \text{ K},\qquad X = X_C = 2 \text{ K}$$
$$P_{max} = \frac{E_{Th}^2}{4R_{Th}} = \frac{(120)^2}{4(0.5 \text{ K})} = \frac{144 \times 10^3}{2 \times 10^3} = 72 \text{ W}$$

3.12 POWER (ac)

In an ac network, the resistive elements are the only components that will dissipate electrical energy. The pure reactive elements simply store the energy in a form that can be returned to the electrical system when required by circuit design. For the total watts dissipated, therefore, it is only necessary to find the sum of the watts dissipated by the resistive elements. The power dissipated can also be determined at the driving point of the network as shown in Fig. 3.79.

The total watts dissipated is determined by

$$P = EI \cos \theta$$ (3.48)

The wattmeter connections are also shown in Fig. 3.79 for determining the total watts dissipated. As for dc networks, the voltage terminals sense the voltage level, while the current terminals reflect the magnitude of I. The meter will include the effect of $\cos \theta$ to indicate the proper wattage level. A dynamometer wattmeter can read the watts dissipated in a dc or ac network. In fact, it can properly indicate the wattage level for any input voltage no matter how distorted it may appear.

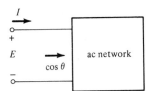

Figure 3.79

Although there is no dissipation of electrical energy by the reactive elements, a level of electrical energy is drawn from the supply and stored in the electric or magnitude fields of the reactive elements. This additional energy can be supplied only through an increase in the current drawn from the supply. Of course, the reactive elements can return the energy to the system but at particular instants of time there is an increase in the supply current due to the reactive elements. This increase in current requires that the supply or generator be designed to handle the increased currents. For fixed voltage levels this increase in current results in an increase in the maximum instantaneous power the supply must be able to provide. Increased current or power rating results in increased costs for the generating equipment and the production of the necessary electrical energy. This difference in demand from that dissipated by the resistive elements is accounted for in the power factor ($F_p = \cos \theta$) of the network. For $F_p = 1$, all the power supplied is dissipated, while for more reactive elements F_p approaches zero and more energy is being stored in the reactive (storage) elements.

The product EI, which is independent of whether dissipation or storage is the result, is called the *apparent* power of an ac system and is measured in volt-amperes. That is, for the network of Fig. 3.79 the apparent power is determined by

$$P_a = EI$$ (volt-ampere: VA) (3.49)

The current I is the current that must be supplied by the power station even if a major portion is diverted to the storage elements. Larger industrial outfits pay for the apparent power demand rather than the average or real power dissipated due to the current requirement above that of the resistive elements. Any power factor less than 1 will certainly result in a wattage demand less than the volt-ampere demand.

On a trigonometric basis, the apparent power and *real, average,* or *dissipated* power are related as shown in Fig. 3.80. The third component of the triangle is similar to the real power equation except for the sine rather than the cosine term. It is called the *reactive* power and is measured in volt-ampere-reactive units. That is,

Reactive power: $\boxed{P_q = EI \sin \theta}$ (volt-ampere-reactive: vars)   (3.50)

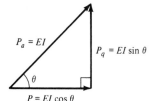

$P = EI \cos \theta$ **Figure 3.80** Power triangle.

The reactive power is a measure of the input power absorbed (but not dissipated) by the reactive elements. For a fixed supply voltage the smaller this component, the less the source current. Peak efficiency of the system is obtained (insofar as the source is concerned) when $P_q = 0$ and $P = P_a$.

For any network, the total reactive power is simply the difference of the capacitive and inductive components as determined by any one of the following equations:

Inductor: $\boxed{P_q = I_L^2 X_L = \dfrac{V_L^2}{X_L} = V_L I_L}$ (vars)   (3.51)

Capacitor: $\boxed{P_q = I_C^2 X_C = \dfrac{V_C^2}{X_C} = V_C I_C}$ (vars)   (3.52)

For a network in which the capacitive vars equals the inductive vars, the net reactive power is zero, and the real and apparent powers are equal.

Since

$$P_T = EI \cos \theta = P_{a_T} \cos \theta = P_{a_T} F_p$$

we find that

$$\boxed{F_p = \dfrac{P_T}{P_{a_T}}}$$ (3.53)

where P_T and P_{a_T} represent the total of each quantity of the system.

EXAMPLE 3.12

The currents and voltages of the network of Fig. 3.81 are indicated. Determine

1. The total power dissipated.
2. The net reactive power.
3. The total apparent power.
4. The power factor of the network.

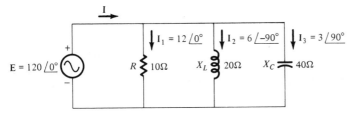

Figure 3.81

Solution:

1. Power is dissipated only by the resistive element:

$$P_T = I^2R = (12)^2 \times 10 = (144)(10) = \textbf{1440 W}$$

2. $$P_{q(C)} = I^2X_C = (3)^2(40) = (9)(40) = \textbf{360 vars (cap.)}$$

$$P_{q(L)} = I^2X_L = (3)^2(20) = (9)(20) = \textbf{180 vars (ind.)}$$

$$P_{q_T} = P_{q(C)} - P_{q(L)} = 360 - 180 = \textbf{180 vars (cap.)}$$

3. $$P_{a_T} = \sqrt{p^2 + p_q^2} \qquad \text{(from right triangle)}$$

$$= \sqrt{(1440)^2 + (180)^2}$$

$$\simeq \textbf{1488 VA}$$

4. $$F_p = \frac{P_T}{P_{a_T}} = \frac{1440}{1488} = \textbf{0.9677 (capacitive, leading)}$$

Note in the solution to Example 3.12 that no consideration was given to the type of network. Each element was treated individually to determine either the real or reactive power. There was no concern (once the currents were known) for the fact that is was a parallel network in the determination of the total watts or reactive power. In addition note that the total apparent power was determined from the total real and reactive power. It cannot be determined by a simple algebraic addition of the real and reactive power to each element.

We shall encounter the concept of apparent power in later chapters when we deal with transformers and ac machinery. At that time the choice of apparent power as a rating rather than real power will be explained as a necessary choice.

3.13 POLYPHASE SYSTEMS

Throughout the analysis of ac networks we have been concerned with the response of a network to a single sinusoidal signal as generated by a single-phase ac generator. There are *polyphase* generators available that can develop two, three, or more phases simultaneously. Each generated voltage has a definite phase relationship with the other phase voltages. For power applications where large voltages are generated, the three-phase generator is employed almost exclusively. The three ac voltages are each 120° out of phase as shown in Fig. 3.82.

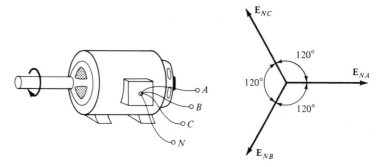

Figure 3.82 Three ϕ generator and the voltages developed.

In the phasor domain,

$$E_{NA} = E_{NA} \underline{/0°}$$
$$E_{NB} = E_{NB} \underline{/-120°} \qquad (3.54)$$
$$E_{NC} = E_{NC} \underline{/+120°}$$

which in the time domain are

$$e_{NA} = \sqrt{2}\, E_{NA} \sin(\omega t + 0°)$$
$$e_{NB} = \sqrt{2}\, E_{NB} \sin(\omega t - 120°) \qquad (3.55)$$
$$e_{NC} = \sqrt{2}\, E_{NC} \sin(\omega t + 120°)$$

and appear in Fig. 3.83.

Note that at any instant of time three voltage levels are available and that each has the same peak value.

The schematic representation of a three-phase generating system appears in Fig. 3.84. The letter N is a shorthand notation for the *neutral* or common terminal of the

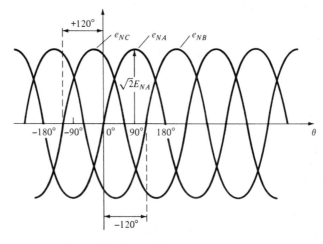

Figure 3.83 Phase voltage of a 3 ϕ generator.

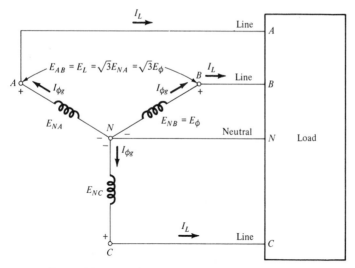

Figure 3.84 Three ϕ, 4-wire, Y-connected generator.

generated voltages. Since the induced voltages appear across the armature *coils* of the generator, the symbol for each phase employs the coil notation. The system of Fig. 3.84 is referred to as a 3ϕ, *Y-connected, four-wire system*. The choice of the terminology Y-connected should be obvious from the schematic representation. There are four lines to the load, resulting in the four-wire label. From the diagram it should be obvious that the line current equals the generated phase current. That is,

$$\boxed{I_L = I_{\phi_g}}$$

(3.56)

for each phase.

It can be shown through the vector addition of any two phases that the voltage from one line to another (such as from A to B) is equal in magnitude to the square root of three times the magnitude of the phase voltage. That is,

$$\boxed{E_L = \sqrt{3}\, E_\phi}$$

(3.57)

The line voltages are all equal in magnitude and have a 120° phase angle between each as shown in Fig. 3.85.

For a three-phase supply, the *phase sequence* can be of particular importance. It is the order in which the phase voltages will appear at the output terminals of the generator of Fig. 3.82. In Fig. 3.85 the phases appear in the order E_{NA}, E_{NB}, E_{NC}, and the phase sequence is said to be *ABC* as determined by the order of the second subscripts. For the only other possibility E_{NA}, E_{NC}, E_{NB}, the phase sequence would be *ACB*. By placing a mark on the phasor diagram (such as the X in Fig. 3.85) and rotating the vectors counterclockwise, the sequence of the first or second subscripts of the voltages will provide the phase sequence. The operation of certain electrical equipment is quite dependent on the phase sequence. The direction of rotation of a shaft of a 3ϕ motor can be reversed by simply changing the phase sequence.

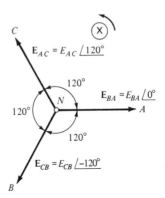

Figure 3.85 Three ϕ line voltages.

In a three-phase system the load can be connected in a Y or Δ configuration. The equations for the Y-connected load as shown in Fig. 3.86 are fundamentally the same as introduced for the generator. For obvious reasons, the system is referred to as *Y-Y-connected*.

The relating equations are

$$E_L = \sqrt{3}\, V_\phi \tag{3.58}$$

$$I_\phi = I_L \tag{3.59}$$

and for a Y-Y system,

$$V_\phi = E_\phi \tag{3.60}$$

Note that Eq. (3.58) is in magnitude form, while Eq. (3.59) and (3.60) are in phasor form.

The load network of Fig. 3.86 is said to be *balanced* since the same impedance appears in each branch. The result is zero neutral current ($I_N = 0$) since each phase

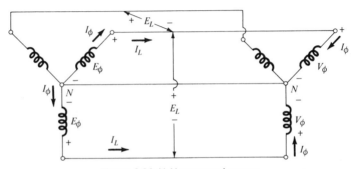

Figure 3.86 Y–Y connected system.

current is equal but out of phase by 120°. The result of the following vector addition
is therefore 0 A:

$$I_N = I_{\phi_1} + I_{\phi_2} + I_{\phi_3} = 0$$

An attempt is always made to *balance* the load as much as possible to reduce
the current I_N to its minimum value so that the line loss $I_N^2 R_{\text{line}}$ is the smallest possible
value.

The obvious advantages of the three-phase system is two available voltage levels
(phase and line), a phase shift between each of 120°, and a zero return current (for
minimum line loss) for balanced loads. For industrial and commercial use, two of the
three phases are provided so that the larger line-to-line voltage of 208 V is available
for heavy machinery, welders, etc. The 120-V outlet voltage for the same plant is
available from each phase. The same is true in the home where 208 V may be necessary
for an air conditioner, while only 120 V is necessary for the the outlets.

EXAMPLE 3.13

For the Y-Y-connected, 3ϕ, four-wire system of Fig. 3.87, determine

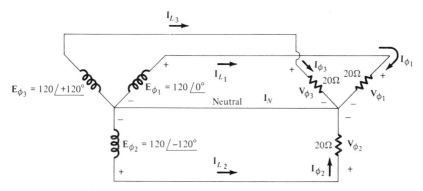

Figure 3.87

1. The load phase voltages.
2. The magnitude of the load line voltages.
3. The phase current of each load element.
4. The vector sum of the three phase currents.
5. The line currents.

Solution:

1.
$$V_{\phi_1} = E_{\phi_1} = 120 \ \underline{/0°}$$
$$V_{\phi_2} = E_{\phi_2} = 120 \ \underline{/-120°}$$
$$V_{\phi_3} = E_{\phi_3} = 120 \ \underline{/+120°}$$

2.
$$V_L = \sqrt{3} \ V_\phi = (1.73)(120) \simeq \textbf{208 V.}$$

3.
$$I_{\phi_1} = \frac{V_{\phi_1}}{Z_1} = \frac{120 \ \underline{/0°}}{20 \ \underline{/0°}} = 6 \ \underline{/0°}$$

$$I_{\phi_2} = \frac{V_{\phi_2}}{Z_2} = \frac{120\;/-120°}{20\;/0°} = 6\;/-120°$$

$$I_{\phi_3} = \frac{V_{\phi_3}}{Z_3} = \frac{120\;/+120°}{20\;/0°} = 6\;/+120°$$

4. $$I_N = I_{\phi_1} + I_{\phi_2} + I_{\phi_3} = 6\;/0° + 6\;/-120° + 6\;/+120°$$

$$= (6 + j0) + (-3 - j5.196) + (-3 + j5.196)$$

$$= (6 - 3 - 3) + j(-5.196 + 5.196)$$

$$= 0 \text{ (balanced load)}$$

5. $$I_{L_1} = I_{\phi_1} = 6\;/0°, \; I_{L_2} = I_{\phi_2} = 6\;/-120°, \; I_{L_3} = I_{\phi_3} = 6\;/+120°.$$

The three phases of a polyphase generator may be connected in a Δ configuration as shown in Fig. 3.88 for reduced line voltage but increased line current (for the same induced voltages and load). In this case it is immediately obvious that

$$\boxed{E_\phi = E_L = V_\phi} \tag{3.61}$$

and through vector algebra it can be shown that

$$\boxed{I_L = \sqrt{3}\,I_\phi} \tag{3.62}$$

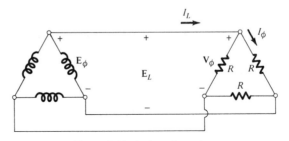

Figure 3.88   Δ–Δ configuration.

EXAMPLE 3.14

For the Δ-Δ-connected, 3ϕ, three-wire system of Fig. 3.89 determine

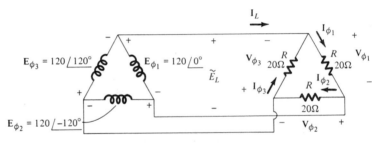

Figure 3.89

1. The line and phase voltages.
2. The load phase currents.
3. The magnitude of the line currents.
4. The relationship between the line current and voltage of a system as compared to a Y-Y system.
5. The power to each load element for the Δ-Δ as compared to the Y-Y system.

Solution:

1.
$$V_{\phi_1} = E_L = E_{\phi_1} = 120 \underline{/0°}$$
$$V_{\phi_2} = E_L = E_{\phi_2} = 120 \underline{/-120°}$$
$$V_{\phi_3} = E_L = E_{\phi_3} = 120 \underline{/+120°}$$

2.
$$I_{\phi_1} = \frac{V_{\phi_1}}{Z_1} = \frac{120 \underline{/0°}}{20 \underline{/0°}} = 6 \underline{/0°}$$

$$I_{\phi_2} = \frac{V_{\phi_2}}{Z_2} = \frac{120 \underline{/-120°}}{20 \underline{/0°}} = 6 \underline{/-120°}$$

$$I_{\phi_3} = \frac{V_{\phi_3}}{Z_3} = \frac{120 \underline{/+120°}}{20 \underline{/0°}} = 6 \underline{/+120°}$$

3. $\qquad I_L = \sqrt{3}\,I_\phi = (1.73)(6) = \textbf{10.4 A}$

4. $\qquad$ Y-Y: $\quad |I_L| = \textbf{6 A,} \qquad |E_L| = \textbf{208 V}$

$\qquad \Delta$-Δ: $\quad |I_L| = \textbf{10.4 A,} \qquad |E_L| = \textbf{120 V}$

Note the increased current but reduced voltage for the Δ-Δ configuration.

5. $\qquad \Delta$-Δ: $\quad P = I^2R = (6)^2 \times 20 = (36)(20) = \textbf{720 W}$

$\qquad$ Y-Y: $\quad P = I^2R = (6)^2 \times 20 = (36)(20) = \textbf{720 W}$

Two other configurations are possible, the Y-Δ and the Δ-Y. In each case, it is only necessary to apply those equations that apply to that configuration for the generator or load. For instance, for a Y-Δ system $E_L = \sqrt{3}\,E_\phi$ but $V_\phi = E_L$ and so on.

The total real, apparent, and reactive power to a load can be determined using the concepts introduced in the previous section. In general,

$$\boxed{P_T = 3I_\phi^2 R} \qquad \text{(watts)} \qquad\qquad (3.63)$$

$$\boxed{P_{q_T} = 3I_\phi^2 X} \qquad \text{(vars)} \qquad\qquad (3.64)$$

$$\boxed{P_{a_T} = 3V_\phi I_\phi} \qquad \text{(VA)} \qquad\qquad (3.65)$$

$$\boxed{F_p = \frac{P_T}{P_{a_T}}} \qquad\qquad\qquad\qquad (3.66)$$

For a Δ- or Y-connected, balanced or unbalanced load, the total power to the load can be determined if the wattmeters are hooked up as shown in Fig. 3.90.

For a three-phase balanced load the total power delivered can be determined using only two wattmeters as shown in Fig. 3.91. Two other possible combinations are possible with two wattmeters. It is only necessary to ensure that both potential coils (*PC*) are connected to a line without a current coil (*CC*) and the current terminals are properly connected in series with the line current to be sensed by that meter. With only two meters, however, it is necessary to know whether the readings should be added or subtracted. The following simple test will make the determination: After ensuring that both meters have an upscale reading, remove that terminal of the potential coil connected to the line without a current coil of the low-reading wattmeter and touch the line that has the current coil of the high-reading wattmeter. An upscale deflection of the low-reading wattmeter will indicate that the reading should be added, while the reverse indicates that their difference is the total wattage.

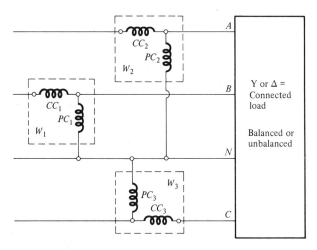

Figure 3.90 Three wattmeter measurement of total power to a 3 ϕ load.

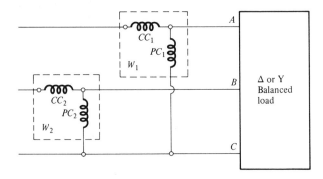

Figure 3.91 Two wattmeter measurement of total power to a 3 ϕ load.

Polyphase wattmeters are available that have the terminals necessary in a single package for measuring the total power to a three-phase load.

Further comment on three-phase systems will be provided when we consider the role of transformers in the distribution of power to the home and industry.

3.14 TUNED (RESONANT) NETWORKS

For the series RLC circuit there is a frequency that if applied will result in maximum power to the circuit and a condition where the energy released by one reactive element is exactly equal to that being absorbed by the other. Under these conditions, the network is said to be in a state of *resonance* or *tuned* to its resonant frequency.

The series RLC configuration of Fig. 3.92 is called a series resonant circuit when this condition of maximum power transfer is satisfied. Note that *both* an inductor and capacitor are required to reach this state. The condition of resonance will occur when

$$\boxed{X_L = X_C} \tag{3.67}$$

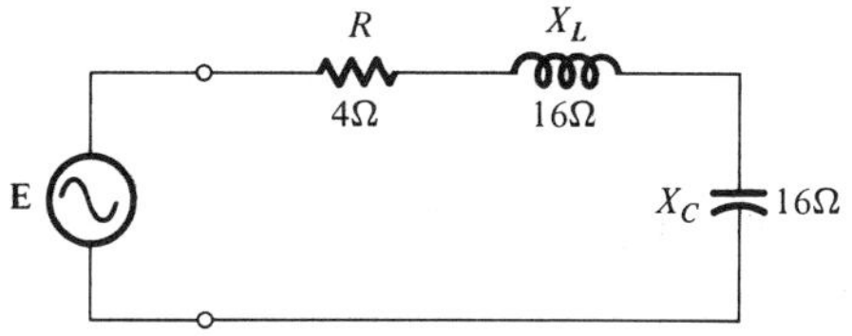

Figure 3.92 Series resonant circuit.

If we substitute for each we find the resonant frequency to be determined by

$$2\pi f_s L = \frac{1}{2\pi f_s C}$$

$$f_s^2 = \frac{1}{4\pi^2 LC}$$

and

$$\boxed{f_s = \frac{1}{2\pi \sqrt{LC}}} \quad \text{(Hz)} \tag{3.68}$$

At resonance, the impedance of the network is determined by

$$\boldsymbol{Z}_T = R + j(X_L - X_C) = R + j0$$

$$\boxed{\boldsymbol{Z}_T = R} \quad \text{(at resonance)} \tag{3.69}$$

Since the net reactance at resonance is zero, the apparent power equals the real power, and the power factor of the network $P_T/P_{a_T} = 1$.

The current I vs. frequency curve appears in Fig. 3.93. Note that since the net impedance is a minimum value (R), the current is a maximum. At frequencies to the

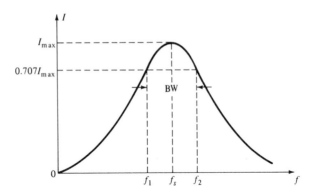

Figure 3.93 Series resonance curve.

left of the resonant value the reactance of the capacitor will increase above that of the inductor, and the net impedance will increase, causing a decrease in I. For frequencies above the resonant value the inductive reactance will be greater than that of the capacitor, causing an increase in impedance and a decrease in I toward zero.

The power delivered at resonance is determined by

$$P_{max} = I_{max}^2 R \qquad (3.70)$$

where $I_{max} = E/Z_T = E/R$.

If the current drops to 0.707 of its peak value as indicated on the curve, the power delivered will drop to

$$P = (0.707 I_{max})^2 R = 0.5 I_{max}^2 R = 0.5 P_{max}$$

or one-half the maximum at resonance. Therefore, as long as frequencies are applied between the *half-power* frequencies, at least half the maximum power will be delivered to the network. The frequencies f_1 and f_2 are called the *half-power, cutoff,* and *band* frequencies. The last term comes from the fact that f_1 and f_2 define a *bandwidth* (BW) as shown in Fig. 3.93. For most applications, it is assumed that the resonant frequency bisects the bandwidth so that f_1 and f_2 are equidistant from f_s.

In equation form,

$$BW = f_2 - f_1 \qquad (\text{Hz}) \qquad (3.71)$$

It can be shown through a sequence of substitutions that the bandwidth is related to the circuit elements in the following manner:

$$BW = f_2 - f_1 = \frac{R}{2\pi L} \qquad (\text{Hz}) \qquad (3.72)$$

Defining

$$Q = \frac{X_L}{R} \qquad (3.73)$$

where X_L is the resonant frequency value, to be the *quality factor* of the network, the following relationship will result:

$$\boxed{BW = \frac{f_s}{Q}} \quad \text{(Hz)} \tag{3.74}$$

The quality factor of a network provides information about the shape of the curve of Fig. 3.93, which is also the *selectivity* curve. The term comes from the fact that one must be selective in his choice of frequencies to ensure that he is within the bandwidth. As the Q factor increases, the peaking curve narrows and the bandwidth decreases as indicated by Eq. (3.74).

EXAMPLE 3.15

For the resonance curve of Fig. 3.94, determine

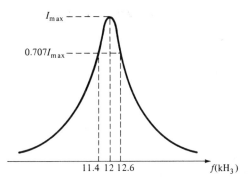

Figure 3.94

1. The resonant frequency.
2. The bandwidth.
3. The Q of the network.
4. The inductance of the network if the resistance of the network is 4 Ω.

Solution:

The solution appears in Fig. 3.95.

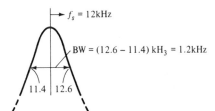

Figure 3.95

1. $f_s = $ **12 kH$_z$**.
2. BW = **1200 H$_z$**.
3. BW $= f_s/Q$ and $Q = f_s/$BW $= 12{,}000/1200 = $ **10**.

4.
$$Q = \frac{X_L}{R} \quad \text{and} \quad X_L = QR = (10)(4) = \textbf{40}\boldsymbol{\Omega}$$

$$X_L = 2\pi f_s L \quad \text{and} \quad L = \frac{X_L}{2\pi f_s} = \frac{40}{(6.28)(12 \times 10^3)} = \frac{40 \times 10^{-3}}{75.36}$$

$$= \textbf{0.53 mH}$$

EXAMPLE 3.16

For the series RLC circuit of Fig. 3.96, determine

1. X_L for resonance.
2. The Q of the network at resonance.
3. The bandwidth.
4. The power delivered at resonance.
5. The power delivered at the HPFs (half-power frequencies).
6. The general shape of the resonance curve.

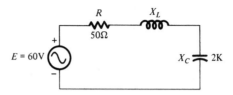

Figure 3.96

Solution:

1. $X_L = X_C = \textbf{2 K}.$
2. $Q = X_L/R = 2000/50 = \textbf{40}.$
3. BW $= f_s/Q = 30{,}000/40 = \textbf{750 Hz}.$
4. $P_m = I_m^2 R = (E/R)^2 R = E^2/R = (60)^2/50 = 3600/50 = \textbf{72 W}.$
5. $P_{\text{HPF}} = P_m/2 = 72/2 = \textbf{36 W}.$
6. See Fig. 3.97.

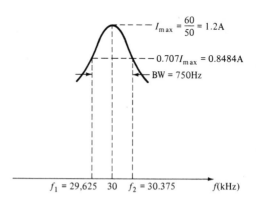

Figure 3.97

In addition to the series resonant circuit there is a parallel resonant network such as shown in Fig. 3.98 that will also develop a frequency curve similar to that obtained for I in Fig. 3.93. However, for this network the curve is that of the voltage v_C vs. frequency. Note that both a coil and capacitor are required to obtain the resonant condition. The resistance R can be either the resistance of the coil or a combination of that resistance and some added resistance chosen to affect the shape of the curve in a particular manner. At the condition of resonance there will again be an equilateral transfer of energy between the reactive elements. This particular resonant circuit is also referred to as a *tank* circuit due to its storage of energy in the reactive elements. As shown in Fig. 3.98 the source applied to the parallel resonant circuit has a constant current characteristic rather than the voltage source appearing for the series resonant circuit. This fact, in combination with the shape of the imped- ance vs. frequency curve for this network will result in the desired frequency curve. As shown in Fig. 3.99 the impedance of the parallel resonant circuit is a maximum at the resonant frequency and drops off to the right and left of this frequency much like the current characteristics of the series resonant circuit. The maximum impedance is given by

$$Z_T = Q_p^2 \cdot R \tag{3.75}$$

where Q_p is in the quality factor of the network and is determined by

$$Q_p = \frac{X_L}{R} \tag{3.76}$$

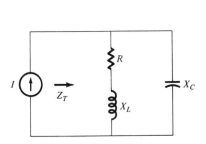

Figure 3.98 Parallel resonant circuit.

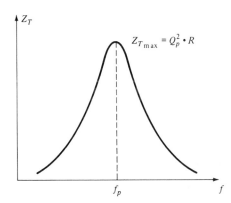

Figure 3.99 Impedance characteristics of a parallel resonant circuit.

For this text our interest will lie in networks in which $Q_p \geq 10$. When this condition is satisfied the related equations take a much simpler form. The majority of practical situations encountered satisfy this condition. For quality factors less than 10 there are numerous excellent references available. Note in the analysis to follow the similarity between the equations for parallel resonance and those provided for series resonance. For example, parallel resonance will occur when

$$X_L = X_C \qquad (Q_p \geq 10) \tag{3.77}$$

and the resonant frequency is determined by

$$f_p = \frac{1}{2\pi\sqrt{LC}} \qquad (Q_p \geq 10) \qquad (3.78)$$

The voltage across the capacitor is determined by

$$V_C = I_s \cdot Z_p$$

where I_s is the constant magnitude of the current source and Z_p is the frequency-dependent impedance of the tank circuit as depicted by Fig. 3.99. Since I_s is a constant, the voltage V_C will have the same shape as the impedance curve as shown in Fig. 3.100.

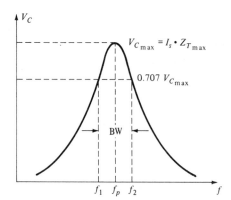

Figure 3.100 Resonance curve for "tank" circuit.

The bandwidth is determined by

$$\boxed{BW = f_p/Q_p} \qquad (3.79)$$

and the half-power frequencies are defined as for the series resonant circuit. This network, therefore, will allow us to *pick* the frequencies that we would like to have maximum output voltage for the next stage. The tuning dial on a radio adjusts the capacitance and therefore the resonant frequency to the station we are interested in listening to. All other stations at the low ends of the curve receive less voltage and therefore power and are not audible.

EXAMPLE 3.17

For the tank circuit of Fig. 3.101, determine

1. X_C for resonance.
2. The impedance at resonance.
3. The resonant frequency if $L = 10$ μH.
4. The cutoff frequencies.
5. The shape of the resonant curve.

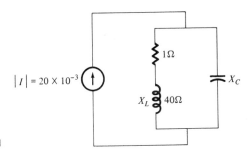

Figure 3.101

Solution:

1. $Q_p = X_L/R = 40/1 = 40 > 10$; therefore, $X_C = X_L = $ **40 Ω.**
2. $Z = Q_p^2 R = (40)^2 \times 1 = 1600(1) = $ **1600 Ω.**
3. $X_L = 2\pi f_p L \Rightarrow f_p = X_L/2\pi L = [40/(6.28)(10 \times 10^{-6})] = $ **637 kHz.**
4. $BW = f_p/Q_p = 637{,}000/40 = 15{,}925$ Hz, and

$$f_1 = f_p - \frac{BW}{2} = 637{,}000 - \frac{15{,}925}{2} = \textbf{621{,}075 Hz}$$

$$f_2 = f_p + \frac{BW}{2} = 637{,}000 + \frac{15{,}925}{2} = \textbf{652{,}925 Hz}$$

5. The resonant curve appears in Fig. 3.102.

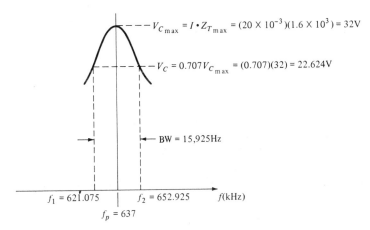

Figure 3.102

3.15 FILTERS

Filters are simply an extension of the tuned networks introduced in the previous section. As the name would imply filters pick out a range of frequencies for passage or blockage. They *filter out* the unwanted frequencies. The first to be described is called the *band-pass* filter since it "passes" a particular range of frequencies. In Fig. 3.103 a series resonant circuit was designed to perform this function. The output is

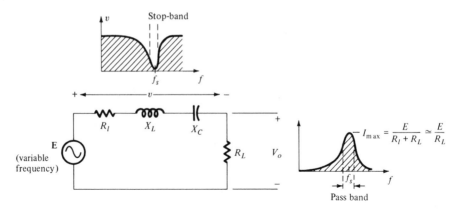

Figure 3.103 Series resonant filter.

taken off the resistor R_L. Usually the resistor R_L is considerably larger than R_l so that at the resonance condition the major portion of the applied voltage will appear across R_L or as V_0. At resonance, the impedance of the series R_l, X_L, and X_C circuit is a minimum, and

$$V_0 = \frac{R_L E}{R_L + R_l} \simeq E$$

At lower frequencies the reactance of the capacitor increases, resulting in an increasing part of E appearing across the resonant circuit. At higher frequencies the reactance of X_L will override, and an increasing part of E will again appear across the resonant circuit. The output frequency curve appears in Fig. 3.103. Only those frequencies near the resonant value will pass to the next stage. If the output were taken off the resonant circuit, we would have a *band-stop* filter where only a certain band of frequencies is not permitted to pass.

A similar discussion can also be applied to the network of Fig. 3.104, which employs a parallel resonant circuit for its selectivity. As a *pass-band* filter, the output is taken across the tank circuit since the impedance of the network is a maximum only near resonance. This will result in $V_0 \simeq E$ since the impedance at the resonant frequency is much greater than that of R. The frequency characteristics appear in Fig.

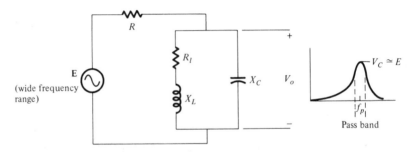

Figure 3.104 Parallel resonant filter.

3.104. Again the band-stop characteristics could be obtained by taking the output off the resistor R.

EXAMPLE 3.18

Sketch the output frequency characteristic for the network of Fig. 3.105.

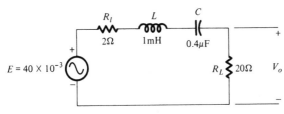

Figure 3.105

Solution:

The resonant frequency is

$$f_s = \frac{1}{2\pi\sqrt{LC}} = \frac{1}{(6.28)\sqrt{(1 \times 10^{-3})(0.4 \times 10^{-6})}}$$

$$= \frac{1}{6.28\sqrt{4 \times 10^{-10}}} = \frac{1}{6.28(2 \times 10^{-5})} = \frac{10^4}{1.256}$$

$$\simeq \textbf{8 kHz}$$

$$\text{BW} = \frac{R}{2\pi L} = \frac{2 + 20}{(6.28)(1 \times 10^{-3})} = \frac{22}{6.28} \times 10^3 = \textbf{3.51 kHz}$$

$$f_1 = f_s - \frac{\text{BW}}{2} = 8 - \frac{3.51}{2} = \textbf{6.245 kHz}$$

$$f_2 = f_s + \frac{\text{BW}}{2} = 8 + \frac{3.51}{2} = \textbf{9.755 kHz}$$

$$V_{\max} = \frac{(20)(40 \times 10^{-3})}{20 + 2} = \frac{800}{22} \times 10^{-3} = \textbf{36.4 mV}$$

The pass-band output appears in Fig. 3.106.

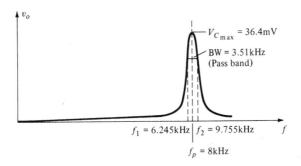

Figure 3.106

PROBLEMS

§ 3.2

1. Sketch the waveforms
   (a) 120 sin α.
   (b) 0.04 sin α.
2. Convert the following:
   (a) $\frac{3}{2}\pi$ rad to degrees.
   (b) 60° to radians.
3. How long will it take a 200-Hz sine wave to pass through 40 cycles?
4. A sinusoidal waveform has a frequency of 10 kHz. Determine the period.
5. Sketch the following waveforms:
   (a) 60 sin 157t.
   (b) 20 × 10⁻³ sin 200t.
6. (a) Determine the frequency of the waveform of Problem 5(a).
   (b) How long will it take for that waveform to pass through five cycles?
7. Sketch the following waveforms:
   (a) $v = 100 \sin(377t + 60°)$.
   (b) $i = 50 \times 10^{-6} \sin(400t - 30°)$.
8. (a) What is the phase relationship between the waveforms of Problem 7?
   (b) Repeat the problem for the following set:

$$v = 4 \sin(\omega t + 60°)$$
$$i = 0.8 \sin(\omega t - 300°)$$

§ 3.3

9. (a) Determine the effective value of each quantity in Problem 8.
   (b) Write the sinusoidal expression for a voltage having an effective value of 220 V.

§ 3.4

10. Determine the average value of the waveform of Fig. 3.107.

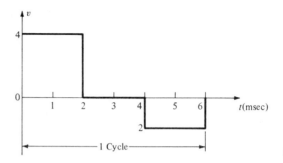

Figure 3.107

§ 3.5

11. Determine the sinusoidal expression for the voltage drop across a 2-K resistor if we pass a current of 10 × 10⁻³ sin 377t through it.
12. Determine the power to the resistor of Problem 11.
13. Calculate the reactance of a 6-H coil at a frequency of 60 Hz.

14. A sinusoidal current, $2 \sin(\omega t + 90°)$ flows through a 10-mH coil. Find the expression for the voltage across the coil.

15. Calculate the reactance of a 5-μF capacitor at a frequency of 60 Hz.

16. A sinusoidal voltage $30 \sin \omega t$ is applied across a 6-μF capacitor. Determine the expression for the resulting current.

§ 3.6

17. Convert the following to the other domain.
    (a) $50 \angle 60°$.   (b) $120 \angle 120°$ (c) $0.04 \angle -60°$.
    (d) $5 + j5$.   (e) $5 + j12$.   (f) $40 - j80$.

18. Perform the following operations:
    (a) $(6 + j6) + (4 - j5)$.
    (b) $(20 \angle 20°) \cdot (60 \angle -30°)$.
    (c) $(30 \times 10^{-3} \angle 10°)/(60 \times 10^2 \angle -50°)$.
    (d) $(120 \angle 0°) + (60 \angle -60°)$.

19. Using phasor notation, determine the voltage across a 5-K resistor if the current through it is $30 \times 10^{-3} \sin(\omega t + 60°)$.

20. Using phasor notation, determine the current through a 200-mH coil if the voltage across it is $40 \sin \omega t$.

21. Using phasor notation, determine the voltage across a 10-μF capacitor if the current is $0.04 \sin 377t$.

§ 3.7

22. For the series ac network of Fig. 3.108, determine

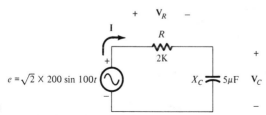

Figure 3.108

    (a) The reactance of the capacitor.
    (b) The total impedance and the impedance diagram.
    (c) The current I.
    (d) The voltages V_R and V_C using Ohm's law.
    (e) The voltages V_R and V_C using the voltage-divider rule.
    (f) The power to R.
    (g) The power supplied by the voltage source e.
    (h) The phasor diagram.
    (i) The F_p of the network.

23. For the series RLC network of Fig. 3.109, determine
    (a) Z_T.
    (b) I.
    (c) V_R, V_L, V_C using Ohm's law.
    (d) V_L using the voltage-divider rule.
    (e) The power to R.

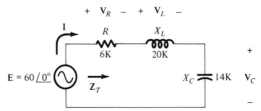

Figure 3.109

(f) The F_p.

(g) The phasor and impedance diagrams.

§ 3.8

24. For the parallel RC network of Fig. 3.110, determine

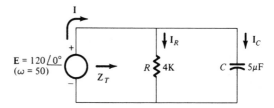

Figure 3.110

(a) Z_T, X_T.

(b) I.

(c) I_R, I_C using Ohm's law.

(d) I_R using the current-divider rule and the results of part (b).

(e) The total power delivered to the network.

(f) The power factor of the network.

(g) The admittance diagram.

§ 3.9

25. Determine the current I for the network of Fig. 3.111.

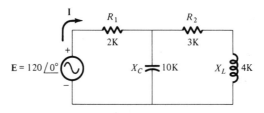

Figure 3.111

26. Determine the total impedance of the network of Fig. 3.112.

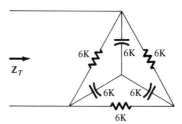

Figure 3.112

27. (a) Is the bridge of Fig. 3.113 balanced?

(b) If not, which element could be changed to achieve this condition?

28. Determine the current *I* in Fig. 3.114 using branch-current analysis.

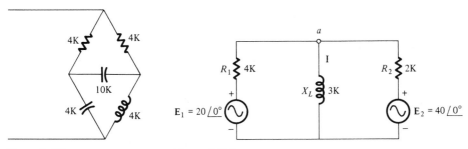

Figure 3.113   **Figure 3.114**

§ 3.10

29. Convert each source in Problem 28 to a current source.

30. Determine the voltage at node *a* (Fig. 3.114) using the results of Problem 28.

§ 3.11

31. Determine the current *I* of Problem 28 using superposition.

32. Find the Thevenin equivalent circuit for the network to the left of the load Z_L in Fig. 3.115.

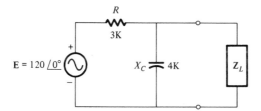

Figure 3.115

33. (a) Determine Z_L for maximum power to Z_L in Fig. 3.115.

(b) Calculate the maximum power that can be delivered to Z_L.

§ 3.12

34. For the network of Fig. 3.116, determine

(a) The current through each branch using Ohms' law.

(b) The total power dissipated (real power).

(c) The net reactive power.

(d) The total apparent power.

(e) The network power factor.

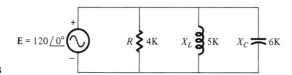

Figure 3.116

§ 3.13

35. A 3φ four-wire Y-Y-connected system has a 4-K impedance coil in each phase of the load. If the line voltage is 220 V, determine
   (a) The generated and load phase voltage.
   (b) The phase currents of the load.
   (c) The line current magnitudes.
   (d) The total power delivered to the load.

36. Repeat Problem 35 for a 3φ, three-wire Y-Δ system.

§ 3.14

37. For the series *RLC* network of Fig. 3.117, determine

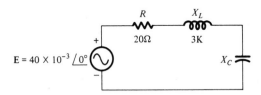

Figure 3.117

   (a) X_C for resonance.
   (b) Q_s.
   (c) BW if $f_s = 5$ kHz.
   (d) The power delivered at resonance and the half-power frequencies.
   (e) *L* and *C* using the information of part (c).

38. For the parallel resonant circuit of Fig. 3.118, determine

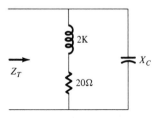

Figure 3.118

   (a) The value of X_C for resonance.
   (b) The impedance at resonance.
   (c) F_p if $C = 0.01$ μF.
   (d) The cutoff frequencies.
   (e) The resonance curve (rough sketch).

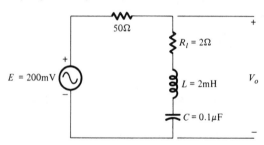

Figure 3.119

§ 3.15

39. For the filter of Fig. 3.119,
    (a) Determine whether it is a pass-band or stop-band filter.
    (b) Calculate the resonant frequency.
    (c) Find the BW and cutoff frequencies.
    (d) Determine V_{max} at resonance.
    (e) Sketch the output waveform (vs. frequency).

4

Magnetics

4.1 INTRODUCTION

The very importance of magnetic effects in a multitude of applications requires that a chapter be devoted to introducing the basic concepts and a few areas of application. The inductor, transformer, mechanical relay, speakers, tape recordings, microphones, etc., all depend on magnetic effects to function properly.

The controlling variables of a magnetic circuit are so closely related to the variables of an electric circuit that the basic concepts of electric circuits can be easily related to magnetic circuits and serve as a guide through the analysis to follow.

There is only one sometimes confusing aspect of magnetic circuits that did not appear for electric circuits that can be cleared up at this point in the development if a moment is devoted to ensuring that the effect of using the proper unit of measurement in an equation is understood. The difficulty arises because three different systems of units are commonly used in relation to magnetic systems: the MKS, CGS, and English. Although each is employed to its fullest advantage in a particular area of application of magnetic circuits, this chapter would become overly cumbersome if an attempt were made to use all three. Rather, the English system, which is frequently applied in industrial applications, will be employed, and a conversion table will have to be consulted if the other units of measurement are encountered. Whichever system is employed, it is absolutely necessary that each quantity in a particular equation have a unit of measurement that is in only one system of units. They cannot be interchanged unless the equation calls for it specifically. Each equation in this chapter will be in the English system. If the units of measurement of a particular problem are all in the MKS or CGS systems, simply convert each to the English using a conversion table and substitute into the equations as they appear in this chapter.

4.2 BASIC PROPERTIES OF MAGNETISM

We have all, at one time or another, enjoyed experimenting with a horseshoe or bar *permanent* magnet, as shown in Fig. 4.1. If two north (or south) poles were brought close to each other, they would make every effort to avoid or repel the other. The reverse would be true if opposite poles were brought together. In other words, *like poles repel, while unlike poles attract*. This phenomenon is demonstrated for each case in Fig. 4.2 using *magnetic flux lines*, which through their density can show the strength of the magnetic field in a particular region. They are continuous lines that leave the north pole of a magnet, return to the south pole, and pass through the magnetic material to the north pole. Flux lines characteristically try to be as short as possible in their continuous path. In Fig. 4.2(a), therefore, the flux lines that pass directly from one magnet to the other will tend to draw the two magnets together. For the unlike poles a buffer action exists between the poles.

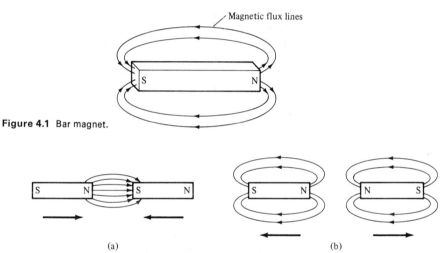

Figure 4.1 Bar magnet.

Figure 4.2 (a) attraction (b) repulsion.

Permanent magnets such as appear in Fig. 4.2 are made of what is referred to as *ferromagnetic materials*. They are materials such as iron or steel that characteristically will permit the setting up of flux lines through them with very little external pressure. Once magnetized their retention properties are such that they exhibit magnetic properties after the external magnetizing agent is removed.

The current through a conductor will establish a circular flux pattern around the conductor such as shown in Fig. 4.3. The direction of the flux lines can be deter-

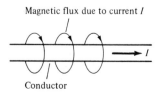

Figure 4.3 Magnetic flux lines surrounding a current carrying conductor.

mined by placing the thumb of the right hand in the direction of current flow and noting the direction of the fingers. This determination is shown in Fig. 4.3. If we wrap a ferromagnetic material with a conducting wire as shown in Fig. 4.4(a), a strong magnetic field can be established through the core. An expanded view of three adjacent windings is shown in Fig. 4.4(b) with the resulting flux patterns for each loop as determined by the technique demonstrated in Fig. 4.3. Note that between the windings there appears to be a canceling affect on the net flux, while above and below the two windings there is a strengthening of the flux in each direction. The net result for the wrapped core of Fig. 4.4(a) is the north pole at the right since the flux patterns are leaving and a south pole at the left since the flux patterns enter. The direction of the resultant flux can be determined by placing the fingers of the right

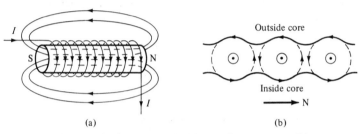

(a)   (b)

Figure 4.4 Electromagnetic (a) complete structure (b) three adjacent windings.

hand in the direction of current flow for the wrappings as shown in Fig. 4.5. The thumb will then point in the direction of the resultant flux. This particular structure, which requires electric current to develop its magnetic characteristics, is called an *electromagnet*. The strength of the magnet can be increased by increasing the current through the conductors. Eventually, however, a condition called *saturation* will be reached where any additional current will have very little effect on the strength of the magnet. Once the magnetizing current is removed the magnetic characteristics of the ferromagnetic core will decrease to a value determined only by the ability of that particular sample to retain the magnetic characteristics—a measure of its *retentivity*. Materials with high retentivities are used to make *permanent* magnets by simply subjecting them to the strong magnetic field such as developed by the coil of Fig. 4.4.

The electromagnetic effect is employed in applications too numerous to possibly list here. However, we shall examine a few of the more interesting areas of usage. A photograph and sectional drawing of a *magnetic relay* appear in Fig. 4.6. When the specified voltage is applied to the coil the electromagnet will be energized by the

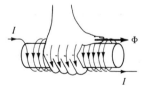

Figure 4.5 Determining the direction of flux developed by an electromagnet.

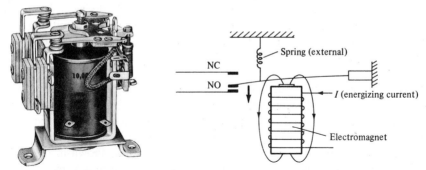

Figure 4.6 Magnetic relay (Courtesy Potter and Blumfield (AMF)).

resulting current. It will draw the arm (armature) down away from the normally closed (NC) contacts to the normally open (NO) contacts. When the energizing voltage is removed the spring will draw the contacts back to the NC terminals.

The internal construction of a circuit breaker appears in Fig. 4.7. The electromagnet at the bottom is sensitive to the service current passing through the breaker. When too high a level is reached the movable arm to the bottom left will be drawn to the electromagnet and the circuit opened to prevent damage in that circuit due to excessive currents. Although it would seem the basic construction of such a device could be quite simple, the additional complexity in appearance in Fig. 4.7 is there to ensure proper operation even when subjected to shock, vibration, or thermal variations.

If the core in the coil is not fixed but free to move, a whole new vista of application

Figure 4.7 Circuit breaker (Courtesy Thomas Sundheim Inc.).

of electromagnetic effects is possible. In this capacity the electromagnetic system is referred to as a solenoid. In Fig. 4.8, the locking bolt on the door can be controlled by the energizing or nonenergizing of the coil. When the proper signal voltage is applied to the coil the magnetizing effect will draw the core back into the coil and permit the door to be opened. When deenergized the spring will draw the core back and lock the door.

The last application of the electromagnet to be briefly described here is in the operation of an electric bell such as shown in Fig. 4.9. When the button is depressed the circuit is completed, and current will flow through the coil, drawing the striker to the gong and sounding the signal, alarm, etc. However, note that once the striker moves toward the sounder, the NC contacts will be open as shown in Fig. 4.9. The electromagnet will then be deenergized and the striker free to return under the action of the spring to the initial position. However, when it does return, current can flow again, and the cycle will continue to repeat itself until the controlling switch is opened.

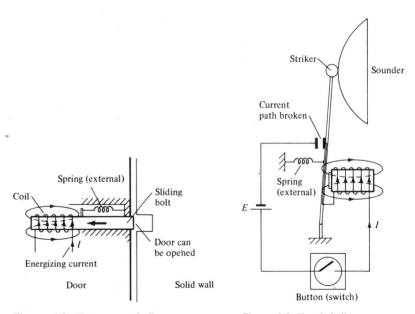

Figure 4.8 Electromagnetically controlled lock.

Figure 4.9 Electric bell.

4.3 MAGNETIC CIRCUITS

Let us now examine in more detail the controlling variables of the simple applications of magnetic effects introduced in the previous section. The magnetic flux established by the permanent magnet or through the electromagnetic core is defined by the Greek letter phi (Φ) and measured in *lines* in the English system. Its properties are very similar to those of current in an electric circuit. In an electric circuit the current will always seek the path of least resistance. The flux in a magnetic system will always seek the path of least *reluctance* ($\Re$). Ferromagnetic materials have very low values of reluctance, while substances such as air, wood, and glass have a very high reluctance. In other words, flux lines will always strive to establish the shortest path possible through the ferromagnetic materials such as iron and steel in the region of magnetic excitation. The reluctance of a sample of material is determined by the following equation, as depicted in Fig. 4.10:

$$\mathcal{R} = \frac{l}{\mu A} \qquad (4.1)$$

where l = length in inches
A = area in square inches
μ = permeability
$\mathcal{R}$ = reluctance in rels

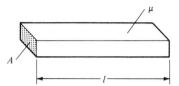

Figure 4.10 Factors affecting the reluctance of a magnetic sample.

The permeability μ is a measure of how easily flux lines can be set up in the material. The relative permeability is given by

$$\mu_r = \frac{\mu}{\mu_0} \qquad (4.2)$$

where μ_r = relative permeability
μ = permeability
μ_0 = permeability of air

It is actually a measure of how good a magnetic material it is compared to air. In the English system the permeability of air is

$$\mu_0 = 3.2 \text{ lines/A-m} \qquad (4.3)$$

For ferromagnetic materials, $\mu_r \geq 100$, while for air, wood, glass, etc., μ_r is for all practical purposes 1. Values of μ_r are not tabulated since they are dependent on the operating conditions of the magnetic circuit.

The fact that magnetic flux lines will always take the path of least reluctance can be put to good use in the shielding of instruments, electronic circuitry, etc., that may be quite sensitive to magnetic effects. By surrounding the piece of equipment by a *shield* of ferromagnetic material such as shown in Fig. 4.11 the magnetic flux will pass through the ferromagnetic material of the shield rather than disturb the sensitive instrumentation inside.

Series, parallel, and series-parallel configurations are also encountered in magnetic systems. However, the majority of the simpler applications of magnetic effects appear in the realm of *series magnetic circuits*. A great deal of literature is available for research into the other areas.

For the series magnetic circuit such as shown in Fig. 4.12 the flux (like the current in a series electric circuit) is the same throughout. The "pressure" (like the applied emf of an electric circuit) that results in the flux in the magnetic circuit is determined by the product of the number of turns, *N*, in the coil and the current through the coil, *I*. It is called the *magnetomotive force* (mmf) of the magnetic circuit and is given the symbol ℱ with the units *ampere-turns*. That is,

$$\mathcal{F} = NI \qquad \text{[ampere-turns (At)]} \qquad (4.4)$$

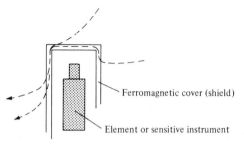

Figure 4.11 Magnetic shield.

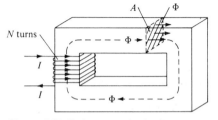

Figure 4.12 Series magnetic circuit.

For any series magnetic circuit the flux established by a particular mmf is determined by the reluctance of the core. The greater the opposition, the less the magnetic flux. The Ohm's law equivalent for magnetic circuits is therefore given by

$$\Phi = \frac{\mathcal{F}}{\mathcal{R}} = \frac{NI}{\mathcal{R}} \qquad \text{(lines)} \qquad (4.5)$$

In Fig. 4.12 the direction of the resulting flux can be found in the same manner as for the electromagnet. Place the fingers of the right hand in the direction of the current through the turns of the coil, and the thumb of that hand will point in the direction of the flux.

The flux per unit area of the core (note Fig. 4.12) is called the *flux density*, has the symbol *B*, and is measured in lines per square inch, as determined by the following equation:

$$B = \frac{\Phi}{A} \qquad (4.6)$$

where $B = \text{lines/in.}^2$
  $\Phi = \text{lines}$
  $A = \text{in.}^2$

Meters are available for measuring the flux density present. A photograph of a

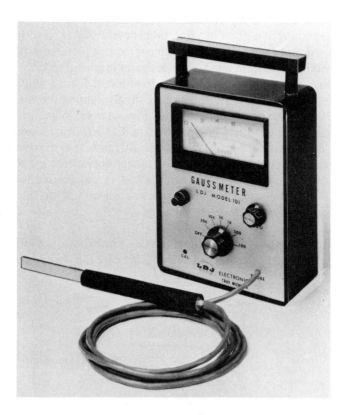

Figure 4.13 Gaussmeter (Courtesy LDJ Electronics, Inc.).

Gaussmeter appears in Fig. 4.13. The conversion factor as provided is the following:

$$1 \text{ gauss (G)} = 6.452 \text{ lines/in.}^2 \qquad (4.7)$$

For example, if a reading of 5×10^2 G were obtained, the flux density in lines per square inch would be the following:

$$\left[\frac{6.452 \text{ lines/in.}^2}{1 \text{ G}} \right] [5 \times 10^3 \text{ G}] = 32.26 \times 10^3 \text{ lines/in.}^2$$

The magnetomotive force *per unit length* required to establish a particular flux in a core is called the magnetizing force (*H*). In equation form,

$$H = \frac{\mathfrak{F}}{l} = \frac{NI}{l} \qquad \text{(At/in.)} \qquad (4.8)$$

The magnetizing force and the flux density are related in the following manner:

$$B = \mu H \qquad (4.9)$$

A curve of B vs. H for a particular magnetic sample appears in Fig. 4.14. Note in the first quadrant that a point of saturation is eventually reached where a further increase in the current I does not result in an appreciable increase in the flux through the core. Something to keep in mind is the design of magnetic systems where a strong flux field is required but limits exist in the current levels. Since a multitude of curves such as shown in Fig. 4.14 can be obtained by simply limiting the variables for each pass in the B-H curve, the singular curve of Fig. 4.15 was obtained from the B-H curves for application purposes. The curve for Fig. 4.15 is called the *normal magnetization* curve, while the cycle of Fig. 4.14 is called a B-H or *hysteresis* curve. Since curves of the type described above are quite important for application purposes, a sophisticated piece of equipment such as shown in Fig. 4.16 was devised to determine the B-H curve for magnetic samples. For a particular value of B the value of μ can be determined from Fig. 4.15 after determining the corresponding values of H and using Eq. (4.9).

As an example in the use of the equations described above, consider the inductor of Fig. 4.17, which has a ferromagnetic core to increase the strength of the magnetic

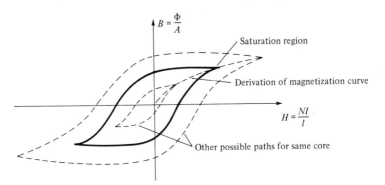

Figure 4.14 Hysteresis curve (cast steel).

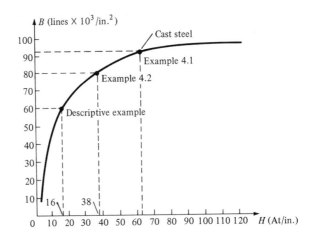

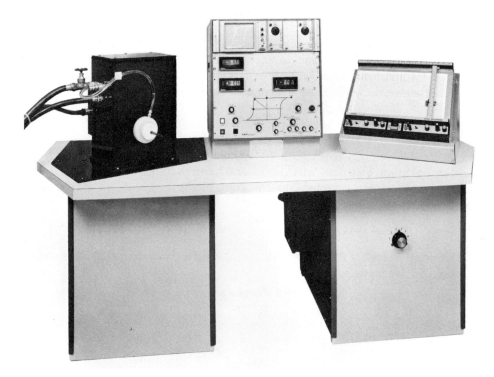

Figure 4.16 B–H meter (Courtesy LDJ Electronics, Inc.).

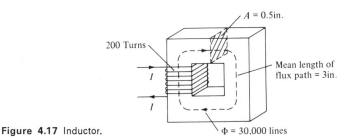

Figure 4.17 Inductor.

field linking the coil and thereby increase its inductance level. The current I necessary to establish a flux of 30,000 lines in the core is to be determined. It would be quite straightforward if we could simply determine the reluctance from the material and dimensions of the core and then plug into Ohm's law equation for magnetic circuits [Eq. (4.5)]. Unfortunately, however, the reluctance cannot be determined since the value of μ is determined by the conditions of operation of the magnetic circuit and cannot simply be read off a chart, as was true for the relative permittivity of the dielectrics discussed for capacitive elements. Rather, we must first apply a law to the magnetic circuit that is very similar to Kirchhoff's voltage law for electric circuits. The law, called *Ampere's circuital law*, states that the algebraic sum of the mmfs around a closed path is zero. The application of the law requires that Eq. (4.8) be written in the form $\mathfrak{F} = Hl$, so that the mmf drops across the branches of the magnetic circuit can be included.

For the series magnetic circuit of Fig. 4.17 the impressed mmf (like the applied voltage E of a series electric circuit) is NI, while the mmf drops (like the voltage drops across the resistive loads of the series circuit) is $H_{abcd}l_{abcd}$. Applying Ampere's circuital law, the algebraic sum of the mmfs must equal zero, and

$$NI - H_{abcd}l_{abcd} = 0$$

or

$$NI = H_{abcd}l_{abcd}$$

The unknown quantities in the equation obtained include I and H_{abcd}. The latter quantity can be determined directly from Fig. 4.15 once B_{abcd} is determined. Therefore,

$$B_{abcd} = \frac{\Phi}{A} = \frac{30,000}{0.5} = 60,000 \text{ lines/in.}^2$$

From the curve we find $H_{abcd} \simeq 16$ At/in. for the material used. Substituting into the equation above,

$$NI = H_{abcd}l_{abcd}$$
$$(200)I = (16)(3)$$

and

$$I = \frac{48}{200} = \textbf{240 mA}$$

The value of μ can then be determined for the ferromagnetic material using Eq. (4.9):

$$\mu = \frac{B}{H} = \frac{60,000}{16} = \textbf{3750} \gg \mu_0$$

The general technique for analyzing magnetic circuits has thus been described. In total, it differs only slightly, due to the need for a B-H curve, from the analysis of electric circuits. In this past example, the applied mmf was to be determined. For a simple series magnetic circuit such as the one just considered, if the impressed mmf were provided, the resulting flux could be determined by simply reversing the sequence of steps. However, for more complex networks such as the series-parallel configurations, the flux cannot be determined directly but must be determined through a "cut and try again" technique. Incidentally, for series-parallel configurations, the dual of Kirchhoff's current law is applicable here in the sense that the total magnetic flux entering a magnetic junction must equal that leaving. That is, for two parallel branches, $\Phi_T = \Phi_1 + \Phi_2$, where Φ_T is the net flux entering the junction.

EXAMPLE 4.1

Determine the flux in the core of the transformer of Fig. 4.18.

Solution:

Since the second winding is left open, the current through the coil is zero, and its mmf is zero. If will not affect the magnitude of the flux Φ.

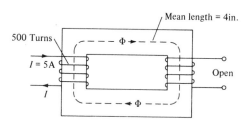

Figure 4.18 Transformer.

Applying Ampere's circuital law,

$$NI = H_{core}l_{core}$$

Substituting,

$$(50 \times 5) = H_{core}(4)$$

and

$$H_{core} = \frac{250}{4} = 62.5 \text{ At/in.}$$

and from the curve of Fig. 4.15 we find

$$B_{core} \simeq 93,000 \text{ lines/in}^2.$$

and

$$\Phi_{core} = B_{core}A_{core}$$
$$= (93 \times 10^3)(0.8)$$
$$\Phi_{core} = \textbf{74,400 lines}$$

In a number of applications involving magnetic effects, air gaps will be encountered. In Fig. 4.19, the air gap is quite evident for the magnetic relay.

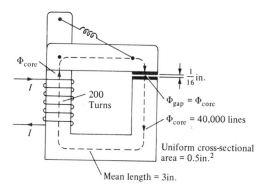

Figure 4.19 Simple relay.

For air gaps, the flux density and magnetizing force are related by

$$\boxed{H_g = 0.313B_g} \qquad \text{(At/in.)} \qquad (4.10)$$

The example to follow will demonstrate that even the smallest air gaps with these high-reluctance characteristics will require the bulk of the impressed mmf to establish a flux of some magnitude in the core and across the gap. In fact, in most situations, the mmf required to establish the flux in the core can be ignored compared to that of the air gap when determining either the mmf required or the flux developed.

EXAMPLE 4.2

For the relay of Fig. 4.19, determine the mmf required to establish a flux of 40,000 lines in the core.

Solution:

Ampere's circuital law will result in

$$NI = H_{core}l_{core} + H_g l_g$$

an equation very similar to that obtained for a series electrical network where $E = V_1 + V_2$.

$$B_{core} = B_g = \frac{\Phi}{A} = \frac{40,000}{0.5} = 80,000 \text{ lines/in.}^2$$

and from Fig. 4.15,

$$H_{core} = 38 \text{ At/in.}$$

$$H_g = 0.313B_g = 0.313(80,000) = 25,040 \text{ At/in.}$$

Substituting into the equation obtained above,

$$(200)(I) = 38(3) + 25,040(\tfrac{1}{16} \text{ in.})$$

$$200I = 114 + 1565 \quad \text{(Note difference in magnitude between two terms)}$$

$$200I = 1679$$

$$I = \textbf{8.395 A}$$

Note the ratio of the mmf for the air gap as compared to that for the core: 1565/114 $\simeq$ 13.72:1, which certainly substantiates the statement above. That is, to assume $NI \simeq H_g l_g$ is certainly a valid approximation in this example.

The basic construction of the D'Arsonval movement as used in both dc and ac instruments is shown in Fig. 4.20. Note in this case that both a permanent horseshoe magnet and an electromagnet are present with two series air gaps between the movable and stationary parts of the movement. The nameplate data of the movement appear in the figure. The 1-K resistance is simply the dc resistance of the windings around the

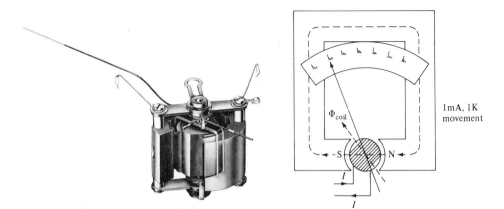

Figure 4.20 D'Arsonval movement (Courtesy Weston Instruments).

movable core. When 1 mA is passed through the movable core a flux will develop through the coil and core that will react with the flux of the permanent magnet and cause the movable core to turn on its shaft. For the full 1-mA rated current the pointer attached to the movable core will indicate a full-scale deflection. For lesser values of current the deflection is that much less on a linear (straight-line) basis.

Since the maximum current is 1 mA, how is an ammeter designed to read higher values such as 1 A? In Fig. 4.21 we find the internal circuitry for a 1-A meter using the movement of Fig. 4.20.

With 1 A applied (the maximum for that scale) we want the full deflection current of 1 mA through the movements as shown in the figure. The remaining current through R_{shunt} is then $1 - 0.001 = 0.999$ A $= 999$ mA; the voltage across the 1-K resistor is $(1 \text{ K})(1 \text{ mA}) = 1$ V, which is also across R_{shunt} since they are in parallel. Therefore

$$R_{shunt} = \frac{V}{I} = \frac{1}{0.999} \simeq 1.001 \ \Omega \simeq \mathbf{1} \ \boldsymbol{\Omega}$$

In other words, high current levels can be read by the sensitive movement by simply diverting that current above its maximum level to the parallel branch.

For voltmeters, a series resistor is added as shown in the 100-V voltmeter of Fig. 4.22 using the same movement. The maximum voltage that the movement can handle is 1 V. The remaining 99 V must appear across R_{series}. Therefore,

$$R_{series} = \frac{V}{I} = \frac{99}{1 \text{ mA}} = \mathbf{99 \ K}$$

Multirange ammeters and voltmeters can be designed using rotary switches that will insert the proper shunt or series resistor into the network.

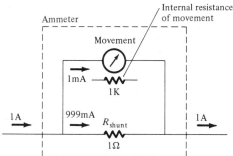

Figure 4.21 Ammeter.

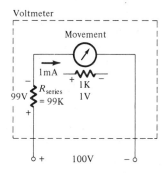

Figure 4.22 Voltmeter.

4.4 TRANSFORMERS

The *transformer* is an electromagnetic device primarily designed to perform one of the three following functions:

1. Step up or step down the voltage or current.
2. Act as an impedance-matching device.

3. Isolate one portion of a network from another.

The core of a transformer, as shown in Fig. 4.23, is laminated (made of thin metallic sheets separated by an insulating material) to reduce the *eddy current* losses in the core. Eddy currents are small circular currents that flow in a core induced by the changing flux through the core. The changing flux induces voltages in the core that set up these currents that result in an additional I^2R loss. By inserting the sheets of insulation the resistance in the path of the currents is so high that their magnitude is reduced significantly.

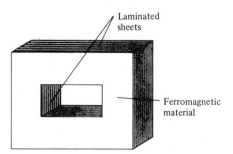

Laminated
sheets

Ferromagnetic
material

Figure 4.23   Transformer core.

A second loss in the core is called the *hysteresis* loss. It is a result of the reversing magnetic field causing the small magnetic alignments in the atoms of the material to reverse their polarities. This reversal motion results in a friction and therefore heating loss in the core. It can be reduced significantly by introducing small amounts of silicon into the core.

Transformers are ac devices. They cannot perform the functions listed earlier with dc inputs. Although the term dc transformer does appear, it is understood that there is an additional amount of circuitry required to perform this function. As shown in Fig. 4.23 at least two windings are required. The winding to which the signal is applied is called the *primary*, while the winding to which the load is connected is called the *secondary*.

The basic operation of the transformer is quite fundamental in nature. An ac voltage is applied to the primary which will induce a sinusoidally varying Φ in the core having the same frequency as the input signal. This sinusoidally varying flux in the core will induce a voltage across the secondary as determined by Faraday's law of magnetic induction, introduced earlier: $e = N\,d\phi/dt$. The result is a sinusoidally induced voltage at the secondary that has the same frequency as the applied signal and a magnitude related to the input by the turns ratio of the coils. That is, the voltages are related by

$$\frac{E_p}{E_s} = \frac{N_p}{N_s} \qquad (4.11)$$

where the indicated voltages are effective values. Solving for E_s, $E_s = (N_s/N_p)(E_p)$, we find that if the secondary turns are greater than the primary turns the secondary voltage will be greater than the primary voltage by the ratio of the turns. A quantity called the *transformation ratio* is defined by

$$\boxed{a = \frac{N_p}{N_s}} \tag{4.12}$$

If a is less than 1, the secondary voltage will be greater than the primary voltage, and we have a *step-up* transformer. If a is greater than 1, the reverse is true, and we have a *step-down* transformer.

The currents of the iron-core transformer are related by

$$\boxed{\frac{I_s}{I_p} = \frac{N_p}{N_s} = a} \tag{4.13}$$

Note that the currents are related by the inverse of the turns ratio, so that if the voltage should increase, the current will decrease and vice versa.

Setting the voltage and current ratios equal to the same turns ratio will result in

$$\frac{E_p}{E_s} = \frac{N_p}{N_s} = \frac{I_s}{I_p}$$

and since things equal to the same thing are equal to each other

$$\frac{E_p}{E_s} = \frac{I_s}{I_p}$$

or

$$\boxed{E_p I_p = E_s I_s} \tag{4.14}$$

which states that the apparent power of the primary must equal the apparent power of the secondary. Or, for resistance loads, the power applied is equal to the power to the load.

EXAMPLE 4.3

For the transformer of Fig. 4.24, determine

1. The transformation ratio.
2. The secondary voltage.
3. The secondary current.
4. The primary current.
5. The power to the load.
6. The power supplied by the source.

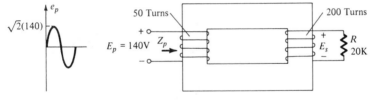

Figure 4.24

Solution:

1. $a = N_p/N_s = 50/200 = \frac{1}{4}$ (step-up transformer).
2. $E_s = (N_s/N_p)E_p = (200/50)(140) = $ **560 V.**
3. $I_s = E_s/R_L = 560/20 \text{ K} = $ **28 mA.**
4. $I_p = (N_s/N_p) I_s = (200/50)(28) = $ **112 mA.**
5. $P_L = E_s I_s = 560(28 \times 10^{-3}) = $ **15.68 W.**
6. $P_s = E_p I_p = (140)(112 \times 10^{-3}) = $ **15.68 W** $= P_L$.

If we divide Eq. (4.11) by (4.13) and perform a few mathematical manipulations, the following will result:

$$\frac{E_p/E_s = a}{I_p/I_s = 1/a} \Longrightarrow \frac{E_p/I_p}{E_s/I_s} = a^2 \Longrightarrow \frac{Z_p}{Z_s} = a^2$$

and

$$\boxed{Z_p = a^2 Z_s} \tag{4.15}$$

which in words states that the impedance "seen" at the primary of a transformer is the transformation ratio squared times the impedance connected to the secondary. Since the impedances are related by a constant (a^2), if a load is capacitive or inductive, it will also appear capacitive or inductive at the primary.

For Example 4.3,

$$Z_p = a^2 Z_L = \left(\frac{N_p}{N_s}\right)^2 R_L = \left(\frac{50}{200}\right)^2 20 = \left(\frac{1}{4}\right)^2 20 = \frac{20}{16}$$

$$= \textbf{1.25 K}$$

If impedance is to be transferred from the primary to the secondary, then it is multiplied by the factor $1/a^2$.

Equation (4.15) can be used to its full advantage in the application of the transformer as an *impedance-matching* device. Consider the following example.

EXAMPLE 4.4

Determine the turns ratio of the transformer in Fig. 4.25 to ensure maximum power to the resistive load and calculate the maximum power.

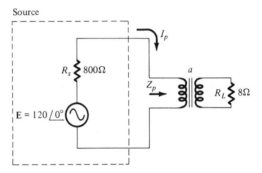

Figure 4.25

Solution:

The maximum power theorem states that maximum power is delivered to the load when its resistance equals the internal resistance (R_{Th}) of the supply. In this case R_L should be 800 Ω. As it stands, with $R_L = 8$ Ω, if the transformer were not present, the power to the load would be

$$P_L = I_L^2 R = \left(\frac{E}{R_s + R_L}\right)^2 R_L = \left(\frac{120}{800 + 8}\right)^2 \times 8 = (0.1485)^2 \times 8$$

$$= \textbf{176.5 mW}$$

However, if we insert the transformer and ensure that $Z_p = 800$ Ω, then the load will "appear" to be equal to R_s and maximum power will be delivered.

That is,

$$Z_p = 800 = a^2 R_L = a^2(8)$$

and

$$a^2 = \tfrac{800}{8} = 100$$

or

$$a = \frac{N_p}{N_s} = \textbf{10}$$

The primary current is then

$$I_p = \frac{120}{R_s + Z_p} = \frac{120}{1600} = \textbf{66.67 mA}$$

and

$$I_s = aI_p = 10(66.67 \times 10^{-3}) = \textbf{666.7 mA}$$

with the power to the load determined by

$$P_L = I_s^2 R_L = (0.667)^2 \times 8 = \textbf{3.559 W}$$

The increase over the power delivered without the transformer is

$$\frac{3.559}{0.1765} \simeq 20{:}1$$

Transformers are rated in kilovolt-amperes rather than kilowatts for the reason indicated in Fig. 4.26. Even though I_s may be well above the rated value and possibly cause severe damage to the secondary circuit, the wattmeter would show zero deflection (ideally speaking, no dc resistance is present in the system) since the load (C) is purely reactive. However, the apparent power rating of $E_p I_p$ or $E_s I_s$ would set a limit on the transformer currents for a particular voltage. In other words, an ammeter would be employed to determine maximum conditions for a fixed voltage rather than a wattmeter.

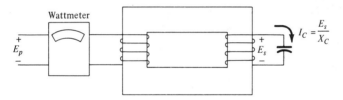

Figure 4.26

The third application of the transformer listed in the introduction to this section was as an isolation device. It will isolate a network insofar as dc and ground connectors are concerned but will permit the passage of ac quantities.

The autotransformer is fundamentally a two-circuit transformer hooked up in a manner that will improve its kilovolt-ampere rating but at the expense of its isolation characteristics. Consider the filament transformer of Fig. 4.27 used to provide the 6 V necessary to heat the filaments of the tubes still appearing in some radio and television receivers.

For the filament transformer we find that the condition of Eq. (4.14) is satisfied. That is,

$$E_p I_p = E_s I_s$$
$$(120)(0.1) = (6)(2)$$
$$12 = 12 \quad \checkmark$$

When used as an autotransformer it has the general appearance of Fig. 4.28. Note the loss of isolation but the increased secondary voltage (now 126 V). The current through each winding must be the same (as indicated on the diagram) so that $I_s = 2$ A but $I_p = 2 + 0.1 = 2.1$ A. The new volt-ampere rating of the device is

$$E_p I_p = (120)(2.1) = E_s I_s = (126)(2) = 252 \text{ VA}$$

Current transformers such as appearing in Fig. 4.29 provide a range of current levels when the specified current is applied to the primary.

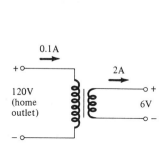

Figure 4.27 Filament Transformer.

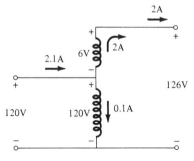

Figure 4.28 Autotransformer.

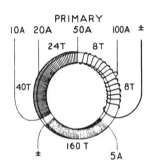

Figure 4.29 Current transformer (Courtesy Weston Instruments).

4.5 OTHER AREAS OF APPLICATION

The *iron-vane* meter movement of Fig. 4.30 can read the effective value of any network without the additional circuitry required for the D'Arsonval movement. Two vanes are connected to a common hinge within a coil that carries the current to be measured. The stronger the current, the stronger the opposite poles induced in the vanes and the greater the deflection. The scale is then properly calibrated to indicate the effective value.

The *electrodynomometer movement* of Fig. 4.31 is the most expensive of the three movements but usually the movement with the highest degree of accuracy and the only one used in the design of a wattmeter. It has a movable coil as shown in the figure that is free to rotate inside a fixed coil. Each coil will sense the current, voltage, or power being measured, and the two fluxes will react to develop a torque on the movable coil to which the pointer is attached. This movement will also indicate the effective value of any waveform.

Electromagnetic effects are the moving force in the design of speakers such as shown in Fig. 4.32. The shape of the pulsating waveform of the input current will be determined by the sound to be reproduced by the speaker at a high audio level. As the current peaks and returns to the valleys of the sound pattern the strength of the electromagnet will vary in exactly the same manner. This will cause the cone of the speaker to vibrate at a frequency directly proportional to the pulsating input. The higher the pitch of the sound pattern, the higher the oscillating frequency between the peaks and valleys and the higher the frequency of vibration of the cone.

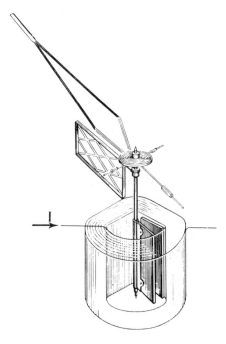

Figure 4.30 Iron-vane movement (Courtesy Weston Instruments).

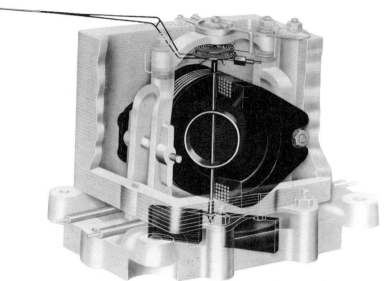

Figure 4.31 Electrodynomometer movement (Courtesy of Weston Instruments).

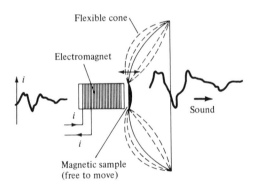

Figure 4.32 Speaker.

A second design used more frequently in more expensive speaker systems appears in Fig. 4.33. In this case the permanent magnet is fixed and the input applied to a movable core within the magnet as shown in the figure. High peaking currents at the input will produce a strong flux pattern in the voice coil, causing that coil to be drawn well into the flux pattern of the permanent magnet. As occurred for the speaker of Fig. 4.32, the core will then vibrate at a rate determined by the input and provide the audible sound.

Microphones such as appear in Fig. 4.34 also employ electromagnetic effects. The sound to be reproduced at a higher audio level will cause the core and attached moving coil to move within the magnetic field of the permanent magnet. Through Faraday's law ($e = N\, d\phi/dt$) a voltage will be induced across the movable coil proportional to the speed with which it is moving through the magnetic field. The resulting induced voltage pattern can then be amplified and reproduced at a much higher audio level through the use of speakers described earlier. Microphones of

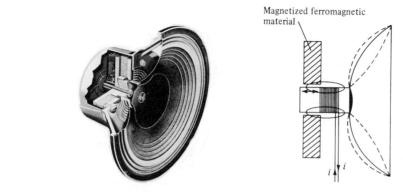

Magnetized ferromagnetic
material

Figure 4.33 Coaxial high-fidelity loudspeaker (Courtesy
Electro-Voice, Inc.).

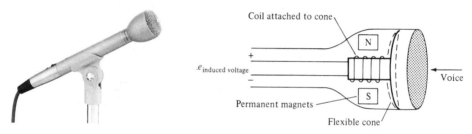

Coil attached to cone

N

+

e induced voltage

−

Voice

Permanent magnets

S

Flexible cone

Figure 4.34 Dynamic microphone (Courtesy Electro-Voice,
Inc.).

this type are the most frequently employed, although other types that employ capacitive, carbon granular, and piezoelectric* effects are available. This particular design is commercially referred to as a *dynamic* microphone.

Magnetic effects play a number of important roles in the telephone handset and base as shown in Fig. 4.35. In the base a ringer coil is intermittently energized as described for the bell in Section 4.3, causing the clapper to repeatedly strike the gongs until the handset is removed from the cradle, opening the bell circuit. The details of the receiver portion are shown in Fig. 4.36. The current through the coil of the electromagnet is controlled by the voice pattern to be reproduced. It will control the varying strength of the electromagnet and cause the vibrating diaphragm to vibrate at the same rate. The vibrating diaphragm will cause the surrounding air to vibrate in the same manner, causing the audible sounds or voice of the sender.

The transmitter or sending unit has the basic construction shown in Fig. 4.37. When the handpiece of the telephone is lifted, current will flow through the carbon granulars. As you speak into the handset the pressure on the diaphragm will vary in accordance with the voice pattern. The greater the pressure on the diaphragm, the more compact are the carbon granulars and the less the resistance encountered for

---

*Piezoelectricity is the generation of a small emf by exerting pressure across certain crystals.

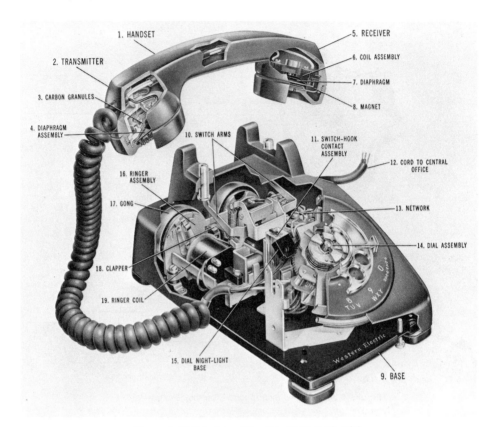

1. HANDSET
2. TRANSMITTER
3. CARBON GRANULES
4. DIAPHRAGM ASSEMBLY
5. RECEIVER
6. COIL ASSEMBLY
7. DIAPHRAGM
8. MAGNET
10. SWITCH ARMS
11. SWITCH-HOOK CONTACT ASSEMBLY
12. CORD TO CENTRAL OFFICE
16. RINGER ASSEMBLY
17. GONG
13. NETWORK
14. DIAL ASSEMBLY
18. CLAPPER
19. RINGER COIL
15. DIAL NIGHT-LIGHT BASE
9. BASE

Figure 4.35 Telephone (Courtesy Western Electric).

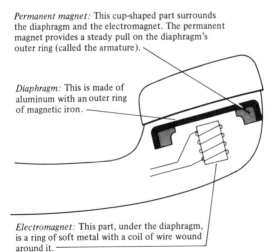

Permanent magnet: This cup-shaped part surrounds the diaphragm and the electromagnet. The permanent magnet provides a steady pull on the diaphragm's outer ring (called the armature).

Diaphragm: This is made of aluminum with an outer ring of magnetic iron.

Electromagnet: This part, under the diaphragm, is a ring of soft metal with a coil of wire wound around it.

Figure 4.36 Telephone receiver (Courtesy Western Electric Co.).

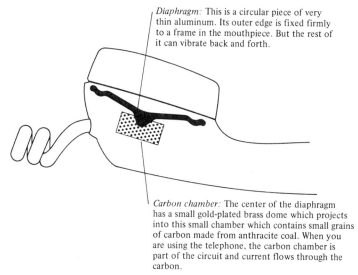

Diaphragm: This is a circular piece of very thin aluminum. Its outer edge is fixed firmly to a frame in the mouthpiece. But the rest of it can vibrate back and forth.

Carbon chamber: The center of the diaphragm has a small gold-plated brass dome which projects into this small chamber which contains small grains of carbon made from anthracite coal. When you are using the telephone, the carbon chamber is part of the circuit and current flows through the carbon.

Figure 4.37 Telephone transmitter (Courtesy Western Electric Co.).

the current through the transmitter. The net result is a variation in current related directly to the voice pattern to be reproduced. The signal is then passed through the wires in the form of a current variation to the receiver set at the other end of the conversation. That current can travel through the lines at the speed of light: 186,000 miles/s. This same principle can be employed in the diagram of carbon microphones, as indicated earlier in this section.

A transformer plays a vital role in the ignition system of an automobile, shown in Fig. 4.38. When the switch is closed current will flow from the battery through the primary of the ignition coil and the closed contact back to the battery through the ground. As the cam shaft turns the contacts will open, and the current through the primary circuit will decrease very rapidly. This will induce a voltage of perhaps 250 V across the primary, as derived by the equation introduced earlier for a coil: $v_L = L \, di/dt$. Recall that this equation indicated that it was the rate of change that determines the induced voltage and not simply its fixed magnitude. For an ignition coil such as shown in Fig. 4.39 the primary is a coil of a few hundred turns of heavy wire and the secondary thousands of turns of fine wire. For a turns ratio of $a = N_p/N_s = 100$ and a primary induced voltage of 250 V the secondary voltage will be 25,000 V. As indicated in Fig. 4.38, this secondary voltage will appear across a spark plug determined by the distributor. For a typical eight-cylinder engine 300 high-voltage surges will be required every second to the spark plugs. The capacitor in the distributor prolongs the contact life by acting as a short across the contacts when the current surge is at its maximum. As the cam shaft continues to turn the contacts will close again, reestablishing a primary flow and another firing cycle.

The *AMP-CLAMP* is an instrument that employs transformer effects to measure current without having to break the circuit. A typical commercially available unit

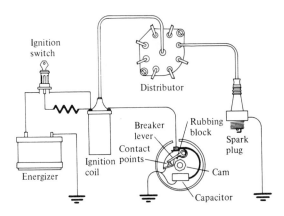

Figure 4.38 Automobile ignition system (Courtesy Delco Remy).

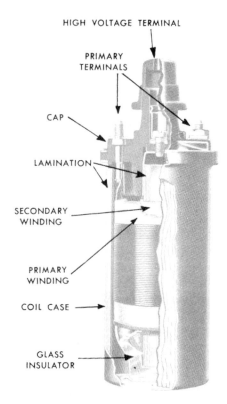

HIGH VOLTAGE TERMINAL

PRIMARY TERMINALS

CAP

LAMINATION

SECONDARY WINDING

PRIMARY WINDING

COIL CASE

GLASS INSULATOR

Figure 4.39 Ignition coil (Courtesy Delco Remy).

appears in Fig. 4.40 that has a 0–5-, 10-, 25-, 50-, 100-, and 250-ac-A scales. Surrounding every current-carrying conductor is a magnetic field such as shown in Fig. 4.41. An ac current in the primary will cause a varying flux pattern directly related to the magnitude of the current through the conductor. This will be sensed by the "secondary"

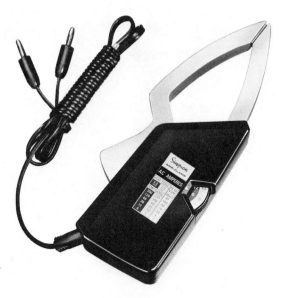

Figure 4.40 Amp-Clamp® (Courtesy Simpson Electric Co.).

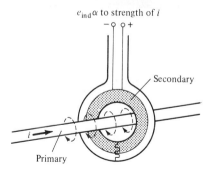

Figure 4.41 Amp-Clamp® (internal view).

winding in the clamp of the instrument, and a reading will be obtained that is calibrated directly to the induced voltage of the secondary.

Magnetic tape such as shown in Fig. 4.42 is in high demand today for the increasing number of recording instruments being used. The eight-track cartridge has become almost as much a part of daily living as the television or radio. The basic recording process is in actuality quite simple. For the single-channel recording head depicted in Fig. 4.43 the current direction through the coil will determine which side of the recording head is north or south. In this case the resulting direction of the flux pattern is shown as established in the magnetic tape. The strength of the current in the coil will also affect the flux strength established in the head and therefore in the tape. A reversal in current direction will reverse the magnetic polarization in the tape as also shown in the figure. The playback process employs a playback head similar in design to the recording head that will be sensitive to the small permanent magnets established on the tape. As the tape passes under the reading head a flux pattern is established in the playback core that will induce a voltage pattern in the coil that can be translated into an audible sound.

Figure 4.42 Magnetic tape (Courtesy Ampex Corporation).

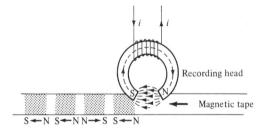

Figure 4.43 Magnetic tape recording head.

The last application of magnetism to be introduced in this chapter is in the field of electronic computers. There is, as most laymen realize, the need for a memory system in every computer to store information. Since the computer circuits operate on binary arithmetic only (two possible levels: 0 or 1), the storage element must be able to store only one of two possible states (that can be retrieved). Since a magnetic core such as shown in Fig. 4.44 can be magnetized in only two directions, if we assign one direction to be a 0 state and the other a 1 state, the small bit of information can be preserved as a magnetic field in the core. A number of wires pass through the core since it is not only necessary to set up the magnetic field in the core but be able to *read out* its state and determine where in a large memory unit such as shown in Fig. 4.45 the core is located.

We have only touched on a few areas of application of magnetism in this chapter. It should be apparent from the diverse areas of application introduced here that magnetism plays a very integral part in the electrical and electronic systems. As we progress through the chapters to follow and deal with, e.g., motors and generators further evidence of the need for magnetic effects will appear with a discussion of sufficient detail to provide at least a surface understanding of the application.

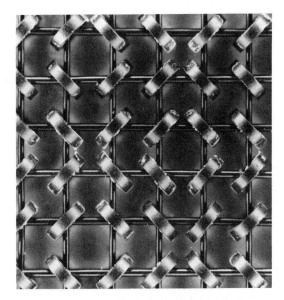

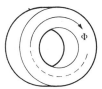

Figure 4.44 Magnetic core.   **Figure 4.45** Magnetic core memory (Courtesy Ampex Corporation).

PROBLEMS

§ 4.3

1. Determine the permeability (μ) of a material with a relative permeability (μ_r) of 250.

2. For the magnetic system in Fig. 4.46, determine
   (a) The magnetizing force.
   (b) The magnetomotive force.
   (c) The flux density.
   (d) The flux Φ (in lines) in the core.
   (e) The direction of the flux Φ in the core.

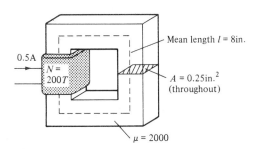

Figure 4.46

3. Determine the current I necessary to establish the flux indicated in Fig. 4.47.

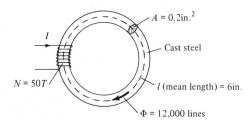

$A = 0.2$in.^{2}

Cast steel

l (mean length) = 6in.

I

$N = 50T$

$\Phi = 12,000$ lines

Figure 4.47

4. Repeat Problem 3 if an air gap of 0.003 in. is cut through the core.

5. Using a 50-μA, 20,000-Ω movement, design
   (a) A 10-A ammeter.
   (b) A 10-V voltmeter.

§ 4.4

6. A transformer with a turns ratio $a = N_p/N_s = 4$ has a load of 2 K applied to the secondary. If 120 V are applied to the primary, determine
   (a) The reflected impedance at the primary.
   (b) The primary and secondary currents.
   (c) The load voltage.
   (d) The power to the load.
   (e) The power supplied by source.

7. (a) If $E = 60$ V, $R_s = 1$ K, and $a = 5$ in the network of Fig. 4.25, determine the load value for maximum power to the load.
   (b) Determine the power to the load under these conditions.

8. A two-circuit transformer has the following measured quantities: $F_p = 240$ V, $I_p = 2$ A, $E_s = 40$ V, $I_s = 12$ A when applied to a 3.33-Ω load. Sketch the autotransformer connection if 280 output volts were desired across the load. What is the new volt-ampere rating of the system? Compare this level to that of the two-circuit configuration.

5

dc and ac
Machinery

5.1 INTRODUCTION

There are four broad areas of coverage to be considered under the given chapter heading: dc generators and motors and ac generators and motors. The basic construction operation and nameplate data along with an area of application or two will be provided for each subject area. The sole purpose of the chapter is to ensure that the reader has sufficient knowledge of each type of machine to choose which to use for the area of application. Typical of most texts devoted solely to dc and ac machinery, we shall begin with dc generators and motors.

5.2 dc GENERATORS

Generators (whether they be dc or ac) are dynamos (any energy-converting device) that convert mechanical energy to electrical energy. As the shaft of the dc generator of Fig. 5.1 is turned at the nameplate speed, a dc voltage is generated at the two output terminals. The basic construction of the dynamo can be broken down into the two basic components appearing in Fig. 5.2: the stationary *stator* and the movable *rotor*. The stationary part has an even number of poles (alternating north and south poles) which are either the permanent magnetic type or induced by a continuous

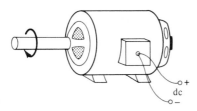

Figure 5.1 Dc generator.

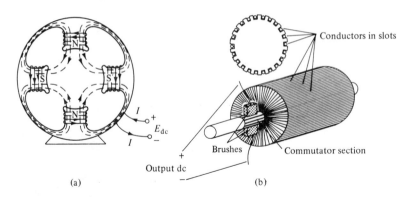

Figure 5.2 Four (4) pole generator (a) stator (b) rotor.

winding around all the poles through which an energizing dc current is passed as shown in the figure. The flux pattern established by the poles reflects the need for alternating field poles. The stator frame (also called the yolk) is constructed of a ferromagnetic material to permit a continuous flux path for the poles with a minimum of reluctance in its path. When the rotor (also constructed of ferromagnetic material in the laminated form discussed for transformers) is inserted within the stator the flux pattern can continue through the rotor with only the air gap between the two structures to severely limit the generated flux. The air gap is typically between $\frac{1}{32}$ and $\frac{1}{8}$ in. depending on the relative size of the machine.

Coils of the type depicted in Fig. 5.3 are set into the slots of the rotor and connected together in a particular manner to develop the nameplate voltage and current rating. They have the shape shown in the figure because of the necessity to overlap the windings at each end of the rotor. There would be insufficient length in the coil for the overlapping if each coil were perfectly rectangular.

Figure 5.3 Rotor winding.

As the rotor turns, under some external force such as hydroelectric power, the conductors of each coil will pass through the flux of the poles of the stator, and a voltage will be induced across the conductors as determined by Faraday's law of electromagnetic induction: $e = N \, d\phi/dt$. The induced voltages of a turn will add as shown in Fig. 5.3, and then the series summation of the coils will result in the generated voltage.

The induced voltages of the windings are connected directly to the small parallel copper sections (insulated from one another), shown in Fig. 5.2(b), called the commutator bars that make up the commutator section. The output voltage taken off the carbon brushes that ride on these sections will show a slight ripple as it rides from one commutator bar to another. However, for all practical purposes, the output is considered pure dc.

For dc generators requiring the dc voltage to energize the field poles there appears to be a contradiction in that dc voltages are required to develop the desired dc output. The generated levels, however, are much greater than that required to develop the desired output. Dc generators can be divided into two basic types: *separately excited* and *self-excited*. The labels imply that the separately excited generator requires a separate dc supply, while the self-excited generator obviously does not. In fact, in the latter case the generated voltage will provide the necessary dc voltage to energize the field windings. This phenomenon is possible only because of an effect referred to as *residual magnetism*. It is that magnetism remaining after the current is removed from the magnetizing coil of an electromagnetic system. When the rotor first begins to turn this residual flux in the core will develop sufficient voltage across the field of the stator to result in a measurable field current. The result is an increase in the field flux and a simultaneous gain in induced voltage. This behavior is cyclic until rated generated voltage is reached and both the generated voltage and limiting field resistance reach a point of common satisfaction. Therefore, as long as the small generated voltage due to the residual flux appears directly across the field windings the dc generator can build up to rated voltage without the need for a separate dc supply.

The generated voltage is directly proportional to the strength of the magnetic field and the speed with which the conductors in the slots of the rotor cut the magnetic field. In equation form,

$$E_g = K\Phi S \qquad\qquad (5.1)$$

where E_g = generated voltage in volts
K = constant determined by construction of dynamo
Φ = main field flux
S = speed in rpm

In other words, the greater the speed of the rotor or magnitude of the flux developed by the poles, the greater the generated voltage.

In general, the *nameplate* on a dc generator will include three important pieces of data, which must be carefully considered in the choice of a machine to perform a particular task. They include the rated terminal voltage of the generator V_t; the rated line current I_L, indicating the maximum current that can be drawn from the supply; and, finally, the speed of rotation of the shaft required to sustain the voltage and current rating.

The output power rating can be determined by the product of the rated output voltage and current:

$$P_{\max} = V_t I_L \qquad \text{(watts)} \qquad\qquad (5.2)$$

For a generator, laboratory dc bench supply, or what have you, the *voltage regulation* (introduced earlier) is an excellent measure of the quality of the supply. In Fig. 5.4(a), the ideal situation appears. The terminal voltage is *unaffected* by the current drawn by the load. In the practical world, however, every supply has some

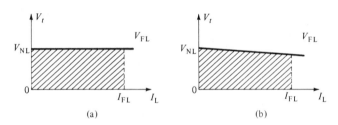

Figure 5.4 Terminal voltage variation with load (a) ideal (b) practical.

internal resistance, as introduced for simple dc batteries. As the load current increases, there will be an increase in the voltage drop across the internal resistance of the supply and the terminal voltage will naturally decrease as shown in Fig. 5.4(b). Certainly, the closer the curve to the ideal, the better since the supply may seldom be operating exactly at its full line current capacity. In equation form the *voltage regulation* is given by

$$\text{Voltage regulation } (VR) \ \% = \frac{V_{NL} - V_{FL}}{V_{FL}} \times 100\% \qquad (5.3)$$

Since the numerator is an indication of the variation in level, the smaller the voltage regulation, the better the characteristic.

There is more than one type of dc generator commercially available. Thus far our interest has been in the *shunt* (parallel-type) generator that has its field coil (of the stator) in parallel with the generated voltage as shown in Fig. 5.5. The generated voltage E_g appears across the schematic representation of the rotor and brushes. The resistance R_A is the resistance of the rotating armature. The resistance R_F and the coil represent the field circuit equivalent. The circuit clearly shows that the generated voltage must be greater than the rated terminal voltage due to the drop across R_A. That is,

$$V_t = E_g - I_A R_A \qquad (5.4a)$$

The current I_A must be sufficiently high to guarantee the proper level of I_F and I_L with

$$I_A = I_F + I_L \qquad (5.4b)$$

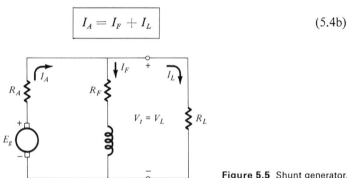

Figure 5.5 Shunt generator.

EXAMPLE 5.1

Determine the following for the shunt dc generator of Fig. 5.6:

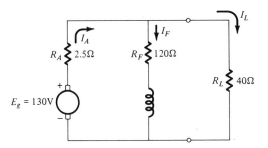

Figure 5.6

1. I_L, I_F, I_A.
2. The voltage drop across R_A.
3. The resistance R_A.
4. The power to the load.

Solution:
1. $I_L = 120/40 = $ **3 A**, $I_F = V_t/R_F = 120/120 = $ **1 A.** $I_A = I_F + I_L = 1 + 3 = $ **4 A.**
2. Kirchhoff's voltage law: $E_g - V_{R_A} - V_t = 0$ or $V_{R_A} = E_g - V_t = 130 - 120$
   $= $ **10 V.**
3. $R_A = V_{R_A}/I_A = 10/4 = $ **2.5 Ω.**
4. $P_L = V_t I_L = (120)(3) = $ **360 W.**

Another type of dc motor that has increased control of the terminal voltage is the *compound* dc generator which has two separate insulated windings around the same field poles as shown in Fig. 5.7.

An electrical schematic of a compound configuration revealing the location of the additional winding is provided in Fig. 5.8. Note that the additional winding

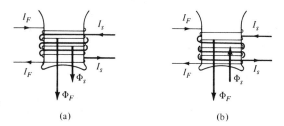

Figure 5.7 Compound pole winding (a)
cumulative (b) differential.   (a)   (b)

is placed in series with generated voltage and therefore has the label *series* winding. The current through the series winding that results in the additional flux is the armature current I_A. The turns of the series winding is far less than that of the main field shunt winding. It is designed solely to act as a vernier (a fine control) on the output terminal voltage (V_t) and not the major provider of flux necessary for generating a sufficiently high voltage.

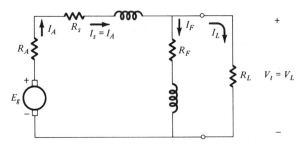

Figure 5.8 Compound dc generator.

If the series field is connected such that its resulting flux aids the flux of the main field such as shown in Fig. 5.7(a), we say we have a compound *cumulative* situation. The greater the line current drawn, the greater will be the armature current I_A and the greater will be the flux in the core and the generated voltage. If the series field flux opposes that of the shunt field flux such as shown in Fig. 5.7(b), it is called a compound *differential* machine. The net effect with increasing line current will be a reduction in net flux and generated voltage such as shown in the characteristics of Fig. 5.9.

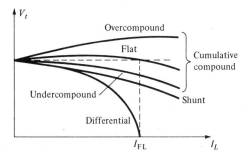

Figure 5.9 Dc generator characteristics.

Note that three possibilities exist for the cumulative compound machine. If the series field is allowed to contribute a significant amount of additional flux, an *overcompound* situation will result. A reduction in the effect of the series field will result in the *flat* response. A still further reduction will result in the *undercompound* situation. If the series coil is bypassed, we return to the shunt field characteristics.

For the overcompound case the voltage regulation will have a negative sign since the full-load voltage is greater than the no-load voltage. Further, the flat response will have 0% regulation and the undercompound a positive regulation. For most compound-wound machines a diverter (a variable resistor) is placed across the series coil to control the amount of current that will pass through the field. For maximum resistance values the overcompound situation would result, and for the 0-Ω setting all the armature current would pass through the parallel path, and the series coil would have absolutely no effect. Although one might wonder why the differential characteristic would be desirable at all, consider that in the far right region we have an almost constant current region for a variation in terminal voltage (like a current source). This characteristic is excellent for applications such as welding and series connection of loads where a fixed current must exist even if the load should vary.

From Fig. 5.8 the following current and voltage equations can easily be developed:

$$I_L = I_F + I_A, \qquad I_S = I_A, \qquad I_F = \frac{V_t}{R_F} \qquad\qquad (5.5)$$

and

$$V_t = E_g - I_A(R_A + R_s) \qquad\qquad (5.6)$$

EXAMPLE 5.2

For the following numerical values given to the elements of the network of Fig. 5.8, determine

$$R_F = 100\ \Omega, \qquad R_A = 0.5\ \Omega, \qquad R_S = 1\ \Omega, \qquad V_t = 240\ \text{V}, \qquad R_L = 20\ \Omega$$

1. The rated line current.
2. The generated voltage.
3. The power to the load.
4. The power delivered by the armature.
5. The total power lost in the armature, series, and shunt field coils.

Solution:

1. $I_L = V_t/R_L = 240/20 = $ **12 A.**
2. $E_g = V_t + I_L(R_A + R_S)$, $I_A = I_F + I_L = (V_t/R_F) + I_L = (240/100) + 12$
   $= 2.4 + 12 = 14.4$ A, and $E_g = 240 + 14.4(0.5 + 1) = $ **261.6 V.**
3. $P_L = V_t I_L = (240)(12) = $ **2880 W.**
4. $P = E_g I_A = (261.6)(14.4) = $ **3767.04 W.**
5. $P_{\text{lost}} = P_i - P_o = 3767.04 - 2880 = $ **887.04 W.**

The third and last type of dc generator to be introduced is the series generator, which has the circuit model of Fig. 5.10. A single field coil is placed in series with the generated voltage rather *than* in the usual parallel position. At small levels of I_L,

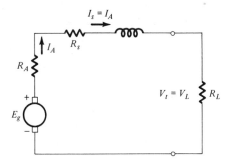

Figure 5.10 Series dc generator.

the current $I_S = I_L$ is small and the flux strength Φ_S of the coil is low. The result is a small generated voltage as shown on the characteristics of Fig. 5.11. As the line current increases, the field strength and generated voltage will increase as shown in the same figure. Due to an effect known as *armature reaction*, which can be researched in many available texts, the curve will drop to zero as shown, and the areas of application indicated for the differential compound machine apply here also. One interesting application of the rising characteristic of the series generator is shown in Fig. 5.12,

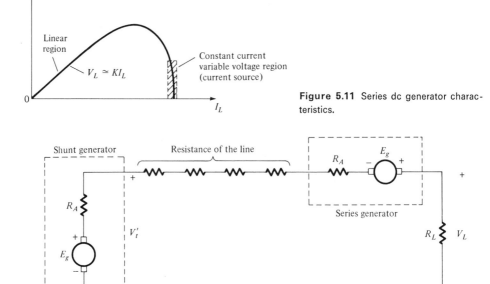

Figure 5.11 Series dc generator characteristics.

Figure 5.12 Series generator in application as a booster.

where it is being used as a *booster*. Some of the generated voltage V'_t is lost across the long transmission line from the generator to the load. The greater the current demand, the greater will be the loss across the lines. To compensate for this loss, a series generator is placed near the load as shown in the figure. Increasing line currents will result in increasing generated voltage across the series booster (characteristics of Fig. 5.11), compensating directly for the increasing line losses so that V_t can be maintained at some relatively constant and sufficiently high magnitude.

For the series generator the following equations are applicable:

$$I_L = I_S = I_A \tag{5.7}$$

$$V_t = E_g - I_A(R_A + R_s) \tag{5.8}$$

5.3 dc MOTORS

The motor is an electrical dynamo that can convert electrical energy into mechanical energy. In other words, as shown in Fig. 5.13, an applied voltage will result in a turning shaft that can do mechanical work.

The network schematic of a dc motor, as shown in Fig. 5.14(a), is simply the reverse of that for the dc generator. Like the generator, there are shunt, compound,

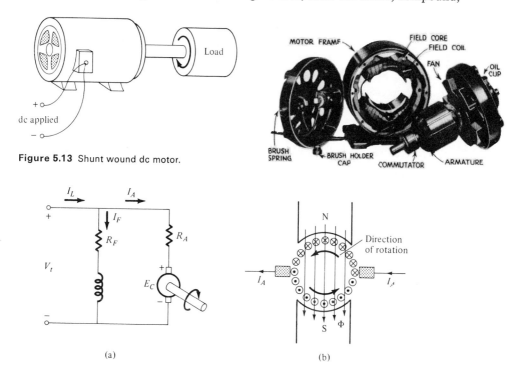

Figure 5.13 Shunt wound dc motor.

Figure 5.14 Dc motor (a) schematic (b) internal construction.

and series dc motors also. The shunt configuration appears in Fig. 5.14(a). The applied voltage will result in a field current I_F and the necessary flux for the poles of the stator as indicated in Fig. 5.14(b). The applied voltage will also result in an armature current I_A through the conductors of the armature as shown in the same figure. The current-carrying conductors in the magnetic field will experience a force in a direction determined by the *left-hand rule*. Simply place the index finger in the direction of the magnetic flux and the second finger at right angles to the index finger in the direction of the current. The thumb, if placed at right angles to the index finger, will give the direction of the force. The force direction for the conductors at the top and bottom of Fig. 5.14(b) are indicated. Note that they are aiding forces in the counterclockwise direction. The reversal in force direction in the lower half is due to the reversal in current direction. The force on the conductors can be determined by

$$F = BIL \times 10^{-8} \tag{5.9}$$

where F = pounds
  B = lines per square inch
  I = amperes
  L = inches

which clearly indicates that the greater the strength of the magnetic field, conductor current, or length of the conductor *in* the magnetic field, the greater the force on each conductor and therefore on the rotating member (rotor) of the machine.

As the rotor turns a voltage will be induced across the rotating conductors of the rotor that will oppose the applied voltage at the input terminals. It is labeled E_C as shown in Fig. 5.14(a) with the subscript C derived from the expression *counter-emf*. It is determined by an equation similar to that for E_S for the dc generator:

$$E_C = K\Phi S \qquad \text{(volts)} \tag{5.10}$$

The greater the speed of rotation or flux of the poles, the greater the counter-emf. For the shunt motor of Fig. 5.14 the following relationships will result:

$$I_L = I_F + I_A \tag{5.11}$$

$$I_F = \frac{V_t}{R_F} \tag{5.12}$$

$$V_t = E_C + I_A R_A \tag{5.13}$$

EXAMPLE 5.3

The dc shunt motor of Fig. 5.14(a) has the following component values:

$$V_t = 120 \text{ V}, \qquad I_L = 8 \text{ A}, \qquad R_F = 100 \text{ } \Omega, \qquad R_A = 1 \text{ } \Omega$$

Determine the counter-emf E_C.

Solution:

$$I_F = \frac{V_t}{R_F} = \frac{120}{100} = 1.2 \text{ A}$$

$$I_A = I_L - I_F = 8 - 1.2 = 6.8 \text{ A}$$

and

$$E_C = V_t - I_A R_A = 120 - (6.8)(1) = \mathbf{113.2 \text{ V}}$$

A quantity of major importance with regard to any type of motor is the *torque* the shaft can develop since it will determine the load the machine can efficiently handle. As depicted by Fig. 5.15 and given by

$$T = F \cdot r \qquad \text{[foot-pounds (ft-lb)]} \tag{5.14}$$

Figure 5.15 Defining torque.

the torque is directly related to the force developed on the conductors of the rotor and their distance from the center of the rotating shaft. In terms of the flux and armature current it is given by

$$T = K\Phi I_A$$ (5.15)

clearly indicating that an increase in magnetic field strength or current through the conductors of the rotor will increase the torque by the same amount.

The nameplate data of any dc motor includes the input voltage and current rating and its power rating. That given power rating is always the rated output power of the motor. It is related to the torque on the output shaft by the following equation:

$$T = \frac{7.04P}{S}$$ (5.16)

where T = foot-pounds
$\quad P$ = watts
$\quad S$ = rpm

Proportionally speaking, the following is true:

$$T \cdot S \propto P$$ (5.17)

which states in words that for a particular horsepower machine there can be some trade-off between torque and speed. In other words, in Eq. (5.17), for a *fixed* power P, the torque available can be increased but only at the expense of speed. Many of us have experienced the effect with the simple $\frac{1}{4}$ in. drill. When the drilling became more difficult the drill could continue to work for us but at a much slower speed. Of course if the torque of the rotating shaft could not handle the load for the horsepower rating of the machine, the drill could easily burn up on us. The solution would perhaps be a $\frac{1}{2}$ in. drill with its usually higher horsepower rating.

The efficiency of a dc motor (or generator) can be determined by the following equation:

$$\eta\% = \frac{P_o}{P_i} \times 100\%$$ (5.18)

For the dc motor the output power is the nameplate rating, and the input power is determined by $P_i = V_t I_L$. For the generator the input power is the power delivered to the shaft, and the output power is $P_{og} = V_t I_L$.

EXAMPLE 5.4

The following nameplate data were provided for a dc shunt motor: 5 hp, 240 V, 18 A, 2400 rpm. Determine

1. Output torque.
2. Efficiency of operation.

Solution:

1.
$$T = \frac{7.04P}{S} = \frac{7.04[(5)(746)]}{2400} = \textbf{10.94 ft-lb}$$

2.
$$P_i = V_t I_L = (240)(18) = 4320 \text{ W}$$

$$\eta\% = \frac{P_o}{P_i} \times 100\% = \frac{(5)(746)}{4320} \times 100\% = \textbf{86.3}\%$$

For the shunt motor the field flux remains fairly constant during load variations since the applied voltage is directly across the field circuit. The equation for torque for the shunt motor can therefore be written as

$$T = K\Phi I_A = \underbrace{KK'}_{K_T} I_A = K_T I_A \tag{5.19}$$

and we find that the torque will vary directly as the armature current. Since I_A and I_L are usually quite close in magnitude, the graph of torque vs. current will include the former as shown in Fig. 5.16. The curve clearly indicates that as we increase our torque demand on the motor the current will increase proportionally. This effect will be noticed in the heating up of the handle as too heavy a load is put on the biting edge of the drill of the example described above in the discussion of torque. The curve will drop off slowly at higher currents due to an effect mentioned earlier called *armature reaction* that reduces the net flux of the field poles by introducing a flux component (developed by the rotating armature) that appears perpendicular to that of the main field winding. As determined by Eq. (5.15), a reduction in net flux will most certainly reduce the available torque.

A second curve of primary importance for dc motors is its speed vs. I_A characteristics—in other words, how the speed varies with change in load. Equation (5.10) can be rewritten in the following manner:

$$\boxed{S = \frac{E_C}{K\Phi}} \tag{5.20}$$

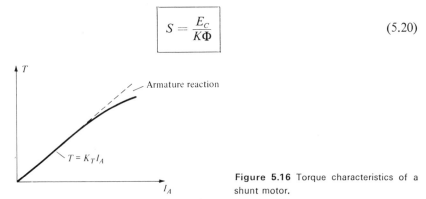

Figure 5.16 Torque characteristics of a shunt motor.

Substituting Eq. (5.13),

$$S = \frac{V_t - I_A R_A}{K\Phi} \tag{5.21}$$

For increasing values of I_A (for fixed V_t and R_A) the numerator will obviously become smaller. The net flux Φ will also decrease slightly due to armature reaction. However, on a whole, the shunt motor is considered to be a fairly constant-speed device as shown in Fig. 5.17. Obviously, therefore, applications that require a fairly constant speed for variations in load should possibly employ a dc shunt-wound motor.

Speed control can be introduced by making R_F a variable resistor since it will control I_F and therefore the flux Φ in the denominator of the equation. Additional control can be instituted by making R_A a variable quantity. Of course variations in V_t will also affect the speed characteristics.

Dc motors, like dc generators, are also manufactured using a second series coil such as shown in the compound configuration of Fig. 5.18. The flux of the added

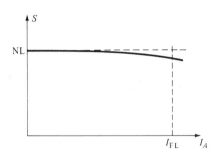

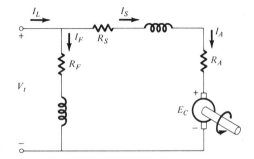

Figure 5.17 Speed characteristics of a shunt motor.

Figure 5.18 Compound dc motor.

series coil will again act as a vernier on the total magnetic flux of the stator poles. As with the dc generator, the series coil is wrapped around the same poles as the main field winding although totally insulated from the other winding. For the cumulative compound $(\Phi_F + \Phi_S)$ machine the three conditions of overcompound, flat, and undercompound can be established depending on the amount of additional series flux. Since $T = K\Phi I_A$ the reduction in net flux associated with changing from an overcompound to undercompound situation will result in the variation in curves indicated in Fig. 5.19. For the differential compound situation $(\Phi_F - \Phi_S)$ the torque will drop off more rapidly as shown in the same figure.

For the compound configuration the equation for speed becomes

$$S = \frac{V_t - I_A(R_A + R_S)}{K(\Phi_F \pm \Phi_S)} \tag{5.22}$$

For the cumulative or differential case an increase in current (I_A) will certainly cause a reduction in the numerator and a decrease in speed. However, Φ_S, which appears in the denominator, is also dependent on the current I_A. For increasing I_A (and therefore Φ_S), in the cumulative case the denominator will continue to get larger and the speed decrease as shown by the set of curves of Fig. 5.20. For the differential case, however, the total flux will decrease, and the speed curve can rise as shown in the same figure. For cumulative or differential machines the effect of armature reaction will be to decrease the net flux and cause an increase in speed.

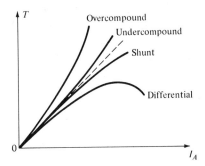

Figure 5.19 Torque characteristics of the compound dc motor.

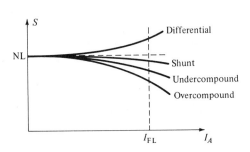

Figure 5.20 Speed characteristics of the compound dc motor.

Speed regulation, defined by

$$\boxed{\text{Speed regulation (SR)}\% = \frac{S_{NL} - S_{FL}}{S_{FL}} \times 100\%} \qquad (5.23)$$

is a measure (like voltage regulation) of how much the speed will vary from no-load to full-load conditions. For the differential case, since the full-load is greater than the no-load value, a negative result would be obtained.

The series dc motor has the network configuration of Fig. 5.21. Since the line current $I_L = I_A$, the flux developed by the series coil is directly related to the I_A, and in equation form, $\Phi = K_1 I_A$. If we substitute for Φ in the torque equation, we find

$$T = K\Phi I_A = \underbrace{K(K_1 I_A)}_{K_T} I_A = K_T I_A^2$$

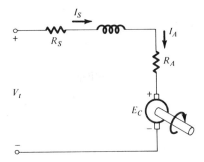

Figure 5.21 Series dc motor.

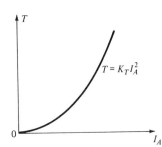

Figure 5.22 Series dc motor torque characteristics.

indicating that the torque will increase exponentially with the line or armature current. The resulting torque characteristic appears in Fig. 5.22. The speed characteristics of the series motor can be determined through

$$S = \frac{V_t - I_A(R_A + R_S)}{K\Phi_S} \qquad (5.24)$$

When $I_L = I_A = I_S = 0$, $\Phi_S = \Phi_{residual}$, resulting in a very small magnitude for the denominator of the equation and a very high speed (usually excessive and must be limited). As $I_L = I_A$ increases, the flux will increase, and the speed will decrease toward zero, as indicated by the characteristics of Fig. 5.23.

One interesting application of the series motor is in the cog train climbing the very steep grade in Fig. 5.24. When climbing the hill a very high torque must be developed, although the speed can be quite low. This condition is described by high levels of I_A on each characteristic. When going down the other side the torque demand is significantly reduced, and the speed can increase substantially. This condition is described by low values of I_A on each curve.

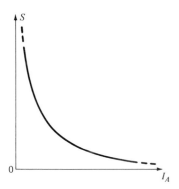

High tourque low speed

Low torque high speed

Figure 5.23 Series dc motor speed characteristics.

Figure 5.24 Application of a series dc motor.

The description of dc motors in this section implies that the dc voltage can be applied directly across the field and armature circuits when starting the dynamo. In actuality, an additional resistive element must initially be inserted in series with the armature circuit at starting, since the counter-emf $E_c = K\Phi S$ is zero (the speed of the rotating shaft is initially zero). The entire applied voltage will appear directly across R_A since

$$V_t = E_c + I_A R_A = 0 + I_A R_A \qquad \text{(at starting)}$$

and

$$I_A = \frac{V_t}{R_A} = \frac{120}{1} \text{ (typically)} = 120 \text{ A}$$

Current levels of 120 A will most assuredly damage the armature circuit. This effect can be easily eliminated by inserting a starting resistance in series with R_A at starting as shown in Fig. 5.25. The starting resistor can be designed to limit the starting current to perhaps 200% of rated line current. For short periods of time the machine can safely handle this level of current. As the machine builds up the counter-emf E_C will increase, and the voltage across the series combination of R_{st} and R_A will decrease. The value of R_{st} can then be reduced in step with the increasing level of E_C until running conditions are established and R_{st} can be eliminated altogether.

The basic design of a four-point starter appears in Fig. 5.26. The handle is manually moved through the contacts of the series resistors until the final position is reached where the energized magnet will hold the arm until the applied emf is removed. Note that in the final rest position the entire starting resistance is out of the armature circuit, while initially it is all in series with R_A.

Technological advances of recent years have resulted in a number of interesting innovations in the design of dc motors and generators. For instance, commutation can now be performed with an electronic package rather than the copper sections introduced earlier, which were a definite maintenance problem. There are now rotor designs where the coils are wrapped near the surface so that the interior of the rotor is hollow. The result is a smaller rotating mass, with less inertia for braking and bearing design. Recently printed circuit techniques have been employed to reduce the overall length of the machine.

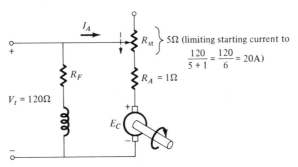

Figure 5.25 The starting resistor.

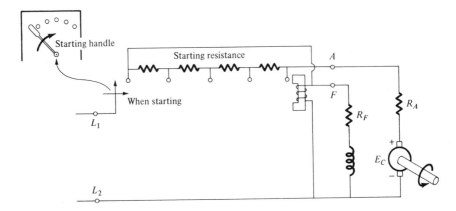

Figure 5.26 Four-point starter.

All in all, however, the basic operation of the vast majority of commercially available dc generators and motors is as described in the past few sections. Only the techniques for performing certain functions are receiving some redesign interest.

5.4 ac GENERATORS

We shall now examine the basic construction and operation of single-phase and polyphase ac generators, which are also referred to as *alternators.* If the commutator sections of the dc generator of Fig. 5.1 were replaced by two slip rings (continuous circular bands) such as shown in Fig. 5.27, a sinusoidal voltage would result at the output terminals as shown. The slip rings are insulated from one another and represent the two ends of the continuous connections of conductors in the rotor. However, the vast majority of larger commercial units are not made in this manner. Rather, the field poles become part of the rotating member, and the voltages are induced across the conductors set in the slots of the fixed stator as shown in Fig. 5.28. The required current for the field poles of the rotor is passed through the brushes to the slip rings. There are a number of reasons for this reversal in approach. The rotating mass in the center is considerably lighter, requiring less concern about such design problems as the end bearings and braking. In addition, if very large generated voltages are taken off the brushes of the construction of Fig. 5.27, heavy sparking and brush

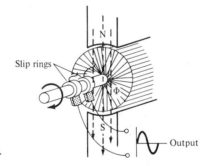

Figure 5.27 Ac generator.

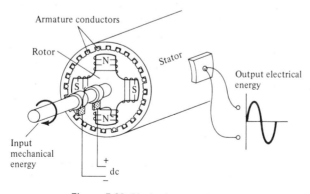

Figure 5.28 Single phase ac alternator.

wear would quickly result. With the design of Fig. 5.28 the high voltages do not pass across a brush connector but are connected directly to the load. The dc voltages across the brushes (to energize the fields) are usually 240 V or less. For polyphase generators such as the 3ϕ type, four slip rings would have to appear on the shaft of the design of Fig. 5.27, requiring a longer frame for the dynamo, while for the other design only two slip rings would be required and an additional two wires would appear at the stator output.

The frequency of the sinusoidal voltage produced by an ac generator is determined by the number of poles on the rotor, the speed at which the rotor is turned, and a factor of 120, as indicated by

$$\boxed{f = \frac{P \cdot N}{120}} \tag{5.25}$$

where P = number of poles of rotor

N = speed of rotor in rpm

EXAMPLE 5.5

Determine the number of poles of the rotor of a 3ϕ generator if a frequency of 60 Hz is generated at a rotor speed of 3600 rpm.

Solution:

$$P = \frac{120f}{N} = \frac{120(60)}{3600} = \frac{7200}{3600} = \textbf{2 poles}$$

Automobiles of today use an alternator rather than the dc generator of years past. Fig. 5.29 is a cross-sectional view of a typical single ϕ generator, the Delco-tron. The dc voltage to energize the field of the rotor is applied through the indicated brushes and slip rings. The output ac off the stator appears across the leads at the bottom left of the figure. They are connected to 6 diode (two-terminal electronic devices, to be discussed in Chapter 6) that converts the generated ac to dc as required by the automobile electrical system.

Single-phase portable generators such as shown in Fig. 5.30 provide the necessary power in remote areas for electrical equipment, power tools, wildlife surveys, etc. This particular unit can also supply dc power, and others are available that can provide the necessary 60 Hz for a cabin in a remote area.

The manner in which the windings of the stator are connected will determine the number of phases to be generated by the alternator. In Fig. 5.31 the stator windings are connected such that three sinusoidal voltages (120° out of phase with each other) are produced by each rotation of the rotor. The frequency is the same for each phase as determined by Eq. (5.25).

The nameplate on a polyphase generator should include, at a minimum, the following information: 3ϕ, 10 kVA, 220 V, 3600 rpm, Y-connected, 60 Hz. The 3ϕ obviously indicates the number of phases generated. The 10 kVA is the total apparent power rating of the generator as determined by $P_{a_T} = \sqrt{3}V_L I_L$, where V_L and I_L are defined by the specified Y connection of phases in Fig. 5.32. The 220 V is always

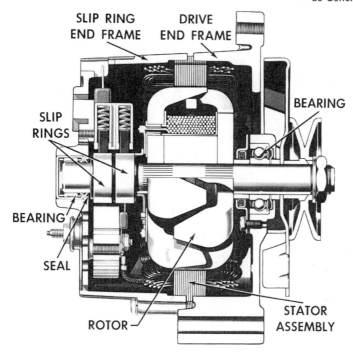

Figure 5.29 Cross-sectional view of the "Delcotron" single-phase ac generator. (Courtesy Delco-Remy, Division of General Motors).

Figure 5.30 Single-phase portable generator (Courtesy NO-BRUSH Corp.). (Courtesy Georator Corporation).

the line voltage (not the per-phase quantity), and the speed is that rpm of the rotor required through some outside means to develop the rated conditions (frequency, voltage, etc.).

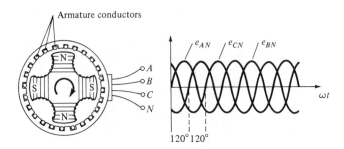

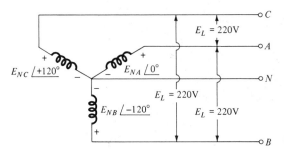

Figure 5.31 Three ϕ alternator.

Figure 5.32 Y-connecter 3 ϕ generator.

The voltage induced across each conductor of the stator can be determined by

$$E_c = 2.22\Phi f \times 10^{-8} \tag{5.26}$$

where Φ = flux in lines

f = frequency generated

For a polyphase generator with Z stator conductors and P phases the generated phase voltage is given by

$$E_\phi = 2.22\frac{Z}{P}\Phi f \times 10^{-8} \quad \text{(volts)} \tag{5.27}$$

EXAMPLE 5.6

Determine the voltage induced per phase of a 3ϕ ac alternator if the stator has 240 conductors, the flux is 1.5×10^6 lines, and the frequency is 60 Hz.

Solution:

$$E_\phi = 2.22\frac{Z}{P}\Phi f \times 10^{-8}$$

$$= 2.22\left(\frac{240}{3}\right)(1.5 \times 10^6)(60) \times 10^{-8}$$

$$= \mathbf{159.84 \ V}$$

In our analysis of dc generators and motors the inductive reactive elements could be ignored (except for their internal resistances) since they simply assumed the dc short-circuit equivalent. For ac dynamos, however, the reaction will have a reactance dependent on the frequency generated that can measurably reduce the terminal voltage of the generator under loaded conditions. The network equivalent for one phase of a polyphase generator appears in Fig. 5.33(b) as derived from the 3ϕ Y-connected generating system of Fig. 5.33(a). R_ϕ is the internal armature resistance, and X_ϕ is the reactance of that phase of the stator windings. The terminal voltage is then determined by the following vector relationship:

$$\mathbf{V}_\phi = \mathbf{E}_g - \mathbf{V}_R - \mathbf{V}_X \qquad (5.28)$$

or

$$\mathbf{V}_\phi = \mathbf{E}_g - \mathbf{I}_\phi\mathbf{R}_\phi - \mathbf{I}_\phi\mathbf{X}_\phi \qquad (5.29)$$

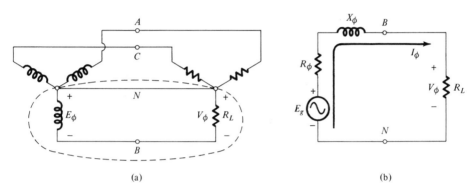

Figure 5.33 Single-phase equivalent of a 3 ϕ alternator.

The vector diagram can quickly be determined by first drawing an arbitrary length current (I_ϕ) vector to act as a reference for the voltages of the closed loop. For a *resistive* load the voltage drop across the load (V_ϕ) is *in phase* with the current through it, and the vector for V_ϕ is drawn parallel to that for I_ϕ. For V_R the same is true, and it is simply added to the end of the V_ϕ vector in the same direction. V_X leads I_ϕ by 90° and is therefore drawn vertically to I_ϕ. Through Eq. (5.28) we see that the generated phase voltage is determined by connecting the beginning of vector V_ϕ to the head of V_X as shown in Fig. 5.34. In equation form, the magnitude of E_g is determined by

$$E_g = \sqrt{(V_\phi + V_R)^2 + V_X^2} \qquad (5.30)$$

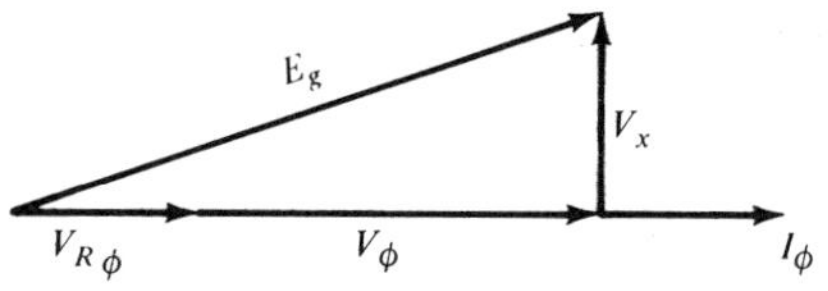

Figure 5.34 Single-phase phaser diagram.

EXAMPLE 5.7

A 3ϕ Y-connected generator has the following nameplate characteristics: 10 kVA, 220 V, $R_\phi = 0.5\ \Omega$, $X_\phi = 4\ \Omega$. Determine the voltage generated per phase to supply rated voltage to a unity F_p load (resistive).

Solution:

$$P_{a_T} = \sqrt{3}\ V_L I_L = 10,000$$

and

$$\sqrt{3}\ (220)I_L = 10,000$$

$$I_L = \frac{10,000}{(1.73)(220)} = \textbf{26.27 A}$$

For a Y-connected system (Fig. 5.32) $I_L = I_\phi$ and $I_\phi = 26.27$ A.

In Fig. 5.35,

$$V_R = I_\phi R_\phi = (26.27)(0.5) = 13.135\ \text{V}$$

$$V_x = I_\phi X_\phi = (26.27)(4) = 105.8\ \text{V}$$

and

$$V_\phi = \frac{V_L}{\sqrt{3}\,(\text{Y-connected})} = \frac{220}{1.73} = 127.17\ \text{V}$$

$$E_g = \sqrt{(V_\phi + V_R)^2 + (V_X)^2}$$

$$= \sqrt{(127.17 + 13.135)^2 + (105.8)^2}$$

$$= \textbf{175.72 V}$$

In other words, in order to provide the rated 127.17 V at the load (resistive) the

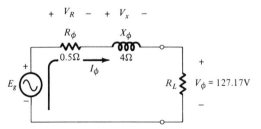

Figure 5.35

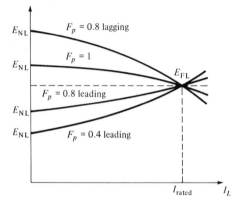

Figure 5.36 Variation in terminal voltage for an ac alternator with change in power factor.

generator would have to develop 175.72 V to compensate for the drops in potential across the internal resistance and reactance of each phase. The analysis here was for a purely resistive load. If the load were inductive to any degree, E_g would have to increase also. For capacitive loads E_g would decrease from that required for a purely resistive load since the series capacitive and inductive reactances will oppose each other and reduce the magnitude of the total impedance. In fact, Fig. 5.36 clearly indicates that the more leading of the power factor, the less the required no-load voltage (E_g) to meet rated conditions, although too low a leading F_p would result in very poor voltage regulation (variation in terminal voltage with load).

5.5 POLYPHASE INDUCTION ac MOTOR

The stator of a polyphase *induction* ac motor is constructed in much the same manner as that of the ac generator except now the 3ϕ voltage is applied to the stator, and the rotor will turn the mechanical load.

The basic construction of the stator and rotor appear in Fig. 5.37 with the photograph of a commercially available unit. The conductors of the stator are connected in such a manner as to develop a resulting direction of flux such as shown in Fig. 5.37(a) when the 3ϕ ac voltage is applied. As the magnitude of the phase components vary in magnitude with time the conductors of the stator are connected such that the resulting flux direction will appear to be rotating around the shaft as indicated in Fig. 5.37(a). The rotating flux pattern is equivalent to having two opposite poles such as in Fig. 5.37(b) rotating around the frame of the machine. That is, the flux pattern would be the same.

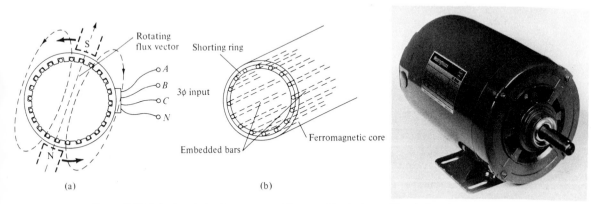

Figure 5.37 Polyphase induction ac motor (Courtesy Westinghouse Electric Corp.).

The speed with which this resultant flux rotates around the stator is determined by the number of poles for which the stator is wound (two for this example) which could be two, four, or more depending on the manner in which the stator windings are connected and on the frequency of the applied 3ϕ signal. In equation form, this *synchronous* speed is determined by

$$N_S = \frac{120f}{P} \qquad (5.31)$$

where f = applied frequency
       P = poles of stator
     N_S = rpm

EXAMPLE 5.8

Determine the synchronous speed of a polyphase induction motor having a stator wound for the four poles and an applied frequency of 60 Hz.

Solution:

$$N_S = \frac{120f}{P} = \frac{120(60)}{4} = \textbf{1800 rpm}$$

The construction of the rotor as indicated in Fig. 5.37 is nothing more than conducting rods embedded in a ferromagnetic core with shorting rings on each end. The bars are skewed to provide a more uniform torque. If the rods and shorting ring were removed from the ferromagnetic core, a cage would result, so-called for no other reason than that its general size is about that of a squirrel cage. For this reason this particular motor is often referred to as a *squirrel-cage* induction motor.

The rotating flux of the stator will cut the embedded conductors in the rotor, induce a voltage across them, and establish a current through the conductors and shorting rings. The current-carrying conductors in a magnetic field will experience a force in a direction determined by the left-hand rule described for dc motors. The torque developed is determined by the current through the conductors and the stator flux. That is,

$$T = K\Phi I_c \qquad (5.32)$$

where Φ = flux in lines
    I_c = conductor current
    K = constant

Since current can only be established in the conductors when they are cut by the rotating stator flux, the speed (N) of the rotor can never equal the speed of the flux vector (N_s). If $N_s = N$, the flux vector could not cut the conductors of the rotor, induce the voltage, and create the necessary current flow. The difference in speed between the stator flux and the speed of the rotor is called the *slip* and is measured in percent or rpm as determined by the following equations:

$$S\% = \frac{N_s - N}{N_s} \times 100\% \qquad (5.33)$$

$$S_{rpm} = N_s - N \qquad (5.34)$$

In terms of N_s, P, and the slip, the speed of the rotor can be determined by

$$N = \frac{120f(1 - S)}{P} \tag{5.35}$$

EXAMPLE 5.9

A four-pole, 60-Hz, 3ϕ induction motor has a full-load speed of 1725 rpm. Determine the slip in percent.

Solution:

$$N_s = \frac{120f}{P} = \frac{120(60)}{4} = 1800 \text{ rpm}$$

$$S\% = \frac{N_s - N}{N_s} \times 100\% = \frac{1800 - 1725}{1800} \times 100\% = \mathbf{4.167\%}$$

The starting torque of a 3ϕ induction motor can be shown to be related to the applied line voltage in the following manner:

$$T_{st} = KV_L^2 \tag{5.36}$$

Equation (5.36) indicates that doubling the applied line voltage will increase the torque by a factor of 4 (the squared value). This high starting torque is an excellent characteristic for use in pumps, compressors, industrial fans, and machine tools.

Near normal operating conditions it can be shown that the running torque is directly related to the slip. That is,

$$T_{\text{running}} = KS \tag{5.37}$$

In other words, if the slip should double (if the difference in speed between N_s and N should increase by a factor of 2), the running torque will increase by the same factor. The complete torque curve for a 3ϕ squirrel-cage induction motor appears in Fig. 5.38. Note the straight-line region determined by Eq. (5.37). The pull-out torque is an indication of the maximum load the motor can handle at any speed. For this machine maximum torque is obtained at a rotor speed s_1.

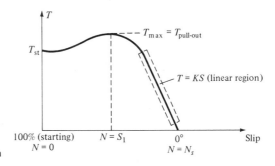

Figure 5.38 Torque characteristics of a polyphase induction motor.

The speed (N) of a motor can be determined directly using a tachometer. The slip can be determined by first calculating N_s from the provided data and substituting into Eq. (5.33).

The polyphase induction motor, unlike the dc motor, can be started with full-line voltage applied. The starting current may be 5–10 times the rated line current, but the machine is designed to handle these surge currents for short periods of time. As the machine builds up to rated speed the line current will drop to rated value. For other loads on the same three-phase system the starting of the 3ϕ induction motor may result in the momentary dimming of the lights, a slowing down of the machinery, etc. If this reaction is objectionable, other techniques, such as using an autotransformer (Chapter 4), may have to be employed to reduce the voltage applied directly to the machine at starting.

Table 5.1 is a list of some commercially available 60-Hz, three-phase squirrel-cage induction motors.

Table 5.1

Polyphase Induction Motors

HP	RPM	VOLTS	AMPERES (I_L)
⅙	1725	240/480	0.9
¼	1725	208	1.2
	1140	240/480	1.3
⅓	1725	208	1.8
½	3450	240/480	1.6
¾	1725	208	3.6
1	3450	240/480	3.0
1½	1725	208	6.0
2	1725	208	6.8
3	3450	240/480	8.9
5	3450	208	15.5

Courtesy of General Electric Company.

In Table 5.1 the length of the $\frac{1}{6}$-hp machine is $9\frac{13}{32}$ in. with a diameter of $5\frac{13}{16}$ in. For the 5-hp machine the length is $14\frac{3}{64}$ in., and the diameter is $6\frac{1}{2}$ in. In addition the weight of the $\frac{1}{6}$-hp motor is only 16 lb, while that of the 5-hp machine is 44 lb—a large unit.

5.6 POLYPHASE SYNCHRONOUS MOTOR

Through the proper design of the rotor it is possible to have a polyphase motor that will run at synchronous speed (the speed of the rotating flux of the stator). For obvious reasons it is referred to as a *polyphase synchronous* motor. Basically, the stator has the same design as for the squirrel-cage induction motor. However, the rotor is now designed to have both the squirrel-cage structure and rotating field poles (electromagnetic or permanent magnet type). The electromagnetic field poles in the rotor can be energized through dc slip rings connected to a dc supply. There are also rotors of a very particular design that will not be examined in this section that actually have the field poles induced through a combination of the shape of the rotor and the stator flux.

As with the squirrel-cage induction motor the speed of the motor is brought near synchronous speed due to the torque developed by the induced current through the rods of the squirrel cage. Near synchronous speed, if the field poles in the rotor are energized, the rotor will jump forward and "lock in" with the rotating poles of the stator as shown in Fig. 5.39. If like poles are encountered at the time the rotating

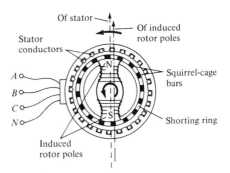

Figure 5.39 Polyphase synchronous motor.

rotor poles are energized, the rotor will slip back until unlike attracting poles are encountered. The slip has then dropped to zero, and the speed of the rotor is then equal to the synchronous value. Namely,

$$N = N_s = \frac{120f}{P} \tag{5.38}$$

where N = speed of rotor.

In an effort to bring the speed due to induced effects as close to synchronous speed as possible the design is usually such as to measurably cut the starting torque to significantly reduce the slip. In fact, it may be cut as much as 50% of full-load value—significantly different from the squirrel-cage type designed to have a very heavy starting torque for its areas of application.

For the polyphase synchronous motor with an external rotor field excitation (I_F) the change in terminal characteristics with field current can provide us with a very efficient method of cutting energy costs in a large industrial plant. At low values of I_F the synchronous machine has an inductive input impedance, and for high values of I_F it is capacitive. For each load condition there is a particular field current that will result in a unity power factor. In general, therefore, no matter what the load on the shaft there is a field current that can be applied to make the machine appear either inductive, resistive, or capacitive.

The capacitive characteristics of the dynamo have important practical application as a *synchronous capacitor*. For a large industrial facility the vast majority of loads (machines, lighting, etc.) has inductive characteristics. As noted by Fig. 5.40

this will result in an apparent power rating far beyond that necessary to supply the total kilowatts (real power) to the load. In large industrial plants, if the manufacturer is paying for the apparent power demand rather than the real power actually dissipated, the introduction of a large reactive power (P_q) component can be a very expensive item. In Fig. 5.41 a synchronous motor operating in its high field current region to establish the highly capacitive characteristics has been attached to a 3ϕ

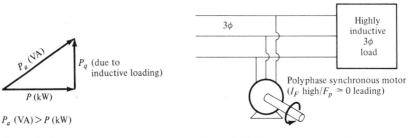

P_a (VA) > P (kW)

Figure 5.40

Figure 5.41 Power factor correction employing a polyphase synchronous motor.

system with a highly inductive load. With no load on the shaft of the motor we can assume as a first approximation that applied voltage lags the line current of the synchronous motor by 90° if we apply a very high field current. For all practical purposes we can therefore assume for application sake that the synchronous motor is nothing more than a large capacitor as shown in Fig. 5.42.

If we assume the machine to be purely capacitive, it will not have a real power component in its power triangle, and its apparent power rating will equal its reactive power component. That is,

$$\textit{Synchronous capacitor:} \quad \boxed{P_{a_s} = P_{q_s}} \tag{5.39}$$

For the power triangle of Fig. 5.40 if we choose a synchronous motor with a VA rating equal to the inductive component of the factory load, the two reactive components will cancel as shown in Fig. 5.43 and the net apparent power (the cost element) will be its minimum value—simply equal to the real power dissipated. It certainly seems strange to apply a motor to a system simply to act as a capacitor and

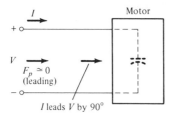

Figure 5.42 Capacitive appearance of a polyphase synchronous motor with no-load and high I_F.

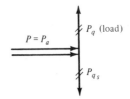

Figure 5.43 Forcing a load-power factor of unity ($F_p = 1$).

have absolutely no load on its shaft, but the practical world considers it a valid energy-conserving technique. Reduced benefits of the capacitive characteristics can still be derived if a load is applied to the shaft of the synchronous motor.

EXAMPLE 5.10

In Fig. 5.41, the total 3ϕ load of the industrial plant is 1500 kW at a lagging (inductive) power factor of 0.6.
1. Sketch the power triangle of the load.
2. Sketch the power triangle of a synchronous motor operating as a synchronous capacitor that will balance the reactive power component of the inductive load. Assume that the synchronous capacitor has pure capacitive characteristics and zero power factor.

Solution:
1.

$$\theta = \cos^{-1} 0.6 = 53.1°, \qquad \sin 53.1° = 0.8$$

From $P = P_a \cos \theta$,

$$P_a = \frac{P}{\cos \theta} = \frac{1500 \text{ kW}}{0.6} = \textbf{2500 kVA}$$

and from $P_a = P_q \sin \theta$,

$$P_q = \frac{P_a}{\sin \theta} = \frac{1500 \text{ kW}}{0.8} = \textbf{2000 kvars}$$

The power triangle for the inductive load appears as in Fig. 5.44.
2. To establish a unity power factor load characteristic the reactive power of the synchronous capacitor must equal the reactive power component of the load. Therefore

$$P_{q_s} = P_{q_L} = \textbf{2000 kvars}$$

and from Eq. (5.39),

$$P_{a_s} = P_{q_s} = \textbf{2000 kVA}$$

The power triangle of the synchronous capacitor appears as in Fig. 5.45

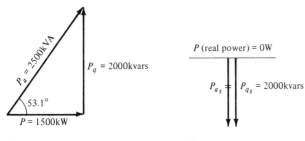

Figure 5.44   Figure 5.45

A great deal of additional information on the construction and analysis of the synchronous machine under varying load conditions is readily available.

5.7 SINGLE-PHASE ac MOTORS

The single-phase outlet is typically all that is available in the home, educational institutions, many small businesses, office buildings, etc. The single-phase ac motor is therefore the most frequently encountered. Since most single-phase outlets have an absolute maximum current rating of 20 A at 120 V, the majority of the single-phase ac motors have power ratings less than $(20)(120) = 2400\ W = 3.22$ hp. The vast majority of the single-phase ac motors are 1 hp or less. A few moments' reflection is all that is necessary to realize how many dynamos of this type appear in the home. The refrigerator, washing and drying machines, dishwasher, sewing machine, electric typewriter, electric heaters with fans, heating system circulator, and even the electric shaver (plug-in) incorporate some type of single-phase ac motor. For most single-phase ac motors, the general characteristics such as starting torque, efficiency, and power factor are not so ideal as for a 3ϕ ac motor if compared on a relative horsepower scale. In fact, the efficiency of some single-phase ac motors can be as unbelievably low as 10%. However, since the power drawn during operation is usually quite small, the loss in watts can be tolerated. Of course, if we consider the number of machines of this type in use today, perhaps a measure of reevaluation is required in times such as today.

There is a wide variety of commercially available single-phase ac motors. However, the series or universal type, the split phase, the capacitor start, shaded pole, hysteresis, and reluctance types are the most frequently encountered. Each will be described briefly here with an application or two.

It is possible to simply apply an ac voltage to a dc shunt-wound motor and obtain an output torque on the shaft. In Fig. 5.46(a), the directions of I_F and I_A are shown for the positive pulse of the sinusoidal input. In the same figure a pole and rotor conductor current direction are shown which clearly indicate that (using the left-hand rule) the torque is clockwise. During the lower half of the sinusoidal input the current in the field and armature will reverse as indicated in Fig. 5.46(b). However, the torque is *still* clockwise, indicating that for the positive and negative regions of the sinusoidal input the torque will always be in the clockwise (or counterclockwise) direction. A reversal in direction can be obtained by reversing either the field or armature circuit but obviously not both at the same time. The shunt motor configuration is seldom

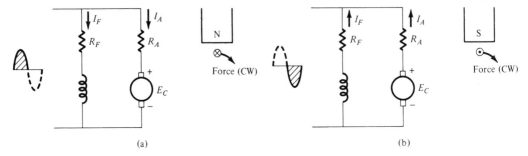

Figure 5.46 Shunt-single ϕ ac motor.

employed for ac application because of the reduction in output torque due to the resulting small main field flux and the increased out-of-phase relationship between the field flux and the armature current. Both effects are introduced by the high inductive reactance of the main field winding not present for dc conditions. The impedance of the field current is thereby increased substantially and the field current reduced (and therefore the field flux) to an ineffective level. This high reactance will introduce a large phase shift between I_F and I_A so that when I_A is its maximum, the flux is not, and the torque $T = K\Phi I_A$ is substantially reduced.

For reasons to be described, the series motor, called the *universal motor*, can display satisfactory characteristics when a single-phase ac voltage is applied. The basic elements of the machine appear in Fig. 5.47. Obviously, the field and armature currents will always have the same direction since they are in fact the same quantity $(I_S = I_A)$. This will result (as for the dc shunt motor) in the torque always having the same direction even though an alternating emf is applied. To reverse direction either the series field coil *or* armature portion of the system must be reversed. Changing both will result in the same direction of rotation,

The effect of the inductance introduced by the turns of the series field is less because there are normally considerably fewer turns in the series field than in the shunt field. This is a necessity, or there would be an intolerable voltage drop across the dc geometric resistance of the series coil, and an insufficient potential level would appear across the armature. The resulting field current and flux are therefore greater, and less phase shift exists between the pole strength and rotor conductor current. The result would be a torque more comparable to an equivalent applied dc voltage and a more efficient system in general.

The characteristics of this *universal* motor are quite similar to those of the dc series motor as shown in Fig. 5.48. Applications that require very high speed but need only low torques to meet the load will work their design around the low-current region. Application where very high torques are needed but speed is not a factor would require the use of the high-current region. As noted on the curves the current must not reach too low a level or the speed could increase to dangerous levels. It is therefore necessary to ensure with resistive elements or other design techniques that the

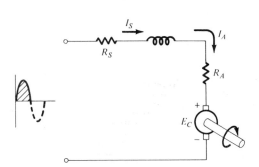

Figure 5.47 Single-phase series ac motor.

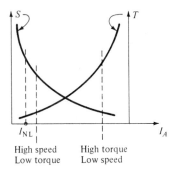

High speed   High torque
Low torque   Low speed

Figure 5.48 Torque and speed characteristics of the single-phase series ac motor.

minimum line current is the no-load value. There are numerous applications for this device in the home that make use of the high-speed, low-torque requirements: electric shavers, sewing machines, food mixers, vacuum cleaners, electric typewriters. The full range of its characteristics are employed in electric power tools (drills, etc.) where heavy loads require the high-torque, low-speed characteristic and lighter loads the high-speed, low-torque characteristic.

Keep in mind that for the discussion above the design is basically dc in the sense that commutator sections and brushes will be present which will result in increased maintenance and cost. Other designs called simply series ac motors are available that work only on ac applied signals with increased simplicity in design (only slip rings) and increased efficiencies. However, the term *universal* is employed to indicate the versatility of the machine in that it can operate on dc or ac applied voltages.

The induction principle as applied in the 3ϕ squirrel-cage induction motor can also be applied to single-phase machines with one important modification: The single phase must be split to appear as two with a phase shift between them of 90°. Hence the terminology for this group of single-phase ac motors—*split phase*. The splitting of the phase is necessary to establish a rotating flux around the stator of the machine. If left as a single phase the magnetic pole alignment would remain fixed. An attempt will not be made here to show why splitting the phase by at least 90° will result in the induced poles rotating about the stator, but we shall investigate how the phase can be split. There are many references for further investigation in this area. The speed of the rotating flux will be determined by the same equation employed for the 3ϕ machine:

$$N_s = \frac{120f}{P} \qquad \text{(rpm)} \qquad (5.40)$$

To split the phase, a second winding is added in parallel with the main field winding as shown in Fig. 5.49 that has considerably less turns than the main field winding. The additional winding will therefore have a much lower inductance level and consequently lower inductive reactance at the applied frequency. This secondary winding is set in the same slots of the stator as the main field winding but is totally insulated from it.

The applied single-phase voltage V_L in Fig. 5.49 appears across both windings and is therefore our reference vector in the phasor diagram for the system in Fig. 5.50. At the applied frequency, the inductive reactance of the main field winding is very high compared to its series internal resistance, and I_F lags V_L by almost 90° as shown in Fig. 5.50. In the parallel branch the inductive reactance is considerably less than its series resistance and I_p lags V_L (since it still is an inductive branch lagging F_p) by only a few degrees as shown in Fig. 5.50. Although I_F and I_p are not exactly 90° out of phase, they have components as shown in the same figure that are exactly 90° out of phase. The flux generated by these component currents in their respective windings will also be 90° out of phase, resulting in the split phase and a flux pattern that will rotate around the stator inducing currents in the conductors of the rotor and establishing the necessary torque to turn the shaft.

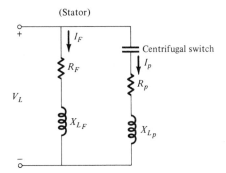

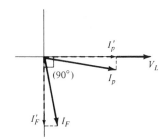

Figure 5.49 Split-phase single-phase ac motor.

Figure 5.50 Phaser diagram for split-phase single-phase ac motor.

Near synchronous speed the necessary torque can be developed solely by the main field winding for reasons shown in Fig. 5.51. With the auxiliary field winding removed the main field will induce currents in the rotor in the direction shown in Fig. 5.51 when the rotor is rotating at or near synchronous speed. The flux pattern developed by these induced currents which appears 90° out of phase with Φ_F is also 90° out of phase with Φ_F on a time scale due to the high inductance of the rotor which causes a phase shift of 90° between the induced voltages across the rotor conductors and the resulting rotor current. The necessary requirement of two flux vectors 90° out of phase to establish a resultant rotating flux vector is therefore present, and the motor can continue to turn without the auxiliary winding. In fact, the auxiliary winding will cut down on the performance characteristics of the motor and should be removed. In the neighborhood of 80% of synchronous speed, a centrifugal switch (a switch whose contacts will open when rotating at a preset speed) will open and remove the auxiliary winding from the system.

The characteristics of the split-phase ac motor as depicted in Fig. 5.52 are quite similar to those of the polyphase induction motor. That is, it is a fairly constant-speed device near synchronous speed, although its slip is more typically 5–6% rather than the 2–3% encountered for the polyphase machines. The starting torque is

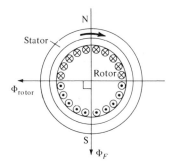

Figure 5.51

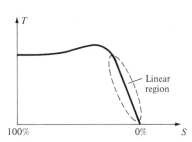

Figure 5.52 Torque characteristics of a split-phase ac motor.

typically about 150% of rated value at a starting current of 5–10 times the rated value. The torque characteristics are quite linear near the synchronous speed with a direct relationship with the slip at these speeds. Since this type of motor is frequently used in washing machines, dishwashers, and oil-burner circulator motors, whenever they are engaged it is possible that a dimming of the lights in the household may be noticeable. The high initial surge current is often the cause of the circuit breaker "popping" if two or more require this high starting current simultaneously on the same circuit in the home. To reverse the direction of rotation of this machine it is only necessary to reverse either one of the windings.

A split-phase motor and its characteristics appear in Fig. 5.53.

Figure 5.53 Split-phase ac motor (Courtesy Westinghouse Electric Corp.).

The 90° phase shift can also be obtained by placing a capacitor in series with an auxiliary winding such as shown in Fig. 5.54. At the applied frequency the capacitive reactance heavily outweighs the inductive reactance of the auxiliary winding so the added parallel branch is effectively capacitive in terminal impedance. The auxiliary winding is necessary to establish the out-of-phase flux resulting from the current through that branch. The phasor diagram of the system appears in Fig. 5.55 with the applied voltage the reference vector as before. The current I_F still lags the voltage across that branch due to the high inductance of the main field winding. However, since the parallel branch is, in total, capacitive, the current I_p will lead the applied voltage by the indicated angle. Again a phase shift of exactly 90° is not usually obtained, but components of I_F and I_p do have this phase relationship and establish the two flux components 90° out of phase. As before, a centrifugal switch will open the auxiliary branch at a speed close to the synchronous.

The increased angle between the parallel branch currents results in improved starting characteristics over that of the split-phase inductive type. The starting torque can now be 2 or 3 times the rated value but with starting currents only 2 or 3 times the

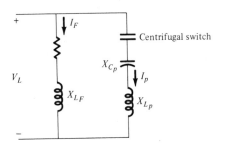

Figure 5.54 Capacitive-start induction motor.

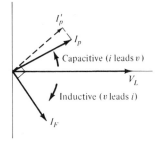

Figure 5.55 Phaser diagram of capacitive-start induction motor.

rated value rather than the 5–10 times encountered for the split-phase inductive machine. In general, however, the capacitive start is a more expensive item than the split-phase variety. Machines of this type are not limited to the fractional horsepower variety due in part to the significantly reduced starting current. A 5–7-hp machine is quite possible and relatively efficient.

The increased starting torque is desirable for application involving pumps, compressors, air conditioners, freezers, refrigerators, and farm equipment. A typical commercial unit appears in Fig. 5.56 with its nameplate data.

The last of the split phase to be considered is the shaded pole, which acts on a principle quite different from the two just described. The poles of the stator are constructed such as shown in Fig. 5.57. A small portion of the pole is separated from the main structure by the indicated cut and a winding of a limited number of turns (often a single heavy copper strap) around the smaller section called a shading coil. The

Figure 5.56 Capacitive-start ac induction motor (Courtesy Westinghouse Electric Corp.).

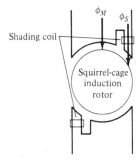

Figure 5.57 Shaded-pole induction motor.

effect of a shading coil is to introduce a time phase shift of 90° between the two poles (of the singular structure). This is accomplished through the current induced in the shading coil by the changing flux of the pole structure. An increasing flux strength through the pole will induce a current in the strap that will develop a flux opposing the increasing flux pattern in the shaded region and severely reduce the flux through the shaded portion. Then as the main field flux reaches its maximum the rate of change of flux is a minimum, and the induced current in the shaded coil is quite small. The resulting opposing flux of the shading coil is therefore quite small, and the flux strength of the shaded region will increase. In other words, when the main field flux is changing at its highest rate the shaded pole flux is a minimum, while when the main field flux is changing at its minimum rate the shaded pole is a maximum. The result is a time delay between maximum flux conditions for each part of the pole, which makes it appear as if the flux pattern is moving clockwise around the stator as required to set the rotor in motion.

The single-phase shaded pole induction motor is one of the least expensive to manufacture due to its relative simple construction, but it is also one of the most rugged and finds application in fans, vending machines, movie projectors, etc. Its horsepower rating is typically less than $\frac{1}{25}$ hp. It must be limited to very small power applications since its efficiency can be as low as 10%.

The principle of operation of the *hysteresis* motor is quite different from those just described. Although the rotating stator field can be established by a polyphase applied voltage, capacitor start, or shaded pole, the rotor as shown in Fig. 5.58 does

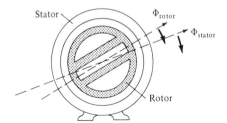

Figure 5.58 Hysteresis ac motor.

not have salient (protruding) poles or dc excitation. It is made of very high-retentivity (holds on to the polarized flux pattern even when the applied mmf is removed) steel that has a large hysteresis loop. The large loop will ensure a lag in the time between the rotor giving up its magnetic characteristics after the applied mmf is removed At starting and while it is building up to synchronous speed the required torque is developed by the eddy currents established in the core and the hysteresis effect just described. In the hysteresis effect the polarized flux pattern of the rotor is always trying to align itself with the rotating stator flux of the stator, causing a torque on the rotor. The lag between the two is due to the high retentivity effect. When synchronous conditions are reached such that the speed of the rotor equals that of the stator speed the eddy current torque is zero, leaving only the hysteresis-induced torque. The crossbars provide a low-reluctance path for the induced polarized flux in the rotor to establish a type of permanent magnet in the rotor that can lock in on the rotating stator flux. A commercially available unit appears in Fig. 5.59 with its characteristics. Note the fixed speed for increase in load. The quiet running of the motor due to the absence of the salient poles and lack of windings and its constant speed characteristic make this motor an excellent choice for applications such as clocks, magnetic tape drives, turntables, and movie projectors.

The limit to which we were able to examine dc and ac machinery is quite obvious from the fact that not one of the areas of application of the various motors was examined in detail in this chapter. However, the salient points of the construction and operation of each of the most frequently encountered dc or ac dynamos were included with an application or two. It was the authors' intention to provide sufficient background material so that the operation and characteristics of any type of machine encountered can be at least partially understood and perhaps applied to perform a particular task.

Figure 5.59 Single-phase hysteresis ac motor (Courtesy Singer Company).

PROBLEMS

§5.2

1. A dc generator develops 120 V at 3600 rpm. At what speed will it generate 160 V if all the other variables of the dynamo remain fixed?

2. If the flux (Φ) of a 240 -V, 2400-rpm generator is reduced by 50%, determine the new generated voltage. The speed remains fixed.

3. Determine the new generated voltage if the speed is doubled and the flux reduced by 25%. Before the change in conditions, the dc generator provided 120 V at 2400 rpm.

4. A dc shunt generator ($R_F = 100\ \Omega$, $R_A = 2\ \Omega$) develops 124 V (E_g) when applied to a 20-Ω load. Determine
   (a) I_L, I_F, I_A.
   (b) V_{R_A}.
   (c) V_L.
   (d) The power to the load.

5. For the cumulative compound dc generator of Fig. 5.7 where $R_F = 100\ \Omega$, $R_A = 1.5\ \Omega$, $R_s = 0.5\ \Omega$, $V_t = 120$ V, and $I_L = 3.5$ A, determine
   (a) I_F and I_A.
   (b) V_{R_A} and V_{R_S}.
   (c) E_g.
   (d) The power to the load.
   (e) The power generated.
   (f) The power lost.

6. For a series dc generator with $R_S = 1\ \Omega$, $R_A = 3\ \Omega$, and $V_t = 60$ V that delivers 240 W to the load, determine
   (a) I_A, I_S, and I_L.
   (b) E_g.

§ 5.3

7. Determine the force on a conductor 4 in. long that is perpendicular to the field if the flux density is 50,000 lines/in.² and the current is 6 A.

8. What is the effect on the counter-emf of a dc motor if the flux is increased by a factor of 1.8 and the speed reduced by 25%?

9. For the dc motor of Fig. 5.14, $R_F = 60\ \Omega$, $R_A = 2\ \Omega$, and the power supplied is 120 V at 5 A. Determine
   (a) I_F and I_A.
   (b) The counter-emf.
   (c) The output power if the machine is 80% efficient.
   (d) The output torque if the speed is 1800 rpm.

10. For a dc motor with the nameplate data 3 hp, 120 V, 20 A, and 3600 rpm, determine
    (a) The output power.
    (b) The input power.
    (c) The efficiency of operation.

11. For the compound configuration of Fig. 5.18, $R_F = 120\ \Omega$, $R_A = 2.5\ \Omega$, $R_S = 0.5\ \Omega$, $E_C = 102$ V, and $V_t = 120$ V. Determine
    (a) I_A and I_F.
    (b) I_L.
    (c) The input power.
    (d) The power out if the machine is 70% efficient.

12. (a) If the armature current is doubled, will the shunt, series, or compound differential machine have the greatest increase in torque?
    (b) Repeat part (a) for the effect on speed.

13. Determine the starting resistor necessary to limit the starting current to 200% of its rated 10-A level if the applied voltage is 120 V and $R_A = 1.5\ \Omega$.

§ 5.4

14. Calculate the frequency generated by a four-pole alternator running at a speed of 1500 rpm.

15. Determine the line current and phase current of a 3ϕ ac generator with the following nameplate data: 5 kVA, 200 V, 2400 rpm, Y-connected, 60 Hz.

16. Determine the number of conductors per phase of a 3ϕ generator that can generate 199.8 V with $\Phi = 3 \times 10^6$ lines and $f = 60$ Hz.

17. Determine the voltage generated per phase of a 3ϕ, Y-connected alternator if $R_\phi = 3\ \Omega$, $X_\phi = 4\ \Omega$, and the machine is rated 5 kVA at 208 V.

§ 5.5

18. Calculate the number of poles of a 60-Hz polyphase induction motor that runs at a speed of 3600 rpm.

19. A polyphase induction motor has a slip of 3%. Determine its running speed if it has four poles and the applied frequency is 60 Hz.

20. Determine the new starting torque of a polyphase induction motor if the applied line voltage is increased by 150%.

21. What is the effect on the running torque if the slip drops 100% (polyphase induction motor)?

§ 5.6

22. (a) Determine the speed of a polyphase synchronous motor with two poles and an applied frequency of 50 Hz.
    (b) What is the slip in percent? (*Think.*)

23. In Fig. 5.41, the total 3ϕ load is 1200 kW at a lagging F_P of 0.8.
    (a) Sketch the power triangle of the load.
    (b) Determine the apparent power rating of a synchronous motor that will result in a unity power factor load.

§ 5.7

24. (a) List four types of single-phase ac motors and briefly describe (in your own words) their mode of operation.
    (b) Which of the above has good torque characteristics?
    (c) Which of the above has a constant-speed characteristic?

6

Basic Electronic Devices

6.1 INTRODUCTION

In recent years the advertising media have added a touch of glamour to the electronics industry through their exposure of the new advances in miniaturization. Although the layman may not be totally familiar with the terminal characteristics of the device, he has become familiar with some of the terminology and how it has effectively reduced the size of electronic systems from the calculator to a large memory computer.

In this chapter we shall introduce the construction, characteristics, and general behavior of those electronic devices in popular use today. The mathematics and physics of each device that determine its terminal behavior will not be covered in detail. But, rather, a surface understanding of its function and application will be the goal. Approximations will be used throughout the discussion to minimize the complexity that often surrounds a basically simple device when the complete complex mathematical relationships of the device are employed.

There are two broad classifications of electronic devices in use today: semiconductor and tube. The latter has become progressively less important in recent years as new advances in the semiconductor devices have necessitated that they replace the old design tube systems. Tubes are still employed, however, in high-power and high-frequency applications and have received some redesign interest (ceramic tubes, etc.), so they will be introduced, but on a limited scale. Semiconductors refer to those devices constructed of semiconductor materials that have characteristics somewhere between those of a conductor and an insulator. The most obvious feature of semiconductor devices is their size as compared to a tube that will perform fundamentally the same function. It would appear that the semiconductor device will eventually be able to surpass the characteristics of the tube in all areas including the high-power and frequency regions and perhaps effectively eliminate the tube from the competitive market.

6.2 DIODES

The diode is a two-terminal device that ideally behaves like an ordinary switch. It has an "on" state in which it appears on an ideal basis to be simply a *short circuit* between its terminals and an "off" state in which its terminal characteristics are similar to those of an *open circuit*. The ideal characteristics of the device appear in Fig. 6.1 with the symbols for the semiconductor and tube devices. Note that for positive values of $v_D(v_D > 0)$ the diode is in the short-circuit state and the current through it limited by the network in which it appears. For the opposite polarity $(v_D < 0)$, the diode is in the open-circuit state and $i_D = 0$. For the semiconductor element, the diode current can only pass through the device in the direction of the arrow appearing in the symbol. For the tube there is no arrow present, and the possible flow direction must be memorized.

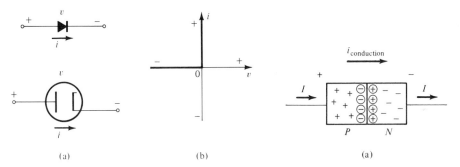

Figure 6.1 Diodes (a) symbol (b) characteristics of the ideal device.

Figure 6.2 Construction and forward bias states of the semiconductor diode.

The construction of each diode appears in Fig. 6.2. In the semiconductor type two semiconductor materials with opposite charge content are placed side by side. The *n*-type material is a semiconductor material such as *silicon* or *germanium* heavily *doped* with negative charges (electrons) as shown in the figure. The *p*-type material is the same semiconductor material heavily doped by other elements so that it will have an excess of positive regions, called *holes*. The holes are in actuality the absence of a negative charge (electron). When the forward bias (term for the application of a dc voltage that establishes the conduction state in the diode) voltage is applied as indicated by the polarities across the device in Fig. 6.2 the holes in the *p*-type material are pressured away from the applied positive potential at the other end of the *n*-type material. The reverse is true for the negative charges in the *n*-type material. They will be attracted to the applied positive potential and pressured away from the negative potential. Conduction is therefore established through the device, and, ideally speaking, the short-circuit state established.

For the reverse polarity across the diode the negative charges of the *n*-type material will be attracted to the positive potential and the positive holes of the *p*-type material to the negative potential. The result is a region between the *n*- and *p*-type materials (at the junction of the two) that is void of any charge carriers. This region is called the *depletion* region due to the depletion of carriers and for all practical purposes represents an open-circuit state between those two points. Thus the two possible states of a semiconductor diode have been described in a very superficial manner. The details of the construction and operation of this two-terminal device require many pages of a test devoted to electronic devices and systems.

The tube diode has a heater that must be energized before the device can exhibit its characteristics. When a preset voltage (often 6.3 V ac-rms value) is applied to the heater terminals, the cathode of the tube as shown in Fig. 6.3 will heat up, and

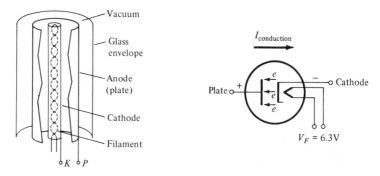

Figure 6.3 Construction and forward-bias state of the vacuum tube diode.

negatively charged electrons will be liberated from its oxide-coated surface to the region between the cathode and plate. If we now apply a positive voltage to the plate, the liberated electrons will be attracted to that surface and the conduction state established. However, if a negative voltage is applied to the plate, the electrons emitted by the cathode will be repelled by the plate and the open-circuit, nonconduction state defined.

If the actual characteristics of each device were perfectly ideal, the application of each in the practical world would be greatly simplified. However, as shown in Fig. 6.4, the actual characteristics of each device are certainly not ideal, although it is immediately obvious that the semiconductor device leans more heavily toward the ideal characteristics than the tube. The slope of the characteristics of each indicates that they are not perfect shorts in the region of conduction but have some resistance associated with each. For the tube it may be 1–5 K, while for the semiconductor device it may typically be 10–30 Ω. Note, however, that the curve of the tube begins at 0 V, while it doesn't start until perhaps 0.7 V positive for the semiconductor diode. For silicon materials this shift (V_F) is typically 0.7 V, while for germanium it is about 0.3 V. It might now be asked why we don't use germanium exclusively since it appears to be closer to the ideal case. The answer lies in the fact that each device has a limited range of operation dependent on its construction and the materials used. In other

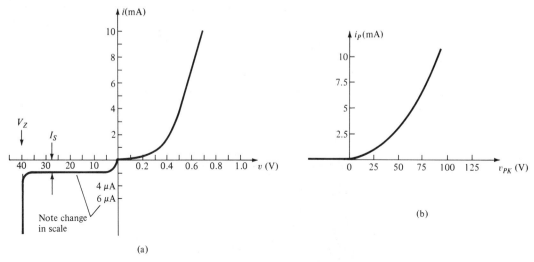

Figure 6.4 Characteristics of the (a) semiconductor and (b) vacuum tube diodes.

words, for each diode there is a maximum current that can pass through the device in its conduction state before it will reach its *burnout* state and its characteristics destroyed. For each manufactured device this limiting current is specified. It has been found that silicon material can handle much higher currents and power levels than germanium and is therefore used in much greater quantities even though its characteristics may appear less "ideal." The offshoot voltage indicates that we must be forward-biased more than this voltage before reaching the conduction state.

Since semiconductor devices will have the limelight in this text (as they rightly deserve), each and every application will employ the semiconductor variety. Further, since the forward resistance is usually quite small compared with the load elements in series with it, we shall approximate the forward resistance of the diode to be 0 Ω. As for the offshoot of 0.3–0.7 V we shall simply remember that the semiconductor diode must be forward-biased by at least this amount before it can assume the approximate short-circuit state. In review, see Fig. 6.5.

Since the ohmmeter section of a VOM has a small internal dc battery voltage, the on or off state of a diode can be determined with this instrument. Placement of the leads across the device in one direction should result in a high current and high ohmmeter indication (low resistance), while the reverse will indicate the opposite.

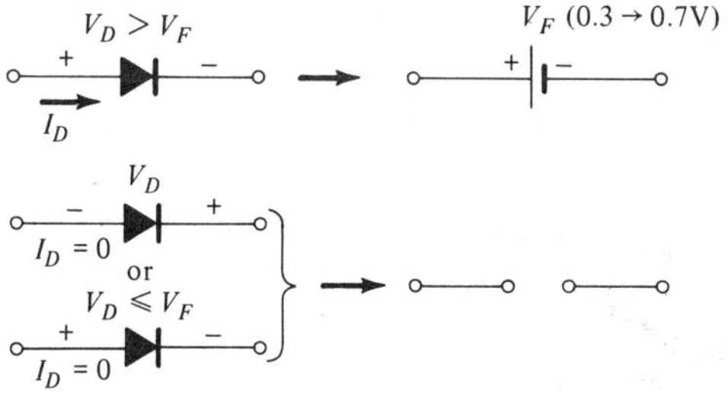

Figure 6.5

A high or very low current (indication) for both directions will indicate a damaged element. The maximum current that can flow through the ohmmeter circuit even with the two leads shorted together is usually not high enough to concern oneself about damaging the diode during the measurement of the forward bias state.

An instrument called the *curve tracer* can display the characteristics of the diode on a screen for comparison and checking purposes. A well-designed commercially available unit appears in Fig. 6.6.

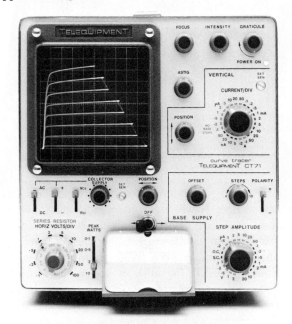

Figure 6.6 Curve tracer.

As indicated earlier, there is a specific current rating for each diode that must not be exceeded. There are some semiconductor devices that can handle hundreds of amperes of current but require the use of heat sinks such as shown in Fig. 6.19 to draw the heat away from the device. A number of semiconductor diodes of different sizes and ratings appear in Fig. 6.7. The direction of diode conduction can be determined by a dot, band, small diode symbol, etc., appearing on the casing of the device as shown in Fig. 6.7.

A pin diagram is required to determine which pin is connected to what element in a diode tube such as in Fig. 6.8. The nameplate data for the device also appear in the same figure.

Before considering other types of semiconductor and tube diodes we shall now examine two relatively simple applications of the diode to give some indication of how its switching action can affect the shape of an input signal. In Fig. 6.9(a) a sinusoidal signal is applied to a network called a *half-wave* rectifier. Since the peak value of the applied voltage is so much higher than the on potential (V_F) of the diode,

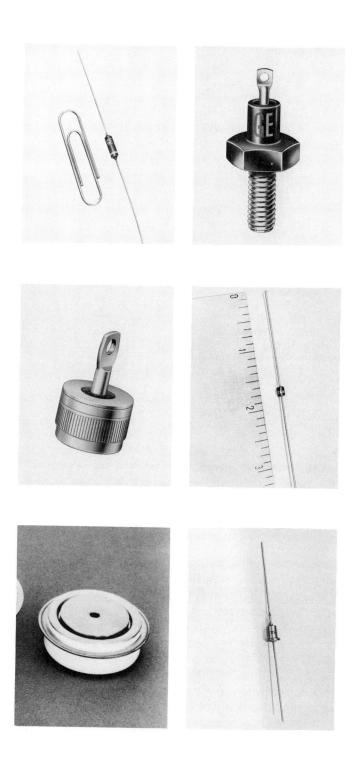

Figure 6.7 Semiconductor diodes.

(a)
35W4
Half-wave rectifier
7-pin miniature type
PIV = 330 V
Peak I_P = 600 mA
Heater voltage = 35 V
(Courtesy Radio Corporation of America)

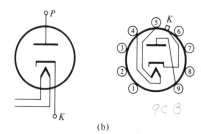

(b)

Figure 6.8 Vacuum tube diode (a) photograph (b) pin diagram.

we shall assume that the diode is perfectly ideal and that $V_F \simeq 0$ V. When the ac signal is positive (above the axis), the polarities across the diode and load resistor R_L are as shown in the figure. The diode is obviously forward-biased and can be replaced by a short-circuit equivalent as shown in the same figure. The positive portion of the input will then appear directly across the load with no loss in potential across the series diode.

However, when the input reverses polarity, the polarities across the elements will be as shown in Fig. 6.9(b) with the diode reverse-biased and the open-circuit representation applicable. The current in the circuit is then 0 A, and the load voltage is $V_L = 0$ V. The resulting waveform for a complete sinusoidal input as shown in Fig. 6.9 is called a *half-wave rectified signal*. The network has converted a sinusoidal signal with zero average value to one with an average value of $G = 0.318V_{peak} = 0.318(120) = 38.16$ V. If the negative portion of the input signal were to be conserved, it would only be necessary to reverse the diode. It might be said that the load resistor has had no effect on the behavior of the network. This may be true for the voltage levels, but keep in mind that it will limit the magnitude of the current in the circuit. If a 500-mA = 0.5-A diode were employed and $R_L = 1\ \Omega$, then $I_D = \frac{120}{1} = 120$ A, and the diode would be immediately destroyed. The load resistor must be at least $V/I = 120/0.5 = 240\ \Omega$ to limit the current to the maximum diode value. Note in Fig. 6.9 that the behavior of the diode was simply that of a voltage-controlled switch—a short circuit for one polarity and an open circuit for the other.

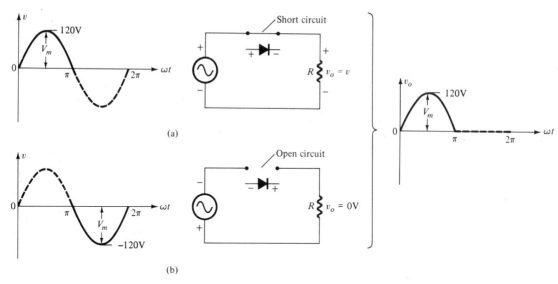

(a)

(b)

Figure 6.9 Half-wave rectification.

The network of Fig. 6.10 is a second diode configuration that finds itself most frequently employed in the first stages of dc power supplies that operate off a sinusoidal ac line. It is called a *full-wave bridge rectifier*. The term bridge was first employed for networks of similar design considered in the dc and ac chapters of the text. The term full-wave indicates that the output is a continuous sequence of pulses rather than having the gaps appearing in the output of Fig. 6.9. During the positive portion of the input sinusoidal waveform, the polarities are as indicated across the elements of the network. You will note that diodes D_2 and D_3 are forward-biased, resulting in the short-circuit equivalent, while diodes D_1 and D_4 are reverse-biased and assume the open-circuit representation. The path of the resulting current through the forward-biased diodes is then obvious and the polarity and waveform across R_L understood.

During the negative portion of the applied signal, diodes will reverse their state, and the conduction path of Fig. 6.11 will result, producing the polarities and output indicated. Note that the polarity across the load resistor remains the same, resulting in a second pulse in the positive region rather than an output of 0 V for the interval. The two pulses in the same time interval will increase the area and therefore the average value. In fact, twice the area is present and $G = 2(0.318V_{\text{peak}}) = 0.636V_{\text{peak}} = 0.636(120) = 76.32$ V. Other techniques are available for obtaining a waveform such as appears in Fig. 6.11 using a different set and number of components. One such configuration that requires only two diodes will appear as a problem at the end of the chapter.

If the diode characteristics are expanded in the reverse bias region as shown in Fig. 6.12, an avalanche region will occur at V_Z that resembles the forward bias state but in the reverse direction. In fact, a closer examination will reveal that at this potential the curve is more vertical than the forward bias region, indicating that it resembles more closely the short-circuit state. For the proper operation of the rectify-

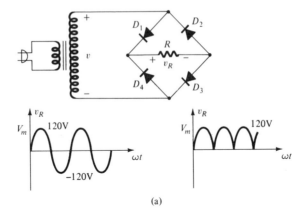

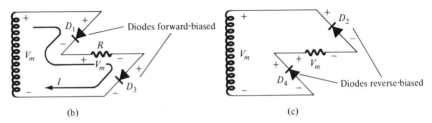

Figure 6.10 Full-wave rectification (a) network (b) conduction path (c) non-conduction path.

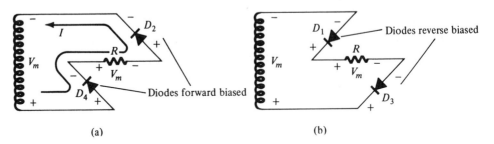

Figure 6.11 (a) conduction path (b) non-conduction path.

ing networks described above the reverse bias must never reach the avalanche value or peak inverse voltage (PIV) that it is most commonly referred to. It is a quantity normally available in manuals or specification sheets. In Fig. 6.9 the peak inverse voltage across each open-circuited diode is simply the negative peak of the input signal. Therefore, a diode must be chosen such that PIV $\geq V_{peak}$. For the bridge configuration, an application of Kirchhoff's voltage as shown in Fig. 6.13 indicates that the peak voltage across the back-biased diode is again the peak value of the sinusoidal input and the condition PIV $\geq V_{peak}$ must be satisfied. Although the input in these examples was a sine wave, other waveforms such as the square wave, triangular

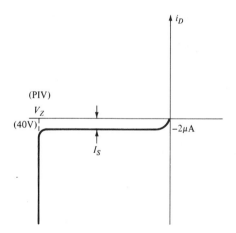

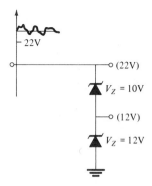

Figure 6.12 Expanded reverse-bias region of the semiconductor diode.

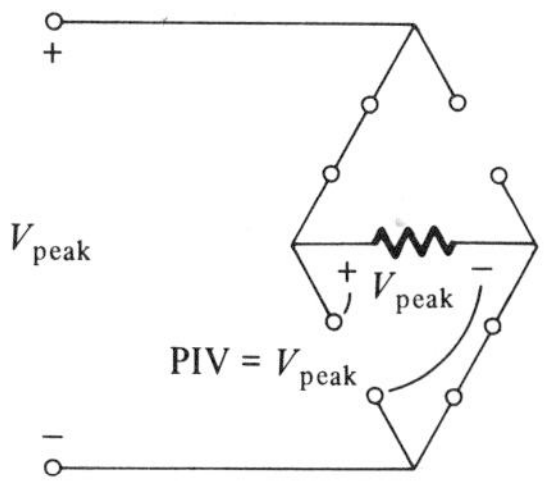

Figure 6.13 Determination of the PIV rating for the bridge rectifier.

waveform, or any nondescriptive waveform can be applied to the same diode configurations with similar results.

There is another type of semiconductor diode called the Zener diode that takes full advantage of this avalanche region (or Zener region as it is often called) as shown by its characteristics in Fig. 6.14. The symbol for the device and its defined polarities appear in the same figure. For V (with the polarity shown) less than V_Z, but greater than 0 V, the device has the characteristics of an open circuit. For applied voltages $V \geq V_Z$; the diode conducts and assumes the short-circuit state indicated by the vertical line. Of course, in this state the short circuit must be associated with the required applied voltage V_Z as shown is Fig. 6.14. For voltages less than $V = 0$ the Zener diode will have the characteristics of a forward-biased semiconductor diode. One frequently employed application of the Zener diode is as a reference potential for other points in the electrical or electronic system. In Fig. 6.15, for example, as long as the applied voltage is greater than 22 V, no matter how distorted the

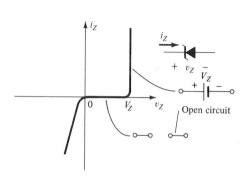

Figure 6.14 Zener diode characteristics.

Figure 6.15 Setting reference voltages using Zener diodes.

riding ripple may be, exactly 12 V will appear at point 1 and 22 V at point 2 since each diode is in its *on* state ($V > V_Z$).

A simple square-wave generator can be developed using two back-to-back Zener diodes as shown in Fig. 6.16. For voltage levels in the positive region of the

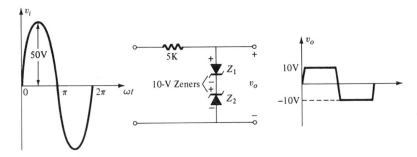

Figure 6.16 Simple square-wave generator.

input signal that are greater than 10 V, Z_1 will be reverse-biased and in the short-circuit equivalent region, while Z_2 will be in the "on" state at its V_Z of 10 V. The remaining input voltage will appear across the 10-K series resistor. The result is a clipping of the top at 10 V. Similarly, the negative portion is cut by 10 V with the

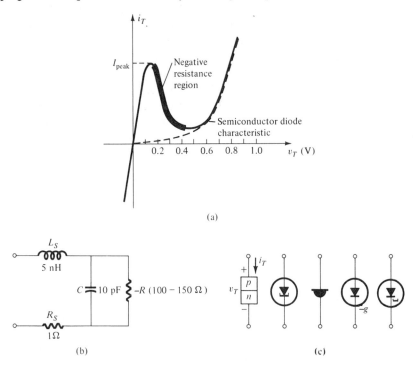

Figure 6.17 Tunnel diode (a) characteristics (b) equivalent circuit (c) symbols.

state of the diodes reversed. The resulting square wave has only slightly sloping size if the peak value of the sinusoid is sufficiently large compared to V_z.

The *tunnel* diode is a two-terminal semiconductor diode whose characteristics appear in Fig. 6.17. The characteristics of a typical semiconductor diode appear as a dashed line on the same set of characteristics for comparison purposes. The region appearing as a heavy dark line is of particular interest since it represents a *negative resistance characteristic*, which means that the current decreases with increase in voltage rather than the reverse, as is true for most resistive loads. The label *tunnel* diode is derived from a tunneling action of the device at a particular potential level. Since the peak voltage that can be applied across the diode may be on the order of millivolts, the ohmmeter section of a multimeter should not be used since the internal dc battery is typically a few volts or more.

Like the junction diode described earlier, the tunnel diode is also a two-layer semiconductor device as shown in Fig. 6.17 with a few of its most frequently employed symbols. The negative resistance has its application in oscillators (sine-wave genera-tors), and its characteristics lend themselves readily to logic switching applications in computers.

The varicap, or varactor diode, as it is sometimes called, is also a two-layer semiconductor diode, but its usefulness lies in the fact that the capacitance between its terminal leads is dependent on the reverse bias voltage applied across the device. In other words, it is a variable capacitance whose capacitance is dependent on the voltage applied across the device. In terms of the reverse bias voltage the capacitance is determined by

$$C_j = \frac{1}{K(V_F + V)^n} \qquad (6.1)$$

where K = constant of semiconductor material
   V_F = knee potential of diode characteristics
   V = applied reverse bias potential
   $n = \frac{1}{2}$ for alloy junctions and $\frac{1}{3}$ for diffused junctions

An equivalent circuit for the device and a few of the symbols frequently employed appear in Fig. 6.18.

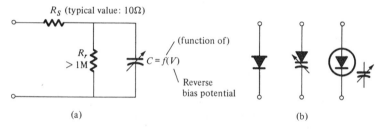

Figure 6.18 Varicap diode (a) equivalent circuit (b) symbols.

It finds application in control systems, variable filters, and FM modulation systems, to be described in a later chapter.

The last of the diodes to be introduced in this chapter is the power variety, which is similar in basic construction to the *p-n* junction type first described but with certain important changes in its construction such as increased surface area at the junction to handle the heavy currents, almost exclusive use of silicon materials, and heat sinks to draw the heat away from the device. The increased size is noticeable in the few power diodes appearing in Fig. 6.19 with their ratings. Consider the very heavy currents that can be handled by these devices of still relatively small size.

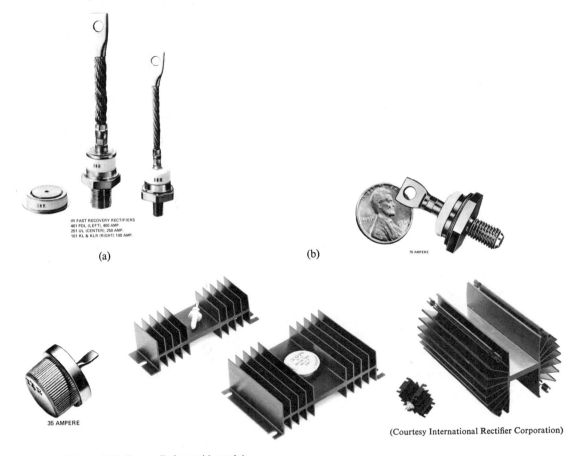

IR FAST RECOVERY RECTIFIERS
401 PDL (LEFT), 400 AMP.
251 UL (CENTER), 250 AMP.
101 KL & KLR (RIGHT) 100 AMP.

(a)

(b)

70 AMPERE

35 AMPERE

(Courtesy International Rectifier Corporation)

Figure 6.19 Power diodes and heat sinks.

6.3 THERMISTORS AND PHOTOCONDUCTIVE DEVICES

The thermistor is a two-terminal device whose terminal resistance is dependent on the body temperature of the device. It has a *negative temperature coefficient*, indicating that the resistance will decrease with an increase in temperature rather than the reverse, as occurred for most commercial resistors introduced earlier. The characteristics of one such device appear in Fig. 6.20. At room temperature the terminal resistance is about 5 K, while at 100°C(212°F—the boiling point of water)its resistance is only 100 Ω. There is therefore a 50-to-1 change in resistance for a temperature span of 80°C. The temperature of the device can be changed by either changing the current

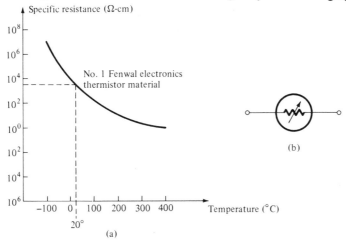

Figure 6.20 (a) Thermistor characteristics and (b) symbol.

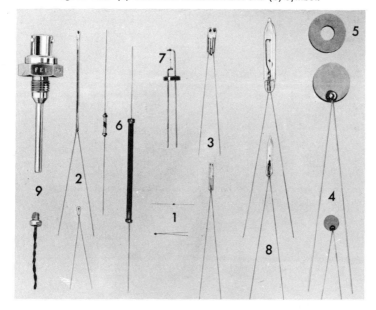

Figure 6.21 Various types of thermistors: (1) beads; (2) glass probes; (3) Iso-curve interchangeable probes and beads; (4) discs; (5) washers; (6) rods; (7) specially-mounted beads; (8) vacuum and gas filled probes; (9) special probe assemblies.

through the device or heating its surface through the surrounding medium or immersion. The symbol for the device is commercially accepted to be that appearing in Fig. 6.20. Thermistors come in all shapes and sizes, as indicated by the photograph of Fig. 6.21.

Photoconductive devices react to the light incident on a particular surface of the element. As the light intensity increases as shown in Fig. 6.22, the resistance decreases, resulting in a negative resistance coefficient. The symbol for the device and a photograph of one commercially available unit appear in Fig. 6.23.

The photodiode is a two-terminal device whose current vs. illumination characteristics (Fig. 6.24) closely resemble that of I_D vs. V_D of a *p-n* junction diode.

The incident light is directed on the junction between the semiconductor layers of the device by the lens in the cap of the diode (Fig. 6.25). The reverse bias region is employed in this device because the reverse saturation current increases almost linearly with increase in illumination as shown in Fig. 6.24.

The energy conversion of a photodiode is reversed in the increasingly popular *LEDs* (light-emitting diodes) used for numerical displays in calculators, timepieces, instruments, etc. Due to the process of *electroluminescence*, radiant light is emitted at a strength dependent on the current through the device as shown by the plot of Fig. 6.26. The symbol for the device and a commercially available unit appear in Fig. 6.27 with its pertinent data.

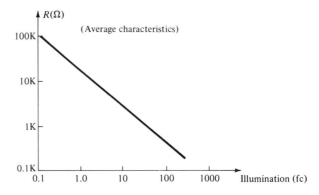

Figure 6.22 Photoconductive cell–terminal characteristics (GE type B425).

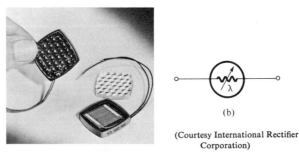

(b)

(Courtesy International Rectifier Corporation)

Figure 6.23 Photoconductive cell: (a) type; (b) symbol.

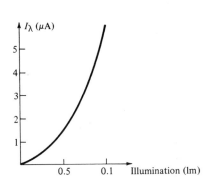

Figure 6.24 Photo-diode characteristics.

(Courtesy General Electric
Company)

Figure 6.25 Photo-diode.

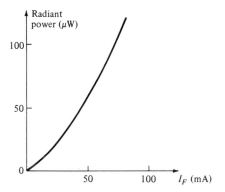

Figure 6.26 LED characteristics.

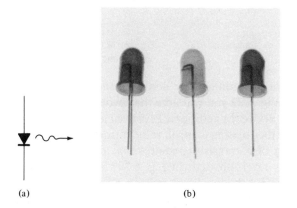

(a)   (b)

Figure 6.27 LED: (a) symbol, (b) device appearance.

6.4 TRANSISTORS AND TRIODES

The transistor (semiconductor) and triode (tube) are each three-terminal devices. They are both very similar in purpose in that they have the ability to amplify (increase in magnitude) the applied signal. Both, however, have different modes of operation and construction and must be considered separately. Since the transistor is the more current and the direction of today's interest, this will be our first area of interest.

As shown in Fig. 6.28, there are three alternating layers of conductor material in a transistor. They also provide the name for each—an *NPN* and a *PNP* transistor. As indicated earlier the direction of this text is to establish some understanding of the terminal characteristics of each device rather than the physics surrounding its internal behavior. We shall therefore progress directly to one of the three configurations that the transistor often assumes and discuss its characteristics and amplifying action.

The *common-base* configuration appears in Fig. 6.29. The label comes from the fact that the *base* terminal is common to both the *emitter* and *collector* terminals.

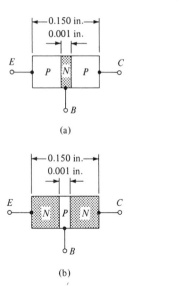

(a)

(b)

Figure 6.28 Types of transistors: (a) PNP; (b) NPN.

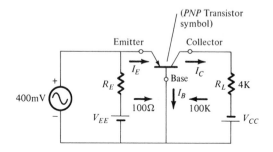

Figure 6.29 Common-base transistor configuration.

The terms used for the terminals are derived from the action of the charged particles in the device. In Fig. 6.29 you will note that two separate dc supplies are required to properly bias the device. In addition a resistive element appears in the input and output circuit for further biasing control, to be introduced in the following chapter. The normal operation of a transistor requires that one *p-n* junction be forward-biased while the other is reverse-biased. In Fig. 6.29 the emitter-base junction is forward-biased, accounting for the low input resistance since current can flow easily through a forward-biased junction. The collector-base junction is reverse-biased, accounting for the high output impedance shown. Since the input junction is forward-biased, the dc voltage from base to emitter is typically $0.3 \rightarrow 0.7$ V, as encountered for the forward-biased diode. When the 400-mV ac signal is applied the emitter current by Ohm's law will be

$$I_E = \frac{400 \text{ mV}}{100 \text{ }\Omega} = 4 \text{ mA}$$

Although Kirchhoff's current law indicates the following for a transistor,

$$\boxed{I_E = I_C + I_B} \tag{6.2}$$

the current I_B is usually so small compared to the magnitude of I_E or I_C that the following approximation is frequently employed:

$$\boxed{I_E \simeq I_C} \tag{6.3}$$

In our example this would dictate that $I_C = 4$ mA also. The result is that

$$V_o = I_C R_L = (4 \text{ mA})(4 \text{ K}) = 16 \text{ V}$$

and the resulting voltage gain is

$$A_v = \frac{V_o}{V_i} = \frac{16}{0.4} = \mathbf{40}$$

In the above example (as with all transistor networks) there was a *transfer* of the input current from a low- to high-resistance circuit. Hence the derivation of the name of the device:

$$\underline{\text{transfer}} + \underline{\text{resistor}} = \text{transistor}$$

In Fig. 6.30 the symbol and biasing arrangements for both the *NPN* and *PNP* devices are provided. Note that the emitter leg is always associated with the arrow in the symbol, while the perpendicular bar is for the base terminal. The arrow is *out* (from the base) for *NPN* transistors and *in* for *PNP* devices.

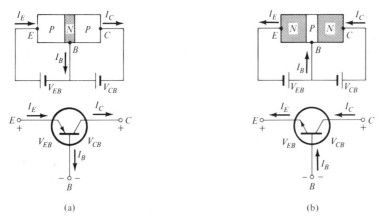

Figure 6.30 Notation and symbols used with the common-base configuration: (a) PNP transistor; (b) NPN transistor.

Two sets of characteristics are required to fully describe a three-terminal device: an input and output set. For the *PNP* transistor the characteristics of Fig. 6.31 apply. It is very seldom that the input or base-emitter characteristics of a transistor are available. Rather the group of curves are looked upon as on with a vertical rise and an average value of 0.5 (for this text) assumed for the transistor base to emitter junc-

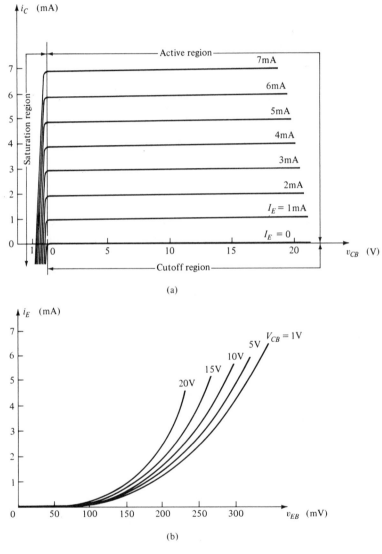

(a)

(b)

Figure 6.31 Characteristics of a PNP transistor in common-base configuration: (a) collector characteristics: (b) emitter characteristics.

tion. From this point on, therefore,

$$\boxed{V_{BE} \simeq 0.5 \text{ V (dc)}}$$ (6.4)

The collector characteristics clearly indicate that the previous statement that $I_C \simeq I_E$ is valid. Simply follow any horizontal time directly over to the collector current axis and it is verified. They also indicate that V_{CB} will have very little effect on the magnitude of I_E since the curves are essentially horizontal. The input current I_E is therefore determined purely by the input circuit.

The collector and emitter currents are related by a factor called *alpha* (α). It is defined by

$$\boxed{\alpha = \frac{I_C}{I_E}}$$ (6.5)

and is typically about 0.98, which is very close to the unity factor employed in Eq. (6.3).

The base and collector currents are related by a factor called *beta* (β) in the following manner:

$$\boxed{\beta = \frac{I_C}{I_B}}$$ (6.6)

Beta is typically $50 \longrightarrow 200$, indicating that the base current is always significantly less than I_C or I_E. Typically I_B is measured in microamperes, while I_C is measured in milliamperes.

The terms α and β are further related by

$$\boxed{\beta = \frac{\alpha}{1 - \alpha}}$$ (6.7)

A two transistor configuration that appears more frequently than the other is called the *common-emitter* configuration. It appears in Fig. 6.32 for the *PNP* and *NPN* transistors. It is important to note that for the common-base and common-emitter configurations a change from a *PNP* to an *NPN* transistor requires that the dc biasing be reversed. As with the common-base configuration, the base-emitter junction is forward-biased, and we shall assume for dc conditions that $V_{BE} \simeq 0.5$ V and remove the need for the input or base characteristics.

The collector characteristics are quite useful, however, and appear in Fig. 6.33. For this set you will note that the input base current determines the horizontal line to be employed and therefore the resulting collector current for a particular value of V_{CE} (the collector to emitter potential). Since the output collector current is related directly to the input base current by the *beta* factor, the transistor is considered a *current-controlled* amplifying device.

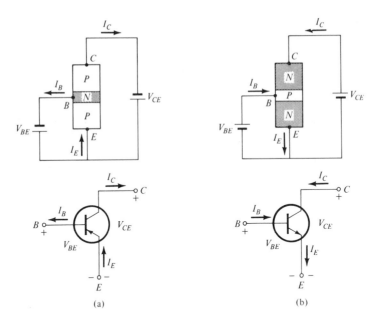

Figure 6.32 Notation and symbols used with the common-emitter configuration: (a) PNP transistor; (b) NPN transistor.

EXAMPLE 6.1

For the characteristics of Fig. 6.33,

1. Determine I_C if $I_B = 40$ μA and $V_{CE} = -10$ V.
2. Determine I_B if $I_C = 3$ mA and $V_{CE} = -8$ V.
3. Calculate β for the results of part 1.
4. Calculate β for the results of part 2.
5. Discuss the results of parts 3 and 4.

Solution:

1. From the characteristics $I_C = $ **4.3 mA.**
2. For a situation such as encountered here where the intersection of the two given points is between two I_B curves an equal number of divisions must be superimposed between $I_B = 20$ μA and $I_B = 30$ μA to give

$$I_B \simeq 27 \text{ μA}$$

3. $\beta = I_C/I_B = 4.3 \text{ mA}/40 \text{ μA} = $ **107.5.**
4. $\beta = I_C/I_B = 3 \text{ mA}/27 \text{ μA} = $ **111.1.**
5. The beta relationship between I_C and I_B is dependent on the point of operation of the device. In some regions the dc amplification factor will be greater than in others.

The remaining common-collector configuration normally appears as shown in Fig. 6.34 rather as one might expect with the collector terminal below the emitter of the transistor. The biasing, etc., is shown for both the *PNP* and the *NPN* in the same figure. In the configuration shown it is referred to as the *emitter-follower,* which

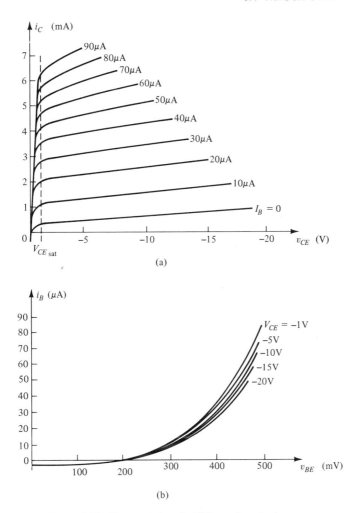

Figure 6.33 Characteristics of a PNP transistor in the common-emitter configuration: (a) collector characteristics; (b) base characteristics.

is employed for impedance-matching purposes, to be discussed in a later chapter in more detail. Briefly, however, the ac output is normally taken off the emitter terminal (the collector is at ground potential), which will *follow* the signal applied to the base potential in both phase and magnitude considerations. The base-emitter junction is forward-biased, and $V_{BE} \simeq 0.5$ V (dc) as before. The output characteristics are essentially those of the common-emitter since the vertical scale of I_E is related to that of the common-emitter characteristics by $I_E \simeq I_C$.

For each transistor configuration, there are maximum ratings of importance: maximum collector to emitter voltage (V_{CE}), maximum collector current (I_C), and maximum power dissipation $P_D = V_{CE}I_C$. Each of these ratings appears on the collector characteristics of Fig. 6.35. The maximum collector voltage and current are

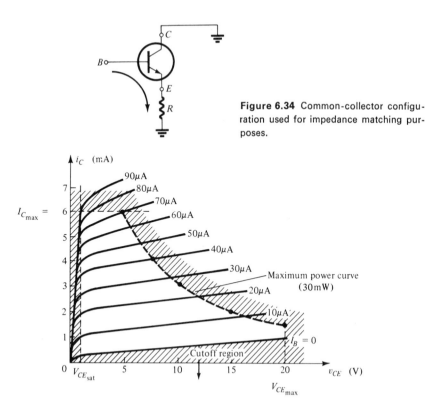

Figure 6.34 Common-collector configuration used for impedance matching purposes.

Figure 6.35 Region of operation for amplification purposes.

simply lines drawn at these values on the characteristics. The power is obtained by simply picking a value for V_{CE} (or I_C) and solving for I_C (or V_{CE}, respectively) using the maximum power rating. For example, if we pick $V_{CE} = 10$ V, then $I_C = P/V_{CE} = 30$ mW$/10 = 3$ mA, and an intersection on the power curve is obtained as shown in Fig. 6.35. Other points can be found and the maximum power curve obtained. Points of operation must now be chosen within this power curve. To avoid distorted (nonlinear) output values of V_{CE} less than the saturation value ($V_{CE_{sat}}$) must not be chosen. Further, the region below $I_B = 0$ is the region of cutoff for the device. The remaining unshaded region must be employed if a nondistorted amplified version of the input is to be obtained.

For increasing wattage ratings the casing must be redesigned to withstand the heat or heat sinks applied as shown in Fig. 6.36(d). For transistors connected directly to a chassis the chassis itself will behave as a heat sink. The terminal identifications of some appear in Fig. 6.37, although every effort is now being employed to label each terminal directly on the casing.

The triode is a three-element vacuum tube with the basic construction and symbol shown in Fig. 6.38. Upon being heated by the filament the cathode will emit electrons from its surface to the region between the cathode and control grid. For 0 V applied to the grid a positive potential plate will result in a flow of charge (I_{PK}) from

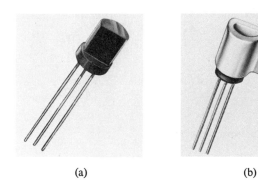

(a)   (b)   (c)

(Courtesy General Electric Company)

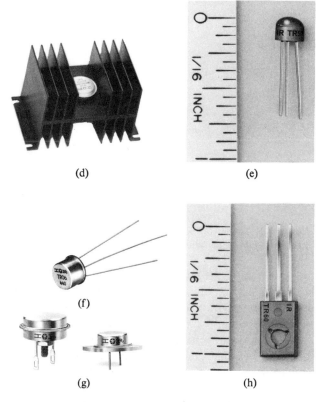

(d)   (e)

(f)

(g)   (h)

(Courtesy International Rectifier Corporation)

Figure 6.36 Various types of transistors.

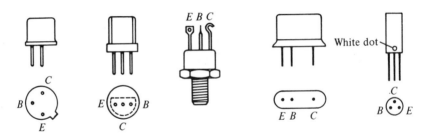

Figure 6.37 Transistor terminal identification.

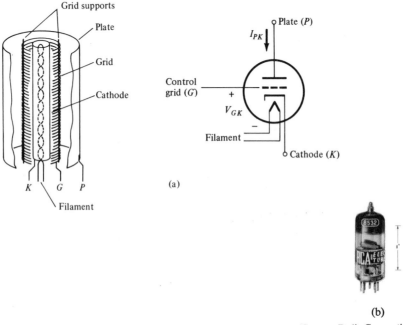

(a)

(b)

(Courtesy Radio Corporation of America)

Figure 6.38 Triode: (a) basic construction and symbol; (b) photograph of a high-mn miniature vacuum tube triode.

cathode to plate as shown by the curve $V_{GK} = 0$ V in Fig. 6.39. Increasing positive potentials on the plate will result in increased plate currents. For $V_{GK} = 0$ the characteristics are essentially those of a vacuum-tube diode. A negative potential applied to the grid (V_{GK}) will result in a large number of electrons being repelled by the structure, preventing them from getting through the grid structure and reaching the plate. There is therefore less plate current (I_{PK}) for the same plate to cathode voltage (V_{PK}), as noted by a vertical line ($V_{PK} = 60$ V) drawn through a few grid to cathode (V_{GK}) lines.

As indicated in Fig. 6.39 the input-controlling variable is a grid voltage which classifies the triode as a *voltage-controlled* device. Since the output signal is measured in volts, it is further categorized as a *voltage amplifier*.

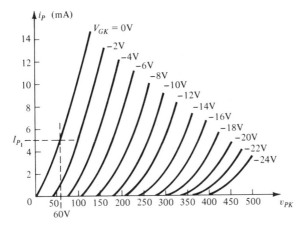

Figure 6.39 Plate characteristics of a vacuum tube triode.

The basic biasing arrangement for the triode appears in Fig. 6.40. Note that the grid to cathode voltage is determined by

$$V_{GK} = -V_{GG} = -4 \text{ V}$$

which replaces our design somewhere on the $V_{GK} = -4$ V curve of Fig. 6.39. The voltage V_{PP} will ensure a positive voltage on the plate of the device less than V_{PP} as determined by the voltage drop across R_L. The path of conduction for the plate current (I_{PK}) is as indicated in Fig. 6.40. The grid current is zero since an open circuit exists between the grid structure and the plate circuit.

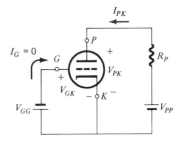

Figure 6.40 Basic biasing arrangement for the vacuum tube triode.

As with the transistor, the triode has maximum ratings of V_{PK}, I_{PK}, and plate dissipation $P_D = V_{PK}I_P$. The region of operation for a triode is shown in Fig. 6.41. The lower region of the characteristics is avoided to minimize the distortion in the output signal. The maximum power curve is determined in the same manner as explained for transistors.

A manual or specification sheet must be consulted to determine which pins of the tube are connected internally to which elements. The pin diagram of a 6J5 triode appears in Fig. 6.42 with additional descriptive information normally available. The interelectrode capacitance is the stray capacitance that exists between the elements of the tube. The effect of these capacitances is negligible at 60 Hz, but at very high

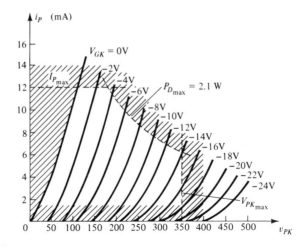

Figure 6.41 Triode region of operation.

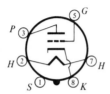

6J5 Triode

Metal and glass octal-type 6J5
used as detectors, amplifiers, or
oscillators in radio equipment

Heater voltage (ac/dc) 6.3V
Heater current 0.3A
Direct interelectrode capacitances (approx.)
    Grid to plate 3.4pF
    Grid to cathode and heater 3.4pF
    Plate to cathode and heater 3.6pF

Figure 6.42 Pin connections and associated data for the
6J5 triode.

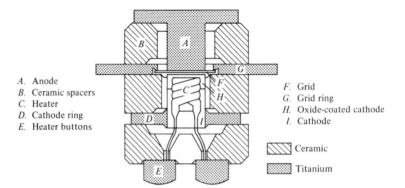

A. Anode
B. Ceramic spacers
C. Heater
D. Cathode ring
E. Heater buttons

F. Grid
G. Grid ring
H. Oxide-coated cathode
I. Cathode

▩ Ceramic

▩ Titanium

Figure 6.43 General construction and pertinent data of the
ceramic triode.

frequencies the capacitive reactance between the elements can be so small that a shorting effect results between the elements. Although most laymen are familiar with the tube construction of Fig. 6.38(a), ceramic tubes are now being developed that have the appearance and design of Fig. 6.43. They provide a more rugged construction and a much higher frequency range of operation.

6.5 FIELD-EFFECT TRANSISTORS (FETs)

In recent years another three-terminal semiconductor device called the *field-effect transistor* has become increasingly important. Unlike the transistor that has two *p-n* junctions, the *junction FET* (JFET) has only one, as shown in the basic construction diagram of Fig. 6.44.

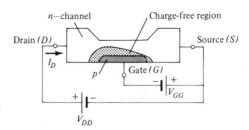

Figure 6.44 Basic construction of the Field Effect Transistor. (FET).

The drain (*D*) and source (*S*) terminals are connected to the *n*-type material, while the gate (*G*) is connected to the *p*-type material. With $V_{GG} = 0$ V and V_{DD} as some positive voltage, a flow of charge (current) is established from drain to source terminals through the narrow *n*-channel. If V_{GG} is adjusted to some negative potential, the *p-n* junction will be reverse-biased, and a depletion region (no charge carriers) will be established at the *p-n* junction; this induced region of noncarriers will reduce the width of the *n*-channel at this point and reduce the drain to source current for the applied V_{DD} voltage. The flow of charge from drain to source can therefore be controlled by the negative potential applied to the gate terminal. This is clearly shown in the *n*-channel JFET characteristics of Fig. 6.45. Note that the more negative the gate potential, the less the drain current due to the increased depletion region and reduced channel width.

In terms of quantities often made available in data manuals and specification sheets the drain current of a JFET is related to the gate to source voltage in the following manner:

$$i_D = I_{DSS}\left(1 - \frac{V_{GS}}{V_P}\right)^2 \qquad (6.8)$$

I_{DSS} is the drain current with V_{GS} set at 0 V. In Fig. 6.45 if we follow the curve for $V_{GS} = 0$ directly over to the i_D axis, we find that I_{DSS} for this transistor is 7.6 mA. V_P is called the *pinch-off* voltage, which is that gate to source voltage that will result in $i_D \simeq 0$ mA. For this device it is -6 V. If we now plot the curve defined by Eq.

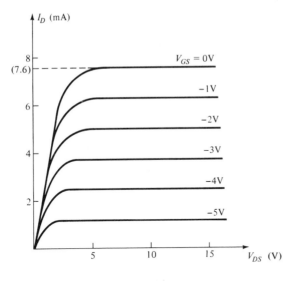

Figure 6.45 *n*-channel JFET characteristics.

(6.8), the *transfer* characteristics of Fig. 6.46 will result. The term transfer is employed because the curve relates an output quantity i_D to an input quantity V_{GS}. Note the absence of V_{DS} in Eq. (6.8). This is an indirect result of the fact that the V_{GS} curves are quite flat and values of V_{DS} have virtually no effect on the drain current level except at very low values of V_{DS}. The transfer characteristics of Fig. 6.46 will find extended use in the design of JFET amplifiers.

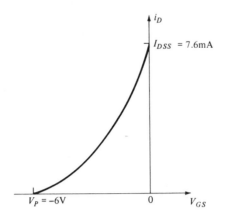

Figure 6.46 JFET transfer characteristics.

The *n*-type and *p*-type materials in Fig. 6.44 can be reversed, and a *p*-channel JFET will result, which requires a reversal in all biasing levels in the network and the characteristics. The symbol and basic biasing arrangement for both *n*-channel and *p*-channel JFETs is provided in Fig. 6.47. Although the basic construction of some commercially available FETs may differ slightly from that appearing in Fig. 6.44, the mode of operation is basically the same for each.

Another type of field-effect transistor is the MOSFET, abbreviation for metal-oxide-silicon FET. The basic construction of the MOSFET devices is somewhat more

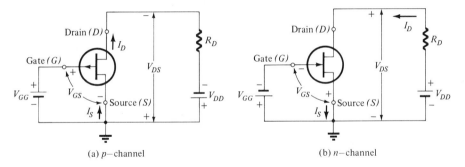

(a) p–channel   (b) n–channel

Figure 6.47 Basic biasing arrangements for the n- and p-channel JFETs.

complex than that of the JFET, but a channel still appears whose width is controlled by the gate potential.

The *enhancement*-type MOSFET has the symbol and characteristics shown in Fig. 6.48 for the p-channel type. For the n-type the arrow is reversed and the polarities reversed on the characteristics. The basic biasing arrangement is the same as for the JFET. The term enhancement is derived from the fact that with *no bias* on the gate ($V_{GS} = 0$) there is no channel for conduction between drain and source terminals and $i_D = 0$, as shown in Fig. 6.48. For increasing negative voltages on the gate the p-type channel will materialize (the channel enhanced) and increase in size so that i_D can increase in magnitude (note Fig. 6.48 for increasing values of V_{GS}). The broken line in the symbol reflects the fact that the channel does not exist without a gate bias.

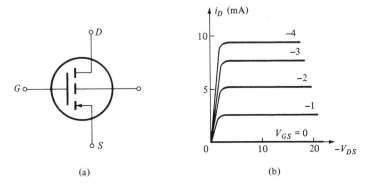

(a)   (b)

Figure 6.48 Enhancement type p-channel MOSFET (a) symbol (b) characteristics.

For the *enhancement-depletion*-type n-channel MOSFET the symbol and charac-teristics appear in Fig. 6.49. Again the arrow is reversed and the polarities on the characteristics changed for a p-channel device. In this case a channel is present with $V_{GS} = 0$ so that a measurable current will flow as shown in Fig. 6.49. For increasing negative voltages on the gate the n-channel is reduced in width and the drain current diminishes. For increasing positive voltages the reverse is true. The continuous bar

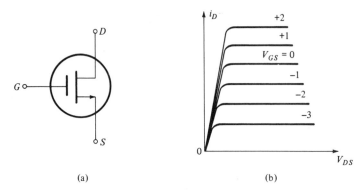

(a)   (b)

Figure 6.49 Enhancement-depletion type *n*-channel MO-
SFET (a) symbol (b) characteristics.

in the symbol connected to the gate terminal reflects the fact that a channel is present
for $V_{GS} = 0$ V. The biasing arrangement for this device is again the same as that
employed for the JFET.

6.6 UNIJUNCTION TRANSISTOR (UJT)

The unijunction transistor is a three-terminal device having the basic construction
of Fig. 6.50. Two contacts B_1 and B_2 are connected to a slab of *n*-type silicon material
with a *p-n* junction formed on the other side between an aluminum rod and the *n*-
type material. The single *p-n* junction is the reason for the terminology *uni*. As shown
in the sketch, the rod is closer to the B_2 connection, resulting in a different slab resistance between the rod and each terminal. However, the resistance between the rod
and B_1 is quite sensitive to the emitter current of the network and may vary from 5 K
down to 50 Ω depending on the magnitude of the emitter current I_E. An equivalent
network for the UJT appears in Fig. 6.51 with the resistance just described appearing
as a variable element. With $I_E = 0$ A the total resistance R_{BB} is defined by

$$R_{BB} = R_{B_1} + R_{B_2}|_{I_E=0} \qquad (6.9)$$

The diode in the equivalent circuit reveals one very important characteristic
of the UJT. For the values of applied emitter voltage V_E not greater than V_1 by the
forward bias voltage of the diode ($V_F \simeq 0.5$ V) the diode is reverse-biased and the
input characteristics of the device are those of an open circuit. However, when sufficient voltage is applied at the emitter terminal the diode is forward-biased. The
major portion of I_E will then flow through R_{B_1} since this resistance decreases rapidly
with the current I_E and current always seeks the path of least resistance. The voltage
V_E will then decrease with increasing values of I_E until a minimum V_E is reached.
Further increase in I_E is then linked with increasing values of R_{B_1}, causing the voltage
V_1 to rise. The result will be a back-biased diode, and the open-circuit input condition will return.

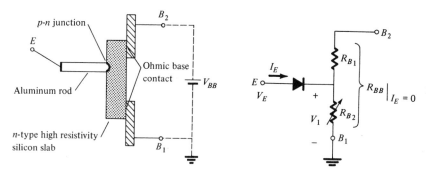

Figure 6.50 Basic construction of the unijunction transistor (UJT).

Figure 6.51 Network equivalent for the UJT.

The characteristics of the device appear in Fig. 6.52 with V_P representing the voltage V_E necessary to initiate conduction in the diode. V_V represents the minimum voltage for which the diode will be held in the conducting state. Beyond this point the increasing resistance is encountered until the back-biased condition for the diode results.

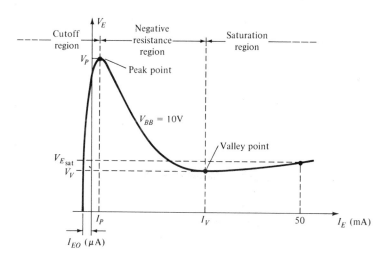

Figure 6.52 UJT characteristics for a fixed $V_{BB} = 10v$.

Note how the resistance decreases (negative resistance coefficient) with increasing values of I_E from V_P to V_V. This negative resistance region will have application in areas such as oscillators (ac signal generators), to be described in a later chapter.

For increasing values of V_{BB} the curve will rise, and increasing values of V_P will be required to *fire* the device. This rise in the characteristics is noted in the characteristics of Fig. 6.53. Note, however, the same relative value of V_V for each curve.

The voltage V_1 is determined by a simple voltage-divider-rule equation when

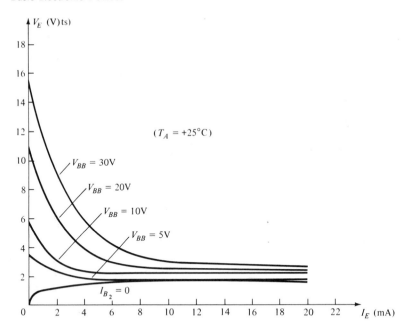

Figure 6.53 Typical static emitter characteristics curves for a UJT.

the diode is reverse-biased. That is,

$$V_1 = \frac{R_{B_1}}{R_{B_1} + R_{B_2}} \cdot V_{BB}\big|_{I_E=0} = \eta V_{BB}\big|_{I_E=0} \qquad (6.10)$$

and

$$\eta = \frac{R_{B_1}}{R_{B_1} + R_{B_2}} \qquad (6.11)$$

The quantity η (Greek letter eta) is typically available on specification sheets. The *firing* potential is then determined by

$$V_P = \eta V_{BB} + V_F \qquad (6.12)$$

A photograph of a commercially available UJT appears in Fig. 6.54 with its characteristics and terminal identification.

EXAMPLE 6.2

For the UJT of Fig. 6.54, determine

1. V_1.
2. V_P.
3. The relative magnitude of R_{B_1} as compared to R_{B_2} for $V_{BB} = 10$ V, $I_E = 0$, and typical values ($V_F = 0.54$).

absolute maximum ratings: (25°C)

Power Dissipation	300 mw
RMS Emitter Current	50 ma
Peak Emitter Current	2 amperes
Emitter Reverse Voltage	30 volts
Interbase Voltage	35 volts
Operating Temperature Range	−65°C to +125°C
Storage Temperature Range	−65°C to +150°C

electrical characteristics: (25°C)

		Min.	Typ.	Max.
Intrinsic Standoff Ratio ($V_{BB} = 10V$)	η	0.56	0.65	0.75
Interbase Resistance ($V_{BB} = 3V, I_E = 0$)	R_{BB}	4.7	7	9.1
Emitter Saturation Voltage ($V_{BB} = 10V, I_E = 50$ ma)	$V_{E(\text{SAT})}$		2	
Emitter Reverse Current ($V_{BB} = 30V, I_{B1} = 0$)	I_{EO}		0.05	12
Peak Point Emitter Current ($V_{BB} = 25V$)	I_P		0.4	5
Valley Point Current ($V_{BB} = 20V, R_{B2} = 100\Omega$)	I_V	4	6	

(Courtesy General Electric Company)

(a)   (b)   (c)

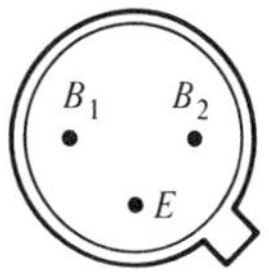

Figure 6.54 UJT: (a) appearance; (b) specification sheet; (c) terminal identification.

Solution:

1. $V_1 = \eta V_{BB} = (0.65)(10) = \mathbf{6.5\ V.}$
2. $V_P = \eta V_{BB} + V_F = 6.5 + 0.5 = \mathbf{7\ V.}$
3.

$$\eta = \frac{R_{B_1}}{R_{B_1} + R_{B_2}} \Longrightarrow 0.65 = \frac{R_{B_1}}{R_{B_1} + R_{B_2}}$$

and

$$0.65 R_{B_1} + 0.65 R_{B_2} = R_{B_1}$$

$$0.35 R_{B_1} = 0.65 R_{B_2}$$

$$\mathbf{R_{B_1} = 1.86 R_{B_2}}|_{I_E = 0}$$

6.7 SILICON-CONTROLLED RECTIFIER (SCR)

The silicon-controlled rectifier is a four-layer semiconductor device having the basic construction and symbol appearing in Fig. 6.55. Even though it has four layers, the SCR has only three external terminals, although the other four-layer semiconductor devices such as the silicon-controlled switch (SCS) and gate turn-off switch (GTO) have four available connections. Priorities do not permit a detailed explanation of the latter two devices, but a number of references are available on each.

As the name implies, the SCR is a *silicon* *rectifier* whose state (open- or short-

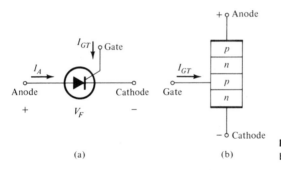

Figure 6.55 (a) SCR symbol and (b) basic construction.

circuit equivalent) is controlled by a third terminal called the gate. In other words, it is not sufficient to forward-bias the diode by greater than V_F ($\simeq 0.5$ V) for the typical semiconductor diode. The gate current, as indicated on the characteristics of Fig. 6.56, will determine the required firing voltage necessary to bring the device into the conduction state. For $I_G = 0$, V_F must be increased to at least $V_{(BR)F^*}$ before the short-circuit state between anode and cathode will result. For increasing values of I_G, the voltage required to fire the device decreases until at I_{G_2} the diode appears to have the characteristics of the two-terminal junction diode. We can look at the behavior of the SCR in another light if we consider a fixed bias of say V_{F_1} (note Fig. 6.56) across the device. For conduction, even though V_{F_1} is greater than the required voltage for a typical semiconductor diode, the current must be increased to I_G before the short-circuit state is set. The holding current (I_H) is that value of current below which the SCR will switch from the conduction to blocking region. The characteristics of this device have widespread application in control systems where

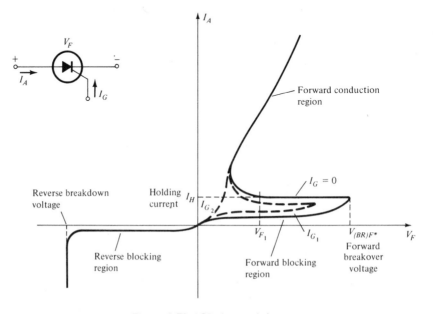

Figure 6.56 SCR characteristics.

a particular function is not to be performed until a gate-connected operation has occurred. A few common areas of application include relay controls, regulated power supplies, motor controls, battery chargers, and heater controls.

There is a wide variety of SCRs commercially available today—from milliwatts to megawatts with currents as high as 1000 A. A set of important gate characteristics for a top-hat SCR appears in Fig. 6.57. For a particular gate current, the range of

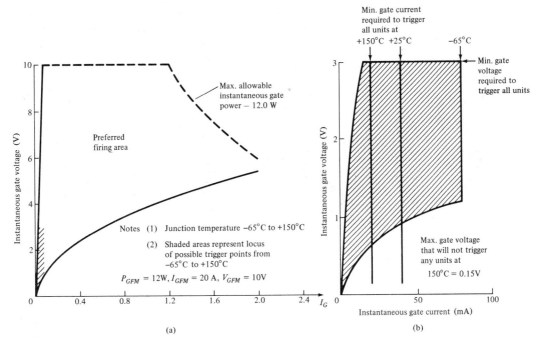

Figure 6.57 SCR gate characteristics (GE series—C38).

preferred firing potentials is indicated. For instance, a gate current of 1.2 A or greater requires a firing potential of $4.2 \longrightarrow 10$ V. As indicated, the maximum gate current is 20 A, the maximum gate voltage is 10 V, and the maximum power dissipation ($P_G = I_G V_G$) is 12 W.

The internal construction of the device appears in Fig. 6.58. Note the very heavy appearance of the device to handle the high currents of this particular SCR. Other types appear in Fig. 6.59 with their terminal identification.

6.8 DIAC AND TRIAC

The DIAC is a two-terminal, five-layer, semiconductor device, constructed as shown in Fig. 6.60. Note that the characteristics in the first and third quadrants are somewhat similar to that obtained for the SCR in the first quadrant. For either voltage polarity across the device in Fig. 6.61 there is a potential V_{BR} above which the voltage decreases

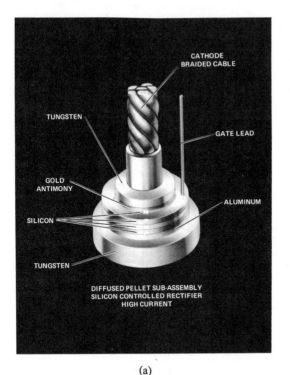

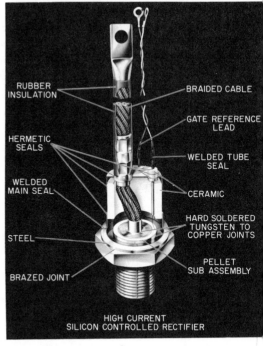

(a)

(b)

(Courtesy General Electric Company)

Figure 6.58 (a) Alloy-diffused SCR pellet; (b) thermal fatigue-free SCR construction.

with increase in current until the short-circuit equivalent region is approached.

Two widely accepted symbols for the device with a photograph of a low-wattage unit appear in Fig. 6.62.

The TRIAC has a set of characteristics very similar to the DIAC, but it has, in addition, a gate terminal that can control the state of the device in either direction. Note in Fig. 6.63 the presence of a holding current I_H not present for the DIAC. The symbol and construction of the device appear in Fig. 6.64. The actual device appears in Fig. 6.65.

6.9 PENTODE

The pentode is a five-element vacuum tube with characteristics, as shown in Fig. 6.66, quite different from those of the triode vacuum tube. Note the small vertical change in I_{PK} with V_{PK} as compared with the large change for the triode. As will be shown, this is an indication of the higher gain of the pentode, as compared to the triode. In addition, for increasing negative voltages, the curves drop vertically rather than to the right as for the triode. The basic construction of the device is shown in Fig. 6.67 with the label applied to each element.

(Courtesy General Electric Company)

(Courtesy International Rectifier Corp. Inc.)

Figure 6.59 SCR case construction and terminal identification.

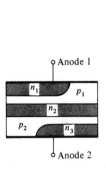

Anode 1

n_1 p_1

n_2

p_2 n_3

Anode 2

Figure 6.60 DIAC construction.

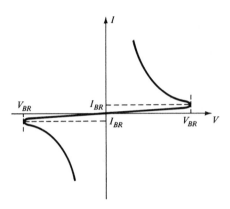

Figure 6.61 DIAC characteristics.

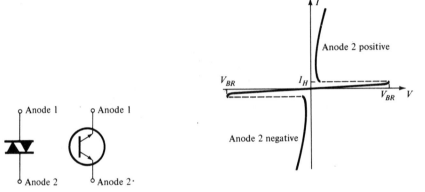

Figure 6.63 TRIAC characteristics.

Anode 1   Anode 1

Anode 2   Anode 2·

Figure 6.62 DIAC symbols and appearance.

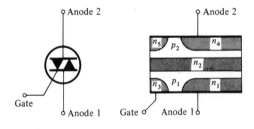

Anode 2   Anode 2

n_5 p_2 n_4

n_2

n_3 p_1 n_1

Gate

Anode 1   Gate   Anode 1

(a)   (b)

Figure 6.64 TRIAC (a) symbol (b) construction.

The symbol for the device and the biasing necessary are shown in Fig. 6.68. Note that the suppressor is connected directly to the cathode and the screen grid at some positive potential. The pin connections for a pentode-triode vacuum tube appear in Fig. 6.69 with a photograph of the device with its ratings.

Figure 6.65 TRIACs.

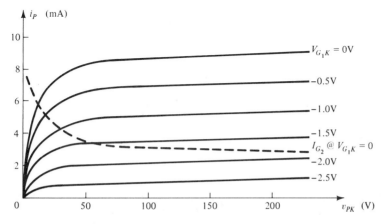

Figure 6.66 Pentode characteristics.

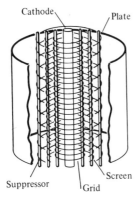

Figure 6.67 Basic pentode construction.

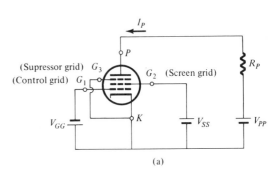

(a)

Figure 6.68 Pentode symbol and basic biasing arrangement.

Heater voltage = 6.3 volts
C_{pk} = 5.0 pf
Plate voltage = 330 max.
Plate dissipation = 3.5 watts

(a)

(Courtesy Radio Corporation of America)

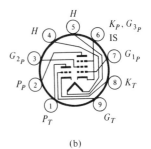

(b)

Figure 6.69 Pentode-triode vacuum tube (a) photograph (b) pin connections.

PROBLEMS

§ 6.2

1. Describe (in your own terms, from memory) the basic operation of the semiconductor junction diode in both the forward and reverse bias regions.

2. Repeat Problem 1 for the vacuum-tube diode.

3. Sketch the output of the diode networks of Fig. 6.70.

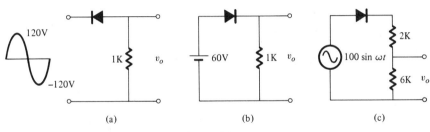

(a) (b) (c)

Figure 6.70

4. Sketch the output waveform of the bridge configuration of Fig. 6.71.

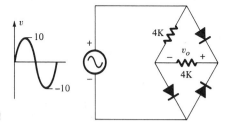

Figure 6.71

5. Sketch a full-wave rectifier network using (a) semiconductor diodes and (b) vacuum-tube diodes.

6. Sketch the output waveform of the full-wave rectifier of Fig. 6.72.

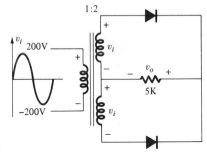

Figure 6.72

7. (a) Sketch the output waveform (V_o) for the configurations of Fig. 6.73.
(b) Sketch the waveform of V_R for each case.

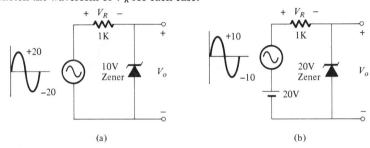

(a) (b)

Figure 6.73

8. Find the resistance of the tunnel diode of Fig. 6.17 in the negative resistance region.

9. For a varactor diode with an alloy junction, $V_F = 0.5$ V, $V = 35.5$ V, determine the junction capacitance if $K = 200 \times 10^7$.

10. What is the current limitation of a 30-V, 10-W-Zener diode?

§ 6.3

11. Find the resistance of the thermistor of Fig. 6.20 at $0°$C. How does it compare to its resistance at $100°$C?

12. Determine the resistance of the photoconductive cell of Fig. 6.22 at 0.5 and 50 fc. Did the relative change in resistance compare closely with the relative change in illumination?

13. Find the resistance of the photodiode of Fig. 6.24 at an illumination of 100 and 1000 fc ($V_\lambda = 40$ V). Compare these values and indicate which more closely resembles the low-resistance forward bias state.

§ 6.4

14. Determine the voltage gain (A_v) for the transistor configuration of Fig. 6.29 if $V_i = 200$ mV, $Z_i = 200$ Ω, and $R_L = 2$ K.

15. Using the characteristics of Fig. 6.31, determine
    (a) I_C if $I_E = 3$ mA and $V_{CB} = 10$ V.
    (b) I_C if $V_{EB} = 200$ mV and $V_{CB} = 10$ V.
    (c) V_{EB} if $I_C = 2.5$ mA and $V_{CB} = 5$ V.

16. For a particular transistor where $\alpha = 0.98$ and $I_C = 2$ mA, determine I_E and I_B.

17. Using the characteristics of Fig. 6.32,
    (a) Determine I_C if $I_B = +20$ μA = and $V_{CE} = +10$ V.
    (b) Find I_B if $I_C = 4$ mA and $V_{CE} = +5$ V.
    (c) Find I_E at $I_B = +30$ μA and $V_{CE} = +10$ V.
    (d) Determine β for the results of part (a).
    (e) Determine α for the results of part (a).

18. Sketch the power curve for $P_{C_{max}} = 25$ mW on Fig. 6.35. $I_{C_{max}} = 6$ mA and $V_{CE_{max}} = 20$ V.

19. For a fixed plate voltage of $V_{PK} = 200$ V, determine the change in plate current if the grid to cathode voltage is changed from -6 to -10 V in Fig. 6.39.

20. For a fixed grid to cathode voltage of $V_{GK} = -14$ V, determine the change in plate current for a change in plate to cathode voltage from $V_{PK} = 350$ to 300 V in Fig. 6.39.

21. For the triode with the characteristics of Fig. 6.41, determine the maximum power curve for $P_{D_{max}} = 20$ W, $I_{P_{max}} = 12$ mA, and $V_{PK_{max}} = 350$ V.

§ 6.5

22. Determine the drain current I_D for the JFET of Fig. 6.45 at
    (a) $V_{GS} = -2$ V, $V_{DS} = 15$ V.
    (b) $V_{GS} = -2$ V, $V_{DS} = 7$ V.
    What conclusion can be drawn from the above results?

23. Using Eq. (6.8), determine the drain current of a JFET at $V_{GS} = -4$ V with $V_P = -8$ V and $I_{DSS} = 8$ mA.

24. Using Fig. 6.46, determine the drain current at $V_{GS} = -2$ and -4 V and compare it to the results of Problem 22.

25. Describe (in your own words, from memory) the difference between a depletion, enhancement, and enhancement-depletion-type FET.

§ 6.6

26. Determine the following for a UJT with $\eta = 0.57$, $V_{BB} = 20$ V, $V_F = 0.5$ V, and $R_{BB} = 10$ K:
    (a) V_1.
    (b) V_P.
    (c) R_{B_1}.

27. (a) Using Fig. 6.53, determine V_P for $V_{BB} = 5$ and 20 V.
    (b) Determine I_E at $V_{BB} = 20$ V and $V_E = 3$ V.

§ 6.7

28. (True or false?) (a) For increasing values of gate current, the required firing potential V_F across an SCR will increase.
    (b) For increasing gate current, at a fixed V_F, the holding current will decrease.

29. Determine the range of firing potentials (using Fig. 6.57) for a gate current of 1 A.

§ 6.8

30. Describe the basic difference between the DIAC and TRIAC semiconductor devices.

§ 6.9

31. Describe the basic difference between the characteristics of a triode and pentrode.

32. Consulting Fig. 6.66, how sensitive is I_P to changes in V_{PK} for a fixed V_{GK}?

33. Determine I_P at $V_{PK} = 100$ V and $V_{GK} = -0.5$ and -2.5 V.

7

Integrated Circuits (ICs)

7.1 INTRODUCTION

In recent years, the integrated circuit (IC) has enjoyed a period of research and development that can only be matched by the explosive interest in the transistor when it was first introduced. The most noticeable characteristic of the IC is its size. It is often hundreds and sometimes thousands of times smaller than the equivalent semiconductor network constructed of discrete (individual) components. What is the IC? It is not a new type of circuit element or system that performs a very specific function. Rather, it is simply the manner in which the network is constructed and packaged. Instead of building a network with 100 individual parts, all the components are both manufactured (diodes, transistors, FETs, resistors, etc.) and connected together simultaneously. That is, layer upon layer of semiconductor, conducting, and insulating materials are applied to the same base (substrate) using specific patterns, which determines where the substance will be deposited or etched away.

Integrated circuits are also treated quite differently from the individual component type of construction in that if an IC of 100 components should fail, the entire package is usually discarded rather than trying to make repairs on the structure. If individual components were employed, it was usually possible to simply replace the defective element.

The effect of this new type of construction technique is quite visible in the reduced size of calculators, computers, instruments, timepieces, televisions, etc.

Integrated circuits fall within the heading of *monolithic, thin (or thick) film,* or *hybrid* types. The construction of each will receive some surface treatment in this chapter. Practical applications of the IC will appear in the chapters to follow.

7.2 MONOLITHIC INTEGRATED CIRCUIT

The term monolithic is the result of the combination of two Greek words, *monos*, meaning "single," and *lithos*, meaning "stone," revealing that this type of IC is a single-structured package. The reduced size of the integrated circuit is best described by the photograph of Fig. 7.1, which clearly indicates that the entire network is now smaller

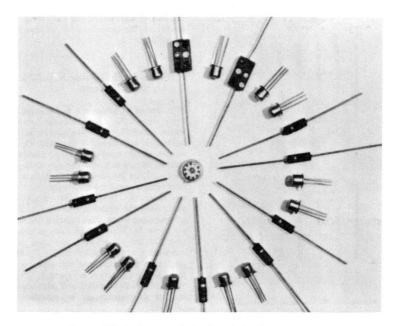

Figure 7.1 An integrated circuit and the discrete components required to build a circuit to perform the same task. (Courtesy Motorola, Inc.)

than some of the individual components used to construct the discrete-type network. As one might expect, a sophisticated sequence of steps is necessary to construct the minute multilayer device. The very basic steps appear in Fig. 7.2.

Since a great deal of time and expense will go into the construction of the integrated circuit, the circuit to be produced must first be of unquestionable design. Once the process of integration is initiated it is virtually impossible to make design improvements.

The second step is laying out the circuit so that the individual components can be constructed simultaneously in the multilayer structure. In other words, if a particular region is to have a transistor, three layers (*PNP* or *NPN*) of semiconductor material will have to be superimposed on the same region. In addition, it is possible that all the transistors in the system will be constructed during the same production steps with their interconnections, if any. It is here that computer techniques can be

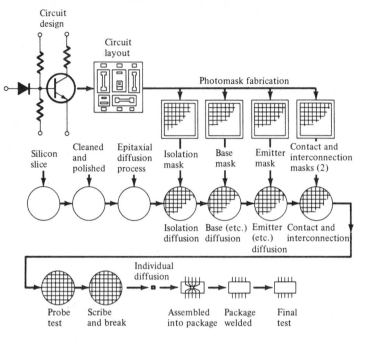

Figure 7.2 Monolithic integrated circuit fabrication. (Courtesy of Robert Hibberd.)

employed to their full advantage to ensure that the layout is the most desirable possible.

When the layout is determined, a mask having the appearance of a film negative with light and dark areas must be made for each layer of the IC to determine the regions in which the substance is to be deposited or etched away. Each mask is then cut with an aristo cutting machine as shown in Fig. 7.3 and then reduced in size using photoreduction techniques as shown in Fig. 7.4. It is then put into a step and repeat machine such as shown in Fig. 7.5 to reproduce the pattern perhaps a hundred times on a final mask depicted in Fig. 7.6. The pattern is reproduced in this manner to permit the production of a large number of ICs at the same time.

Figure 7.3 Aristo handcutting of the mask pattern. (Courtesy Motorola, Inc.)

Figure 7.4 Photo-reduction of the mask pattern. (Courtesy Texas Instruments, Inc.)

Figure 7.5 Step and repeat machine for the placement of a large number of the reduced mask pattern on a single production mask. (Courtesy Texas Instruments, Inc.)

Figure 7.6 Final mask. (Courtesy Motorola, Inc.)

As indicated in Fig. 7.2, the semiconductor silicon must be prepared before the process can continue. An ingot of pure silicon material such as appears in Fig. 7.7 is manufactured using the Czochralski technique described in detail in most texts devoted solely to semiconductor devices. The ingot is then sliced as shown in Fig. 7.8 into wafers as thin as $\frac{1}{1000}$ in. if necessary.

The wafers are then polished and cleaned as depicted in Fig. 7.9.

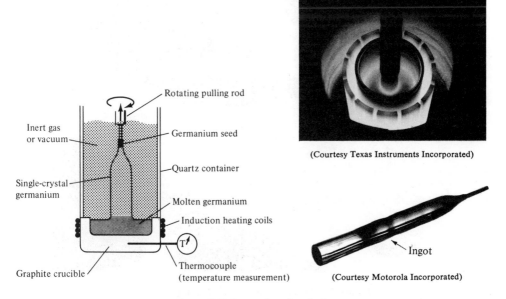

Rotating pulling rod

Inert gas or vacuum

Germanium seed

Quartz container

Single-crystal germanium

Molten germanium

Induction heating coils

Graphite crucible

Thermocouple (temperature measurement)

(Courtesy Texas Instruments Incorporated)

Ingot

(Courtesy Motorola Incorporated)

Figure 7.7 Grown Junction diode.

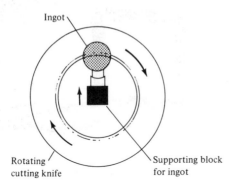

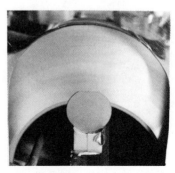

Figure 7.8 Slicing the single-crystal ingot into wafers. (Courtesy Texas Instruments Incorporated)

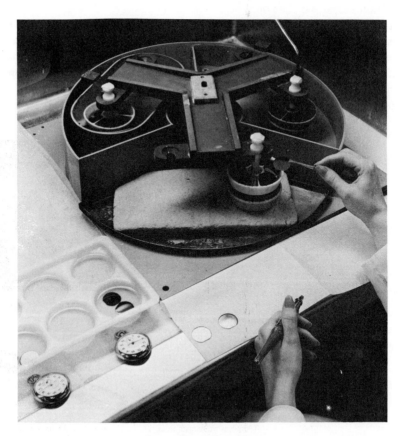

Figure 7.9 Polishing apparatus. (Courtesy Texas Instruments, Inc.)

The next series of steps includes the diffusion of the various doping elements into the silicon wafers. For the *p*-type substrate, an *n*-type region is diffused into the *p*-type silicon material as shown in Fig. 7.10. This is followed by the depositing of an SiO$_2$ insulating layer as shown in Fig. 7.11.

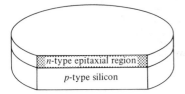

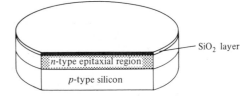

Figure 7.10 *p*-Type silicon wafer after the *n*-type epitaxial diffusion process.

Figure 7.11 Wafer of Fig. 7.10 following the deposit of the SiO$_2$ layer.

The surface is then coated with a material called photoresist as shown in Fig. 7.12, followed by the placement of the first mask as shown in Fig. 7.13.

Ultraviolet light is then applied to expose those regions not covered by the masking pattern as indicated in Fig. 7.13. An applied chemical solution will then remove the unexposed regions of the photoresistive material. Another solution will then remove the SiO$_2$ layer not covered by the photoresist material. Next, the remaining

Figure 7.12 Applying the photoresist; the wafer is spun at a high speed to insure an even distibution of the photoresist. (Courtesy Motorola, Inc.)

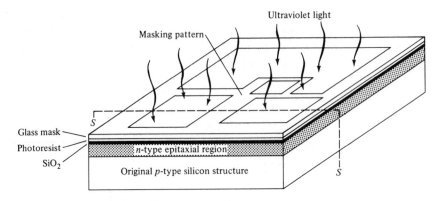

Figure 7.13 Placement of the first mask and application of the ultraviolet light.

photoresist can be removed. The three steps are described by the sequence of drawings appearing in Figs. 7.14 and 7.15.

The process can then be repeated (etching, SiO_2, etc.) to produce the cross section appearing in Fig. 7.16 for a transistor in the structure. Each of the important regions touches the surface, permitting the use of a metalized interconnecting pattern. The purpose of each diffusion process is somewhat described by the applied label in Fig. 7.2.

Following the above sequence of diffusion operations, the wafer will appear as

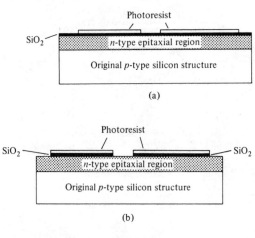

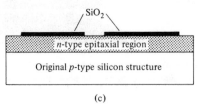

Figure 7.14 (a) Cross-section (*s-s*) of the chip in Fig. 7.13 following the removal of the unexposed photoresist. (b) Cross-section following the removal of the uncovered SiO_2 regions. (c) Cross-section following the removal of the remaining photoresist material. (Courtesy Motorola, Inc.)

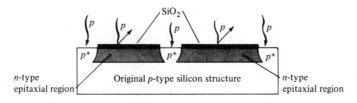

Figure 7.15 Cross-section of Fig. 7.14 following the isolation/diffusion process.

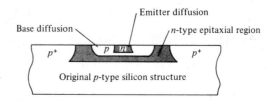

Figure 7.16 Cross-section of a transistor after the base and emitter diffusion cycles.

shown in Fig. 7.17 with the relative size of the elements indicated in the attached sketch.

The wafer is then ready for testing by the rather sophisticated equipment appearing in Fig. 7.18.

Before packaging, the wafer must be scribed and broken into individual chips as shown by the sequence of photographs in Fig. 7.19.

The ICs are then packaged using any one of the techniques depicted in Fig. 7.20

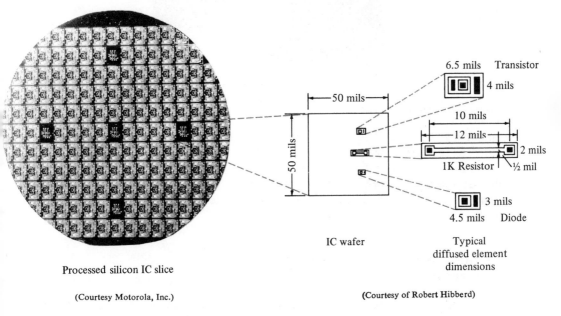

Figure 7.17 Processed monolithic IC wafer with the relative dimensions of the various elements. (Courtesy of Robert Hibberd)

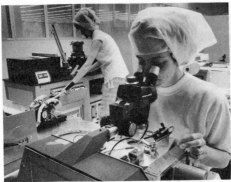 

(Courtesy Autonetics, North American Rockwell Corp.) (Courtesy Texas Instruments, Inc.)

(Courtesy Texas Instruments, Inc.)

Figure 7.18 Production testing.

and then given a final test. Note the relative size of the dual-in-line package as compared to the size of the external pins necessary for connecting it into the system.

The terminal identification and internal circuitry of a commercially available dual-in-line package appears in Fig. 7.21. The placement of the ground terminal is such that the four logic units are totally isolated and can be employed in four distinct areas in the total electronic system.

7.3 THIN- AND THICK-FILM INTEGRATED CIRCUITS

Thin- and thick-film ICs are quite similar in appearance, characteristics, and properties. However, they are both quite different from the monolithic structure just described. In the thin- or thick-film IC only the passive elements such as resistors and

(a)   (b)

(Courtesy Autonetics, North American Rockwell Corp.)   (Courtesy Texas Instruments, Inc.)   (Courtesy Motorola, Inc.)

Figure 7.19 (a) Scribing and (b) breaking of the monolithic water into individual chips.

capacitors are formed using the film techniques that result in the pattern appearing in Fig. 7.22. The active devices such as the transistor, FET, diodes, etc., are added as discrete elements to the surface of the structure. There is an obvious increase in size with this process and, due to the variety of steps required, an increase in cost. However, increased flexibility is often available with the thin- or thick-film IC as compared to the monolithic structure. The difference between the thin- and thick-film ICs is basically a matter of how the passive elements and the metallized conduction patterns are formed on the substrate.

7.4 HYBRID INTEGRATED CIRCUITS

The hybrid IC, as one could probably guess, is a combination (a hybrid) of the two techniques described above (monolithic and film). The monolithic or film process is first employed to develop the necessary electronic packages with any number of electronic elements tied together to perform a particular task. This is followed by bringing these packages together with a metallized conduction pattern as shown in Fig. 7.23.

For the film or hybrid structure, data sheets such as provided for the monolithic structure are available. The area of application will determine which of the three techniques is applied most frequently. For some, the monolithic structure is the only possibility due to size and cost requirements, while for other applications perhaps the flexibility and possibility of some repair or replacement on the individual packages of the film or hybrid approach are desirable.

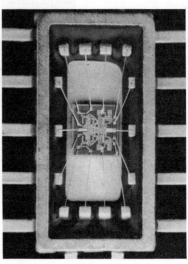

(a)

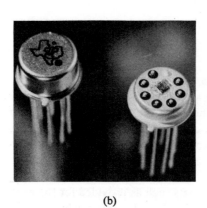
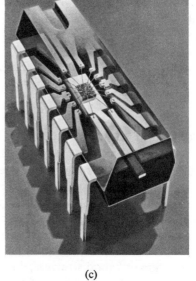

(b)   (c)

(Courtesy Texas Instruments, Inc.)

Figure 7.20 Monolithic packaging techniques: (a) flat package; (b) TO (top-hat)-type package; (c) dual in-line plastic package.

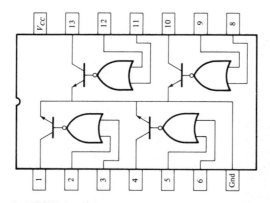

Figure 7.21 Quad 2-input or power driver in a dual-in-line package. (Courtesy Sprague Electric Company)

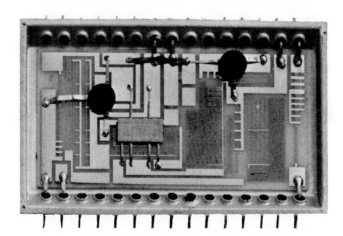

Figure 7.22 Thin film integrated circuits. (Courtesy Autonetics, North American Rockwell Corp.)

Figure 7.23   Hybrid integrated circuits. (Courtesy Texas Instruments, Inc.)

8

Electronic Circuits

8.1 INTRODUCTION TO ELECTRONIC CIRCUITS

Electronic circuits have progressed rapidly from using vacuum tubes to transistors to integrated circuits. Along with this change there has been a related change in the analysis of the circuits using these components with emphasis on different analytical techniques. Vacuum tubes are regarded as essentially voltage amplifiers, bipolar junction transistors are current amplifiers, field-effect transistors are again considered voltage-amplifying devices, and, finally, integrated circuits are considered from the operation of the complete circuit. In time many new devices may be developed and used so that it is most important to concentrate on the basics of electronics and apply these to whatever device is currently used most.

Electronic circuits perform a number of important functions—with amplification being one of the most important. Electronic circuits also provide signal wave shaping (e.g., in radio, hi-fi, radar, television, power supplies) and load matching (e.g., signal pickup from a microphone or other transducer, driving signal into speaker). In this chapter we shall cover some of the basic electronic signal-shaping and amplifier circuits, and then in later chapters we shall apply these circuits to various important applications.

Some basic signal wave shaping is performed by diode circuits as covered in Section 8.2. These include rectification, clipping, and clamping operations. In Section 8.3 we shall discuss ideal voltage and current amplifiers and then practical amplifier characteristics before the chapter gets into the details of how practical amplifier circuits are constructed using BJT and FET devices.

8.2 DIODE WAVE-SHAPING CIRCUITS

A basic electronic wave-shaping component is the two-terminal diode, which is ideally a *short circuit when forward-biased* and an *open circuit when reverse-biased.* Figure 8.1(a) shows a simple circuit application of a diode. During the interval of time that the ac signal is positive the diode is forward-biased, acting as a short circuit as shown in Fig. 8.1(b). The output voltage is seen to be the same as the input during this interval of time. During the time that the input signal is negative, as shown in Fig. 8.1(c), the diode is reverse-biased, acting as an open circuit with no current

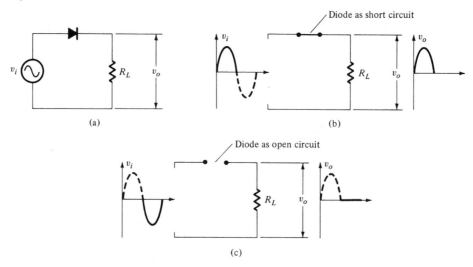

Figure 8.1 Action of ideal diode in rectifying circuit.

flow through the diode and consequently no voltage developed at the output. The overall action of the diode wave-shaping circuit is to pass the positive half-cycles of the input and provide zero output during the negative half-cycles of the input—this action being referred to as rectification (half-wave rectification in this specific circuit). A few different circuit configurations of diode rectifying circuits are shown in Fig. 8.2. The diode in the circuit of Fig. 8.2(a) conducts during the positive half-cycle as shown by the output waveform. Reversing the diode, as in Fig. 8.2(b), results in diode conduction only during the negative half-cycle. The first two connections show the ac signal coupled in through a transformer, while the last two are directly connected to the circuit. Rectifying action will occur with either input connection. In Fig. 8.2(c) the diode is connected so that it conducts (acts as a short circuit) during the positive half-cycle of input signal. When it is a short circuit, however, it results in the output voltage being 0 V. During the negative half-cycle of the input signal it is reverse-biased, acting as an open circuit with the output voltage appearing as shown in Fig. 8.2(c). With the diode element open, or removed from any effect on the circuit action during this negative half-cycle, the output voltage follows the input voltage but is smaller in amplitude due to the voltage-divider action of resistors R_1 and R_2.

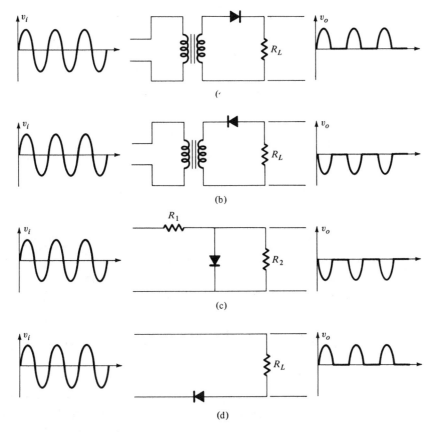

Figure 8.2 Diode rectifying circuits.

In Fig. 8.2(d) the diode conducts during the positive half-cycle of the input signal and is essentially the same circuit as that of Fig. 8.2(a).

Diode Clipping Circuits

The rectifying action of the diodes in Figs. 8.1 and 8.2 can also be referred to as *waveform clipping* since the resulting action of the circuit is to clip off a part of the signal. When the input is sinusoidal and half the signal is clipped off, as in power supply applications, the circuit action is called *rectification*, and the circuit is a *rectifier*. This clipping action can also be done with other signal waveforms. In addition, the clipping level need not be 0 V but can be any other selected voltage value. We shall next consider the circuit action which results in clipping various wave-shape signals at different clipping levels and for different polarity clipping action.

The general form of a clipping circuit includes a resistor, a diode, and a battery (for other than 0 V clipping levels) as shown in the clipping circuit of Fig. 8.3(a). Two different input signals (a sinusoidal and a pulse type) are considered to demonstrate the action of the clipping circuit. For the diode polarity shown the diode will

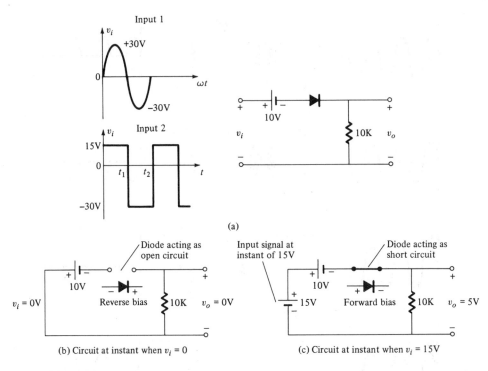

Figure 8.3 Circuit example of waveform clipping.

conduct only when the input signal goes positive, and, in addition, for the battery polarity chosen the input must go more positive than the battery 10-V reference level. As will be demonstrated the output will occur only during the time the input signal goes above a 10-V reference level, the output being 0 V otherwise. The circuit acts as a clipper, passing only those voltage swings that go above a reference $+10$-V level in this sample circuit. Other circuits will be shown to clip signals below reference levels. To consider some detail of how the clipping circuit operates Figs. 8.3(b) and 8.3(c) show the diode action for the conditions of diode reverse-biased and diode forward-biased, respectively. For any voltage of value $v_i < 10$ V, the ideal diode considered is shown to be reverse-biased, and $v_o = 0$ V. For example, with $v_i = 0$ V [Fig. 8.3(b)] the 10-V battery is connected so that the diode is reverse-biased, acts as an open circuit, and allows no current flow through the 10-K resistor so that $v_o = 0$ V. In fact, any voltage of value v_i *less than* 10 V (for the present circuit) will still result in the diode being reverse-biased so that $v_o = 0$ V.

For any value of $v_i > 10$ V the diode is forward-biased, and $v_o = v_i - 10$ V. For example, at the instant of time the input voltage is 15 V [Fig. 8.3(c)] the ideal diode is a short circuit and $v_o = 15 - 10$ V $= 5$ V across the 10-K resistor. The output voltage follows any voltage signal *above* 10 V (in the present circuit) as shown in the output waveforms of Fig. 8.4.

Considerable variations in the form of clipping circuits will be found in practice. In all cases it is important to determine under what signal conditions the diode acts

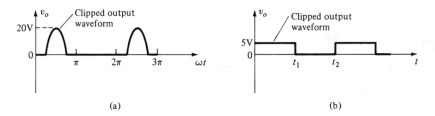

Figure 8.4 Output waveforms from the clipping circuit of
Fig. 8.3.

as an open circuit and as a short circuit. A fairly comprehensive summary of clipping
circuits and resulting waveforms is shown in Fig. 8.5. As practice, cover up the output waveform for a few examples and see if you can determine what the output
waveform should look like.

Diode Clampers

A *diode clamping circuit* operates only on pulse-type signals and *does not remove
or clip off any of the input signal.* It acts to *shift the level* of the signal above or below
the original level. A diode clamping circuit requires at least three elements: a diode,
a capacitor, and a resistor. The clamping circuit may also contain a battery for shifting the input signal to other than a 0-V clamping level. The values of components R
and C must be chosen so that the circuit time constant $(\tau = R \cdot C)$ is large enough,
compared to the period $(T = 1/f)$ of the input signal. This condition is necessary to
ensure that the voltage across the capacitor does not change significantly during the
interval of time that components R and C affect the output waveform.

A typical clamping circuit to consider is shown in Fig. 8.6(a). The input shown
is a pulse-type signal varying between $+5$ V and -10 V at a 1-kHz rate. The action
of the present clamping circuit is to shift the complete signal down to the 0-V level
so that it does not go above 0 V. Such a signal can be considered as clamped below the
0-V reference voltage level.

It is usually advantageous when examining clamping circuits to first consider
the conditions that exist when the input is such that the diode is forward-biased. At
the instant the input signal switches to the $+5$-V level the circuit acts as shown in Fig.
8.6(b). The signal will remain at the $+5$-V level during the interval from 0 to t_1 (one-
half the cycle for the signal considered in this example). The period of signal v_i is
$T = 1/f = 1/1000 = 1$ msec, and the time interval of the signal remaining at $+5$ V
is $T/2$ or 0.5 msec, for the present example. The output voltage is taken across the
diode, which acts as a short during the interval that it is forward-biased. When the
input signal goes positive causing the diode to become forward-biased the output
becomes clamped at 0 V, and the resistor R is shorted out momentarily as shown in
Fig. 8.6(b). The input $+5$-V signal results in capacitor C being at 5 V as shown in the
figure. We shall see shortly that the action of the capacitor will be to set the voltage
shifting level. The presence of the capacitor also prevents the clipping action discussed

SIMPLE SERIES CLIPPERS

POSITIVE

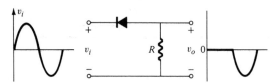

NEGATIVE

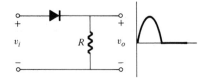

BIASED SERIES CLIPPERS

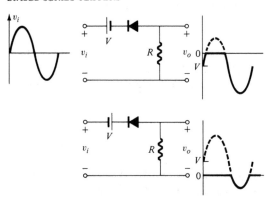

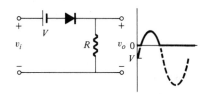

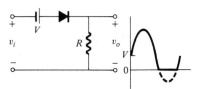

SIMPLE PARALLEL CLIPPERS

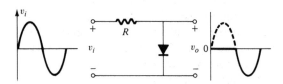

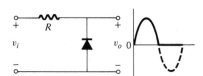

BIASED PARALLEL CLIPPERS

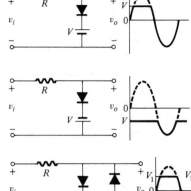

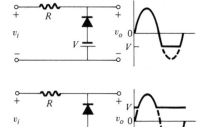

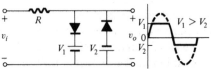

Fig. 8.5 Clipping circuits.

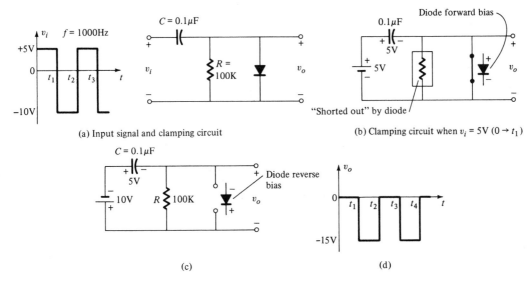

(a) Input signal and clamping circuit   (b) Clamping circuit when $v_i = 5V$ ($0 \rightarrow t_1$)

(c)   (d)

Fig. 8.6 Clamping circuit example.

previously, so that the full input voltage signal swing will still occur in the output—but between different voltage levels.

When the input signal switches to -10 V the circuit results in the diode being reverse-biased (open circuit) as shown in Fig. 8.6(c). With the input now at -10 V and the capacitor charged to 5 V the net output voltage is (for the present example) -15 V—the sum of capacitor and input voltages. At this point the time constant of the RC circuit becomes important. During the time the diode is open circuit and the input is constant at -10 V the capacitor can discharge through resistor R. The discharge time constant is

$$\tau = RC = 100 \times 10^3 \times 0.1 \times 10^{-6} = 10 \text{ msec}$$

Since it takes approximately five time constants for the capacitor to fully discharge, 50 msc in the present example, and the input signal remains constant at -10 V for only 1 msec (for $f = 1$ kHz), we can reasonably assume the capacitor voltage does not change appreciably in the 1-msec time interval. The output voltage is therefore

$$V_o = -10 \text{ V} - 5 \text{ V} = -15 \text{ V}$$
$$\underset{\text{supply}}{\uparrow} \qquad \underset{\text{capacitor}}{\uparrow}$$

The resulting output waveform, v_o, is shown in Fig. 8.6(d). As indicated, the output is clamped below 0 V but is otherwise a duplicate* of the input signal, both in amplitude swing, and frequency. For all clamping circuits the voltage swing of the input and output waveforms will be the same. This was not true for clipping circuits.

*If the condition $\tau \gg T$ does not hold true, then the output waveform will not be a duplicate of the input.

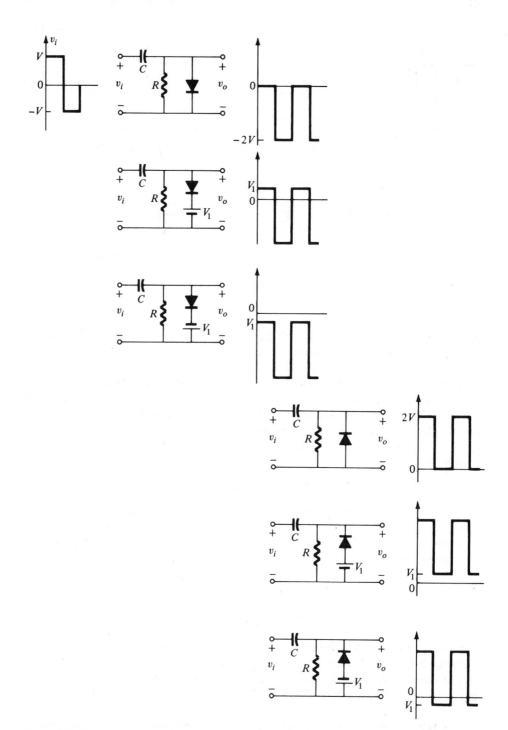

Figure 8.7 Clamping circuits.

A number of clamping circuits and their resulting output waveforms are shown in Fig. 8.7. The addition of a battery allows the clamping level to be set at other than 0 V.

8.3 AMPLIFIER BASICS

Amplifiers are placed in many different categories—voltage or current amplifier, small-signal or large-signal (power) amplifiers—(1) by operating class: class A for small signal (output conducts for full 360° signal cycle), class B for only half-cycle conduction, class AB for output signal swing between 180° and 360°, and class C for less than 180° operation as in tuned circuits; (2) by method of coupling stages together: capacitor-coupled, direct-coupled, transformer-coupled; (3) by range of operating frequency handled: low-frequency, audio frequency, high-frequency, intermediate frequency, ultrasonic frequency; and (4) by area of application: e.g., hi-fi, rf amplifier, audio amplifier, microphone amplifier. These amplifier circuits can be built using bipolar transistors, FETs, or integrated circuits, for example. It would be helpful at this point to consider some aspects of a basic voltage amplifier and a basic current amplifer and then compare these to practical amplifiers. This type of consideration will then give us an overview of the main amplifier features to later investigate for the various circuits considered.

Ideal Voltage Amplifier

Two representations of ideal voltage amplifiers are shown in Fig. 8.8. In Fig. 8.8(a) a voltage source provides an input signal V_i to the amplifier resulting in an output signal V_o applied to the load resistor R_L. The output signal is larger than the input signal by the amplification factor A_v, where

$$A_v = \frac{V_o}{V_i}$$

Both ac and dc signals can be amplified. The circuit shown reflects the action of the amplifier circuit on an input ac signal and does not necessarily include all the details of dc power supply or other bias factors.

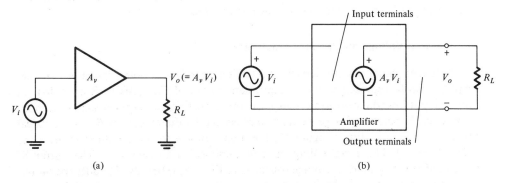

(a)   (b)

Figure 8.8 Ideal voltage amplifier.

Another similar representation of an ideal amplifier is shown in Fig. 8.8(b). The input signal is applied to the amplifier input terminals, which appears ideally as an open circuit or has very high input impedance. The input signal nevertheless affects the amplifier, causing a signal to appear at the output terminals, as an amplified version of the input signal ($V_o = A_v V_i$). This output signal is then applied to another circuit represented by the load resistance R_L. For the ideal voltage amplifier the output voltage is the same for any value of load resistance.

Ideal Current Amplifier

In similar fashion an amplifier could provide current amplification as shown in Fig. 8.9(a). A current I_i applied to the amplifier results in an amplified current I_o to load R_L. Figure 8.9(b) shows the input of the current amplifier ideally as a short

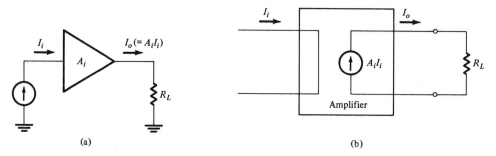

Figure 8.9 Ideal current amplifier.

circuit. Although the internal circuitry of the amplifier is not seen on this block representation, it should be understood that the input current I_i results in an amplified output current, I_o, which is the current amplification of the circuit, A_i times larger than the input current. This is represented by the ideal current source $A_i I_i$, which provides the output current to the load. For this idealized circuit the output current to the load is the same regardless of the value of load resistance.

It might be noted that FET and vacuum-tube devices act as voltage amplifiers, while BJTs act essentially as current amplifiers. In both cases, however, the resulting circuit departs from the ideal conditions just covered.

Practical Voltage and Current Amplifiers

Actual amplifier circuits depart from the ideal conditions noted above. A practical amplifier can be represented by the equivalent circuit of Fig. 8.10(a). Rather than an infinite input impedance (open circuit), the practical voltage amplifier has some input impedance, hopefully of large value, as represented by the resistance R_i. For the ideal voltage source providing a signal V_s, the input voltage to the amplifier input is the same ($V_i = V_s$). This input voltage is then amplified to a value $A_v V_i$. The output of the amplifier is not an ideal voltage source as in Fig. 8.8(a) but also includes the output impedance of the actual circuit, represented by the output resistance R_o. If the output

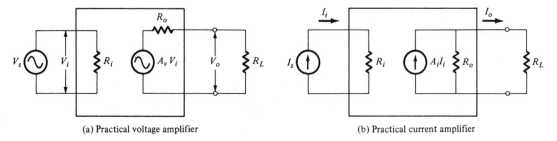

(a) Practical voltage amplifier   (b) Practical current amplifier

Figure 8.10 Practical amplifier circuits.

voltage of the amplifier is measured with no load connected across the output terminals, the voltage is the ideal value $A_v V_i$. In a practical amplifier the output voltage drops down (becomes smaller) as the load draws more current from the amplifier circuit. This reduced voltage is accounted for in the circuit model by the internal amplifier output resistance, R_o. The exact amount of voltage reduction at the output depends on both the value of amplifier output resistance and the amount of the load resistance.

EXAMPLE 8.1

If an amplifier, having a voltage gain of 100 and output resistance of 10 K drives a load of 20 K, what is the resulting output voltage V_o for an input signal of 10 mV.

Solution:

The voltage developed by the ideal output voltage source is $A_v V_i = 100 \times 10$ mV $= 1$ V. Using the voltage-divider rule, the output voltage V_L is calculated to be

$$V_L = \frac{R_L}{R_o + R_L}(A_v V_i) = \frac{20 \text{ K}}{10 \text{ K} + 20 \text{ K}}(1 \text{ V}) = 0.67 \text{ V}$$

Example 8.1 shows that while an ideal amplifier would provide an output of 1 V to any load resistance, the practical amplifier output will be less (0.67 V, in this example), and the values of amplifier output resistance and load resistance are both important factors in the overall operation of the circuit.

A practical current amplifier [see Fig. 8.10(b)] has some input resistance, rather than ideally being a short circuit. When driven by an ideal current source, however, the input current, I_i, is equal to the signal current, I_s. This input current signal is amplified by the circuit, resulting in an output current $A_i I_i$. This amount of current does not appear at the output for all values of load (R_L) as with the ideal amplifier, the amplified current dividing between the circuit output resistance and the load resistance. The larger the value of load resistance, the smaller the amount of output current.

When a source signal is provided by a practical source having source resistance the effect of the practical amplifier having input resistance will result in a value of V_i less than V_s due to the voltage division by source and input resistance. Figure 8.11 shows a practical amplifier circuit driven by a practical voltage source.

While an ideal amplifier would provide an output amplified by the full gain, a practical amplifier provides a gain value which is reduced due to the loading

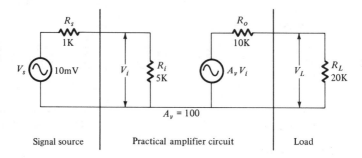

Figure 8.11 Practical voltage source driving practical voltage amplifier circuit.

of the source by the amplifier input resistance and by loading of the amplifier output by the load resistance. Similar loading occurs with current amplifiers for both input and output.

It is important to understand what an ideal amplifier is and then consider what practical factors occur in real circuits to provide other than ideal operation. The amplifier circuits covered next all reflect some aspects of ideal and practical operation, as we shall see.

8.4 BIPOLAR JUNCTION TRANSISTOR (BJT) AMPLIFIER CIRCUITS

Bipolar junction transistors are used in a large variety of applications and in many different ways. It would be impractical to study each area and application of these circuits. Instead, one studies the more fundamental properties and aspects of these devices so that enough is known to carry over this knowledge to slightly different or even greatly different applications.

To use BJT devices for voltage or current amplification, or as control (on or off) elements, or in any other application, it is necessary to first *bias* the device. The usual reason for this biasing is to turn the device on, and in amplification circuits, to place it in operation in the region of its characteristic where the device operates most linearly. Although the purpose of the bias network or biasing circuit is to cause the device to operate in this desired *linear* region of operation (which is defined by the manufacturer for each device), the bias components are still part of the overall application circuit: amplifier, waveform shaper, etc. We can analyze the operation of the overall circuit as well as separately consider the action of the bias components.

Operating point: Since the aim of biasing is to achieve a certain condition of device current and voltage called the *operating point* (or *quiescent* point), we shall first consider a number of various operating points. Figure 8.12 shows a BJT characteristic with four indicated operating points. The biasing circuit may be designed to set the device operation at any of these points within the *operating region*, which is the area of the graph within the limits of maximum power, current, or voltage of that particular device.

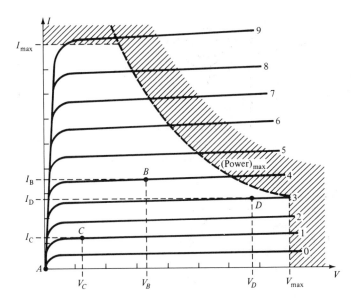

Figure 8.12 Various operating points on BJT device static characteristic.

If no bias were used, the device would initially be completely off, which would result in the current of point A, namely, zero current The undesirable aspect of operating point A is that it will respond only to that part of the input signal that causes the device to conduct. Thus, a sinusoidal voltage going both positive and negative around 0 V would be amplified only during the signal swing time when the BJT is driven to conduct. To allow the device to operate on both positive and negative signal swings the device must be biased at an operating point such as, for example, point B. Both the voltage and current which the BJT device is operating at can be caused to either increase or decrease depending on the signal swing of the input signal. Point C, for example, would allow a large current and voltage variation in only one polarity direction, with little in the opposite direction. Point D will also limit current and voltage swing because of the maximum voltage and power restrictions of the BJT device. Thus, point B is generally the most suitable for small-signal amplifier operation.

We shall next consider the three basic BJT circuit connections—common-base (CB), common-emitter (CE), and common-collector (CC)—emphasizing both dc biasing and ac operation for each type.

BJT Small-Signal Amplifier:
Common-Base (CB) Connection

The bipolar junction transistor can be operated as an amplifier device in a number of ways. Figure 8.13 shows a complete amplifier circuit with dc biasing and operation as an *ac* amplifier. The input ac signal, V_i, is coupled to the amplifier

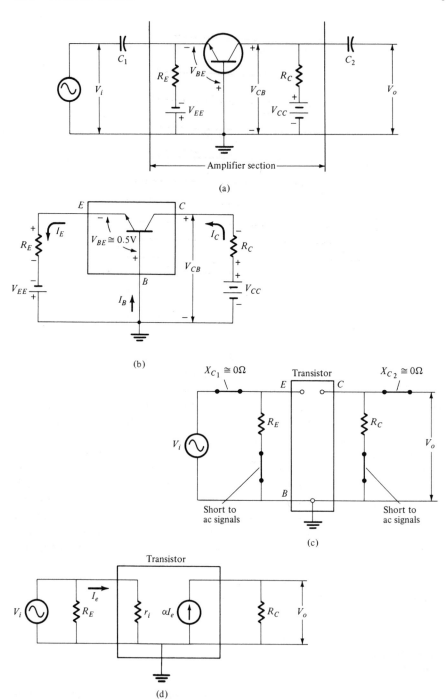

Figure 8.13 Common-base amplifier circuit.

through capacitor C_1, which serves to block any dc voltage from the signal source from affecting the amplifier stage dc bias and vice versa. Capacitor C_2 serves a similar purpose between the amplifier output and the load driven by the amplifier. In the present circuit the input is applied to the transistor emitter-base with the output developed across the transistor collector-base. When the base terminal is common to both input and output the circuit is referred to as a *common-base* (*CB*) circuit configuration, or *CB* amplifier. Analysis of the circuit is usually done separately for dc bias operation and ac amplification operation.

Dc bias. Since the two capacitors in Fig. 8.13 block dc between input and output of the amplifier, the dc voltages and currents for the amplifier circuit can be determined using the partial circuit shown in Fig. 8.13(b). To operate any BJT as an amplifying device requires *biasing the base-emitter junction of the transistor in forward bias and the base-collector junction in reverse bias.* This is true regardless of the type of biasing or transistor configuration. A forward-biased base-emitter junction is essentially a diode junction which has a relatively fixed voltage drop of about 0.3 V for germanium and 0.7 V for silicon devices. As a generalization we shall assume 0.5 V throughout the text (unless otherwise noted). Recall that a diode maintains essentially this fixed voltage drop across the diode junction when forward-biased over a very large range of current. From Fig. 8.13(b) we can now easily calculate the emitter bias current since both V_{EE} and V_{BE} voltages are fixed. The voltage drop $I_E R_E$ and voltage rise V_{EE} must equal the voltage rise V_{BE} so that

$$V_{BE} = 0.5 \text{ V} = -I_E R_E + V_{EE}$$

which can be solved for I_E as

$$I_E = \frac{V_{EE} - V_{BE}}{R_E} = \frac{V_{EE} - 0.5 \text{ V}}{R_E} \tag{8.1}$$

We see that the emitter dc bias current can be selected or adjusted by battery V_{EE} and resistor R_E. Usually V_{EE} is restricted to some available battery or power supply used with the rest of the circuitry so that practical setting of I_E is obtained by selecting R_E. Having set the value of I_E we have also approximately set the value of I_C since $I_C = \alpha I_E$, where α is typically 0.9 to nearly 1 and typically above 0.95. We then have I_C as

$$I_C \simeq I_E \tag{8.2}$$

The voltage drop across R_C is calculated to be

$$V_{R_C} = I_C R_C \tag{8.3}$$

and the voltage from collector to base is

$$V_{CB} = V_{CC} - I_C R_C \tag{8.4}$$

Since V_{CC} is usually already set by the available circuit supply voltage and I_C is determined by the biasing of the base-emitter junction, the value of V_{CB} is set by selection of resistor R_C.

An example should help to emphasize the use of these equations.

EXAMPLE 8.2

Calculate the dc bias currents I_E and I_C and the dc bias voltage V_{CB} for a CB circuit as shown in Fig. 8.13(a) for values $R_E = 2.4$ K, $R_C = 4.9$ K, $V_{EE} = 3$ V, $V_{CC} = 12$ V.

Solution:

Using first the fact that the base-emitter junction must be forward-biased we calculate I_E using Eq. (8.1), and then I_C using Eq. (8.2):

$$I_E = \frac{V_{EE} - 0.5}{R_E} = \frac{3 - 0.5 \text{ V}}{2.4 \text{ K}} = 1.04 \text{ mA} \simeq I_C$$

The voltage drop across the collector resistor is then [Eq. (8.3)]

$$V_{R_C} = I_C R_C = 1.04 \text{ mA} \times 4.9 \text{ K} = 5.1 \text{ V}$$

and V_{CB} is [from Eq. (8.4)]

$$V_{CB} = V_{CC} - I_C R_C = 12 - 5.1 = 6.9 \text{ V}$$

The transistor is thus biased to operate at $I_C \simeq 1$ mA and $V_{CB} = 6.9$ V, which should be a point within the linear operating range of the transistor chosen. Note that the circuit resistors and voltage supply set the operating point without regard to the actual transistor parameters. The transistor used, however, must operate as an amplifying device at the bias point selected by the circuit elements.

Ac operation. Having provided the necessary dc bias so that the transistor operates as an amplifier in its linear operating region, we shall next investigate the circuit and device factors which determine the amount of ac voltage amplification and other important ac circuit values. These include circuit ac input impedance, output impedance, and ac voltage and current gains.

Figure 8.13(c) shows the capacitors replaced by short circuits, as the capacitors used usually have very much smaller ac impedance than the other circuit elements for the frequency range of interest.* The voltage supplies also act as short circuits for the ac signal as their impedance is typically fractions of an ohm. A simplified *ac equivalent circuit* can thus be obtained as shown in Fig. 8.13(d). The ac input voltage is applied directly across the transistor emitter-base, which looks like a practical forward-biased diode of resistance r_i. Actually, the value of this base-emitter resistance is dependent on the dc emitter current from the bias setting and can be expressed as† (for common-base, at room temperature)

$$r_e \simeq \frac{V_T}{I_{EQ}} = \frac{25 \text{ mV}}{I_{EQ}}$$

$$= \frac{25 \times 10^{-3}}{1 \times 10^{-3}} = 25 \ \Omega \tag{8.5}$$

*For a midfrequency of 1000 Hz and $C = 10$ μF the value of X_C would be

$$X_C = \frac{1}{2\pi f C} = \frac{1}{6.28 \times 1000 \times 10 \times 10^{-6}} = 15.9 \ \Omega$$

which is negligible compared to say $R_E = 2400 \ \Omega$.

†A more accurate value of base-emitter resistance includes the diode junction resistance and an internal base ohmic resistance of 2 Ω, so that

$$r_e = \frac{25 \text{ mV}}{I_{EQ}} + 2$$

The ac emitter current is then calculated as

$$I_e = \frac{V_i}{r_e} = \frac{10\text{ mV, rms}}{25\ \Omega} = 0.4\text{ mV, rms} \tag{8.6}$$

Since $I_c \simeq I_e$, the output ac voltage is

$$V_o = I_c R_C = I_e R_C = 0.4\text{ mA, rms} \times 4.9\text{ K} = 1.96\text{ V, rms} \tag{8.7}$$

The ac voltage gain V_o/V_i is

$$A_v = \frac{V_o}{V_i} = \frac{1.96\text{ V}}{10\text{ mV}} = 196$$

This is a sizable voltage gain from a single-stage amplifier. The circuit could be said to have a current gain of

$$A_i = \frac{I_o}{I_i} = \frac{\alpha I_e}{I_e} = \alpha \simeq 0.95$$

or a gain of nearly 1.

The ac input impedance of the transistor, in this example, is r_e, and the ac input impedance of the full circuit is r_e in parallel with R_E or

$$R_i = r_e \| R_E = \frac{r_e R_E}{r_e + R_E} = \frac{25 \times 2.4\text{ K}}{25 + 2.4\text{ K}} \simeq 25\ \Omega$$

The ac output impedance of the circuit is merely the value of R_C, or 4.9 K in this example.

A *CB* amplifier stage can provide good voltage gain, be driven by a low impedance source, and drive into a high-impedance load (any load about 10 times larger than R_C).

BJT Small-Signal Amplifier: Common-Emitter (*CE*) Connection

A simple common-emitter (*CE*) amplifier circuit such as that shown in Fig. 8.14 will be considered next. Although the circuit shown can operate and amplify small ac input signals, more practical circuit versions are generally used such as those considered later. An input ac signal, V_i, is coupled through capacitor C_1, which blocks any dc signal component of the input from affecting the present amplifier circuit

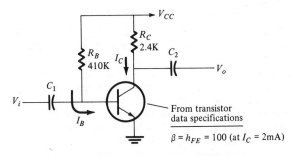

Figure 8.14 BJT common-emitter (CE) amplifier.

(and vice versa). The amplified ac signal from the BJT collector is similarly coupled out through capacitor C_2. The values of resistors R_B and R_C are important in setting the dc bias of the BJT. They will also affect the ac amplification of the circuit, as we shall shortly see. The ac input signal varies the base current set by the dc bias. This variation in base current then causes a variation in the collector current from the level set by the dc bias circuit and subsequently results in a variation of the voltage at the collector—the output voltage.

Dc bias. To simplify the circuit analysis it is possible to separately consider the dc and ac operations. Figure 8.14(a) shows the part of the circuit which has an effect in setting the dc bias voltages and currents. In this circuit configuration the base current is fixed or set to a specific value by resistor R_B and supply voltage V_{CC}. Notice that the dc base bias current I_B is set by the fixed voltage across R_B, which is the difference between the supply voltage and the base-emitter voltage drop when forward-biased— 0.5 V, in general. We can thus calculate I_B as

$$I_B = \frac{V_{CC} - V_{BE}}{R_B} = \frac{9 - 0.5 \text{ V}}{410 \text{ K}} = 20.7 \ \mu A \tag{8.8}$$

Since the collector dc bias current is h_{FE} times the base current, the collector current I_C is calculated to be*

$$I_C = h_{FE}I_B = 100 \times 20.7 \ \mu A = 2.07 \text{ mA} \tag{8.9}$$

The collector voltage measured with respect to ground is

$$V_{CE} = V_{CC} - I_C R_C$$
$$= 9 - (2.07 \times 10^{-3})(2.4 \times 10^3) = 9 - 4.97 = 4.03 \text{ V} \tag{8.10}$$

The transistor is therefore biased to operate at the quiescent point or operating point of

$$I_{C_Q} = 2.07 \text{ mA}, \qquad V_{CE_Q} = 4.03 \text{ V}$$

The values of both R_B and R_C, as well as the transistor parameter h_{FE}, affect the operating point. To use different transistors (having different values of h_{FE}) in the present circuit it would be necessary to adjust, say, R_B for each transistor used to obtain the desired operating point specified above. This individual adjustment of dc bias values is most undesirable when manufacturing many circuits, and later circuit designs will show more suitable bias circuits.

Ac operation. We are concerned mainly with the ac voltage gain of the present amplifier circuit. It is also often necessary to know the input and output resistance of the circuit so that it can be properly connected with other circuits. An ac equivalent

---

*It is important to realize that although the value of V_{BE} for a forward-biased base-emitter remains fairly constant, the value of h_{FE} may vary considerably from the typical value of 100 specified for a particular type of transistor. If the actual transistor used had a current gain, h_{FE}, of 100, then the above collector current value will be correct. If the current gain value were only 80, then I_C would be only 1.6 mA in the present example. This dependence of the value of I_C on h_{FE}, a device parameter which cannot be set to an exact value in manufacturing a very large number of transistors, is the practical reason for not using the circuit of Fig. 8.14.

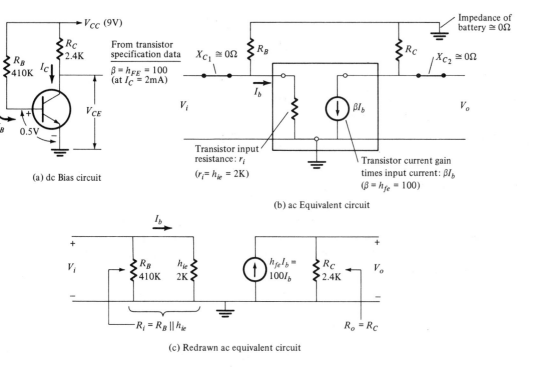

(a) dc Bias circuit

(b) ac Equivalent circuit

(c) Redrawn ac equivalent circuit

Figure 8.15 Dc bias and ac equivalent circuits.

circuit can be obtained from the circuit of Fig. 8.14 as shown in Fig. 8.15(b). Capacitors C_1 and C_2 are chosen so that their capacitive impedance at the lowest frequency of interest is very small (compared with other circuit impedance values).

To analyze the amplifier circuit it is also necessary to represent the transistor by equivalent components which will closely reflect the actual action of the transistor. Insofar as the input at the base is concerned the transistor appears as a resistance from base to emitter (ground in this circuit) whose resistance value, r_i, specified in the manufacturer's information as h_{ie}, may be obtained by either approximation or directly from the manufacturer's specifications.* A clearer version of the ac equivalent circuit can be redrawn as in Fig. 8.15(c). This is the same circuit as shown in Fig. 8.15(b). The circuit input impedance is seen to be that of resistors R_B and h_{ie} in parallel, so that

$$R_i \cong h_{ie} = 2 \text{ K}$$

*It is possible to express the transistor input resistance value, h_{ie}, as

$$h_{ie} = r_i = \frac{v_{be}}{i_b}\bigg|_{Q\text{-point}} = \frac{V_T}{I_{B_Q}} = h_{fe}\frac{V_T}{I_{C_Q}} = h_{fe}r_e$$

where $V_T = 25$ mV at room temperature. For $h_{fe} = 100$ and $I_{C_Q} = 1$ mA

$$h_{ie} = r_i = 100 \times \frac{25 \text{ mV}}{1 \text{ mA}} = 2.5 \text{ K}$$

Note that at larger collector currents the value of h_{ie} would be proportionately less.

The circuit output impedance is simply the value of R_C so that

$$R_o = 2.4 \text{ K}$$

The voltage gain of the circuit can be calculated as

$$A_v = \frac{V_o}{V_i} = \frac{(-h_{fe}I_b)R_C}{I_b h_{ie}} = -\frac{h_{fe}R_C}{h_{ie}} = -\frac{R_c}{r_e} \tag{8.11}$$

which is calculated to be

$$A_v = -\frac{100(2.4 \times 10^3)}{(2 \times 10^3)} = -120$$

The minus sign indicates a 180° phase reversal of the output signal (with respect to the input signal), and the value of 120 is the voltage gain of the circuit from input to output. This value of gain applies only to small signals (micro- or millivolts of input). If the resulting output voltage were near the supply voltage value, say a few volts, then the small signal operation assumed in the equivalent circuit of Fig. 8.15(b) would no longer hold true. For a small signal of, say, 100 μV, rms the ac output voltage of the present circuit would be

$$V_o = A_v V_i = -120 \times 100 \ \mu\text{V} = -12 \text{ mV, rms}$$

dc Bias Circuit Independent of h_{FE}

A more practical dc bias connection for a small-signal amplifier is that of Fig. 8.16. We can again determine the circuit dc bias voltages and currents first and then calculate the ac voltage gain and input and output impedances.

Dc bias. To simplify the analysis of the circuit a sketch of the part of the network which sets the value of the base voltage is drawn in Fig. 8.17(a). In a typical design of a circuit such as that of Fig. 8.16 the transistor parameters and the value R_E are chosen so that the equivalent dc impedance seen looking into the transistor from the base, R_i is much larger than the value of R_2. If one then neglects the loading effect

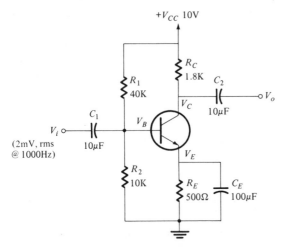

Figure 8.16 BJT small signal amplifier circuit with beta-independant bias.

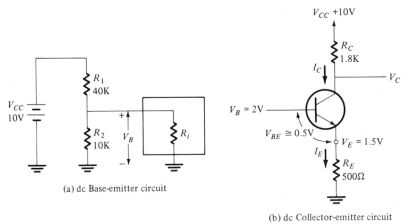

(a) dc Base-emitter circuit

(b) dc Collector-emitter circuit

Figure 8.17 Partial circuits of Fig. 8.16 for dc bias calculations.

of R_i, the base voltage V_B can be calculated as the voltage component across R_2,
and using the voltage-divider rule

$$V_B = \frac{R_2}{R_1 + R_2} V_{CC}$$

$$= \frac{10\ \text{K}}{10 + 40\ \text{K}} (10\ \text{V}) = 2\text{V}$$

(8.12)

Referring to Fig. 8.16, the dc voltage at the emitter, V_E, can be calculated since it is
less than V_B by the fixed base-emitter voltage drop V_{BE} for the conducting transistor
(typically 0.5 V):

$$V_E = V_B - V_{BE} = 2 - 0.5 = 1.5\ \text{V}$$

The value of emitter current I_E can then be calculated using Ohm's law:

$$I_E = \frac{V_E}{R_E} = \frac{1.5\ \text{V}}{0.5\ \text{K}} = 3\ \text{mA}$$

Since the collector current is approximately equal to the emitter current, we know
that $I_C = 3$ mA and can then calculate the collector voltage using Kirchhoff's voltage
law:

$$V_C = V_{CC} - I_C R_C = 10 - 3\ \text{mA} \times 1.8\ \text{K} = 4.6\ \text{V}$$

The voltage across the collector-emitter, which is also a device bias voltage, can be
calculated:

$$V_{CE} = V_C - V_E = 4.6 - 1.5 = 3.1\ \text{V}$$

Thus, the device is biased to operate at

$$I_{C_Q} = 3\ \text{mA}, \qquad V_{CE_Q} = 3.1\ \text{V}$$

which should be somewhere in the center of the device operating region for small-
signal amplification operation.

The three capacitors shown in Fig. 8.16 have no effect on the dc bias currents and voltages as calculated above and are essentially part of the ac operating circuit considered next.

Ac operation. Calculation of ac input and output impedances and circuit voltage or current gain values can all be obtained using the ac equivalent circuit of Fig. 8.15(a). At the frequency of 1000 Hz the capacitors are very low impedances and therefore essentially act as short circuits. At 1000 Hz the ac impedance of the 10-μF coupling capacitors is $X_C = 16\ \Omega$, which is small compared with the 10-K, 40-K, or even 1.8-K resistor values. Similarly, the 100-μF capacitor in parallel with R_E has an impedance at 1000 Hz of 1.6 Ω, which acts as a short circuit across the 500-Ω emitter resistor *for ac operation*. Since the dc supply is also a short circuit (or low impedance) for ac considerations the ac equivalent circuit of Fig. 8.18(a) can be redrawn as shown in Fig. 8.18(b). The bias resistors R_1 and R_2 are seen to be a parallel equivalent resistance of 8 K. The transistor can be represented by a number of equivalent circuit configurations. One simple equivalent of the transistor is shown as a base-emitter resistor, r_i (also referred to as h_{ie} in hybrid parameter nomenclature), and a current source dependent on I_b and the transistor current gain (also called h_{fe} in hybrid parameter nomenclature).

Ac input impedance. We can calculate the circuit ac input impedance as the parallel combination of 8 K (from the dc bias resistors) and 2 K (of the transistor) to obtain

$$R_i = \frac{8\ \text{K} \times 2\ \text{K}}{8\ \text{K} + 2\ \text{K}} = 1.6\ \text{K}$$

Ac output impedance. The circuit ac output impedance is that of the collector resistor,

$$R_o = 1.8\ \text{K}$$

as shown in Fig. 8.18(c).

Voltage gain. The circuit ac voltage gain can be obtained after first calculating I_b:

$$I_b = \frac{V_i}{2\ \text{K}} = \frac{2\ \text{mV, rms}}{2\ \text{K}} = 1\ \mu\text{A, rms}$$

I_c is calculated to be

$$I_c = h_{fe}I_b = 200(1\ \mu\text{A}) = 200\ \mu\text{A, rms}$$

Finally, the output voltage V_o is

$$V_o = -I_c R_o = -200\ \mu\text{A}(1.8\ \text{K}) = -0.36\ \text{V, rms}$$

The circuit voltage gain can then be calculated:*

$$A_v = \frac{V_o}{V_i} = -\frac{0.36\ \text{V}}{2\ \text{mV}} = -180$$

*This voltage gain can be directly expressed in terms of the circuit elements as

$$A_v = \frac{V_o}{V_i} = -\frac{I_c R_o}{I_b h_{ie}} = -\frac{h_{fe} I_b R_o}{I_b h_{ie}} = -\frac{h_{fe} R_o}{h_{ie}} = -\frac{R_o}{r_e}$$

where

$$r_e = \frac{h_{ie}}{h_{fe}} \simeq \frac{V_T}{I_{EQ}} = \frac{25(\text{mV})}{I_{EQ}(\text{mA})}$$

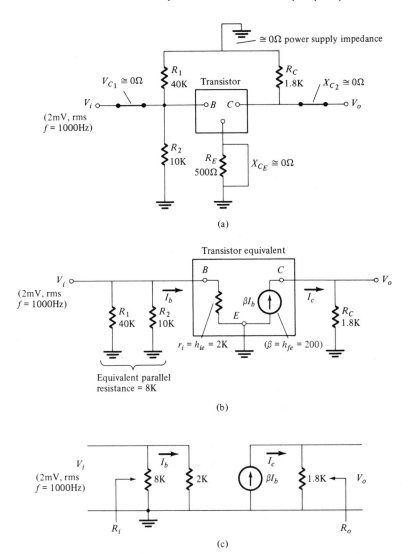

Figure 8.18 Partial circuits of Fig. 8.16 for ac calculations.

A summary of the overall circuit operation is given in Fig. 8.19(a), where the gain is specified as −180. The values of input impedance or output impedance are of concern when connecting a source as signal input or connecting the amplifier output to a load. For example, if the present amplifier is driven by a signal source of 10 mV at 1000 Hz, with the source having an impedance of 1000 Ω, and no output load is present, the circuit connection is that of Fig. 8.19(b). The voltage *actually* appearing as input to the amplifier, marked V'_i, can be calculated using the voltage-divider rule:

$$V'_i = \frac{1.6\,\text{K}}{1\,\text{K} + 1.6\,\text{K}} V_i = \frac{1.6\,\text{K}}{2.6\,\text{K}} (2\,\text{mV}) = 1.2\,\text{mV, rms}$$

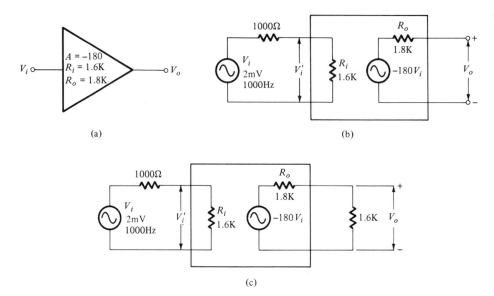

(a)   (b)

(c)

Figure 8.19 Overall amplifier operation.

Thus, the presence of a source impedance in this example results in a reduced amplifier input voltage, from 10 to 6.15 mV. The amplifier circuit then operates on this voltage, producing an output voltage of

$$V_o = -180(V_i') = -180(1.2 \text{ mV}) = -0.216 \text{ V, rms}$$

Compare this output voltage to 0.180 V, which results when no source impedance is present.

If the output of the amplifier is then connected to a load, say another amplifier stage having 1.6-K impedance, the resulting value of V_o is further reduced to [see Fig. 8.16(c)]

$$V_o = \frac{1.6 \text{ K}}{1.8 \text{ K} + 1.6 \text{ K}}(-180V_i') = \frac{1.6 \text{ K}}{3.4 \text{ K}}(-0.216 \text{ V}) = -0.102 \text{ V, rms}$$

It should be noticed that the largest gain would result if the source impedance is lowest and output impedance highest, or, conversely, if the amplifier input impedance is highest and output impedance lowest. Ideally, then, an amplifier should have infinite input impedance and zero output impedance so that any source and load values would result in the maximum amplifier voltage gain.

BJT Common-Collector (Emitter-Follower) Amplifier

Another popular form of amplifier circuit is the common-collector or emitter-follower circuit as shown in Fig. 8.20(a). Notice that the output is taken from the transistor emitter terminal. Intuitively we should expect that any voltage applied to the transistor base will appear as the same signal at the transistor emitter since the

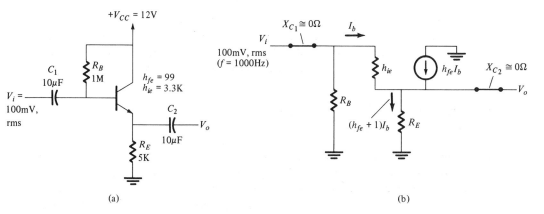

(a)   (b)

Figure 8.20 Emitter-follower circuit.

signal is coupled through the forward-biased base-emitter, which acts as an ac short circuit (or very low impedance). The circuit name emitter-follower suitably denotes that the output ac voltage taken from the emitter *follows* or is the same as the input signal applied to the base.

The dc bias condition can be calculated to be $I_C = I_E = 0.77$ mA, $V_{CE} = V_C - V_E = 12 - 3.8 = 8.2$ V.

The ac operation can be analyzed using the ac equivalent circuit of Fig. 8.20(b) so that the circuit voltage gain is $A_v = 1$, and the current gain of the circuit is

$$A_i = \frac{I_o}{I_i} = \frac{I_e}{I_b} = \frac{(h_{fe} + 1)I_b}{I_b} = h_{fe} + 1 = 100$$

Thus, the circuit ideally provides a current gain with unity voltage gain ($A_v \simeq 1$).

Two important features of the emitter-follower circuit are its input and output impedances. The input ac impedance, seen looking into the base of the transistor, can be calculated as follows:

$$R_i = \frac{V_i}{I_b} = \frac{V_i}{V_i/[(h_{fe} + 1R_E]} = (h_{fe} + 1)R_E \tag{8.13}$$

In the present circuit this results in $R_i = (99 + 1)5\,K = 0.5$ M. This is quite high, which is an important feature of the emitter-follower circuit. Actually, the circuit input impedance seen looking into the full circuit is somewhat lower because of R_B; in this case the resistance resulting from resistor R_B in parallel with the equivalent resistance R_i is the circuit ac input resistance, R_{in}:

$$R_{in} = R_B \| R_i = 1\,M \| 0.5\,M = 0.33\,M \tag{8.14}$$

The output impedance of the circuit is dependent on the transistor parameters h_{ie} and h_{fe} and any input source impedance (R_s) and can be shown to be

$$R_o = \frac{h_{ie} + R_s}{1 + h_{fe}} = \frac{3.3\,K}{100} = 33\,\Omega \tag{8.15}$$

which is seen to be quite low.

The emitter-follower thus acts as a very good buffer amplifier with high input impedance (so that it doesn't load down any source) and provides the same as the input signal but from a very low output impedance (to act as a very good driver for any external load). The emitter-follower circuit provides an increased voltage to a load by its favorable impedance matching rather than by direct voltage amplification.

8.5 FIELD-EFFECT TRANSISTOR (FET) AMPLIFIER CIRCUITS

Both junction FET (JFET) and metal-oxide semiconductor (MOSFET) devices will be considered in amplifier circuits. The FET has a number of advantages over the BJT as an amplifying device. One main advantage of the FET is its high input impedance (resulting in almost infinite input impedance). Others are the smaller (than BJT) size for manufacturing large-scale integrated circuits and the relative simplicity in constructing FET devices. Although most FET circuits, especially IC circuits, use MOSFET devices, discrete element circuits also use JFET devices, which are considered first.

JFET Operating Parameters

A number of necessary device operating values can be obtained from either the device characteristic or device specifications. Figure 8.21 shows a *drain* or output characteristic for a typical JFET. Two device parameters of great interest in dc biasing are the values associated with the $v_{GS} = 0$ V curve of the characteristic. Recall that it is possible to pass current through the JFET channel until pinch-off occurs. On the drain characteristic pinch-off occurs when the characteristic curve bends to the horizontal showing that no further drain current flows (above a saturation value dependent on the value v_{GS}) with increased v_{DS} voltage. This pinch-off condition is shown as the knee of the curve, and for the $v_{GS} = 0$ V curve it is equal to the device pinch-off voltage, V_P. It can be seen that at more negative gate-source

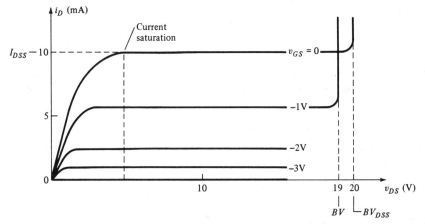

Figure 8.21 FET drain characteristic.

voltages (negative for an *n*-channel or positive for a *p*-channel JFET) pinch-off occurs at lower values of v_{DS} since the gate-source already has started to pinch off due to the gate-source bias voltage. The value V_p of interest, however, is that voltage required to fully pinch off the channel, and it can be obtained as either the knee of the curve for $v_{GS} = 0$ V or more clearly as the zero current condition of the device transfer characteristic (Fig. 8.22).

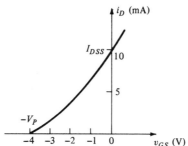

Figure 8.22 FET transfer characteristics.

A second important device parameter is the value of drain saturation current at $v_{GS} = 0$ V, called I_{DSS} (current in Drain to Source for Saturated condition, at $v_{GS} = 0$ V). The drain current will be decreased from this value as the gate-source voltage becomes more negative (for an *n*-channel JFET) as shown in Fig. 8.21. This saturation current value is also clearly shown on the device transfer characteristic (Fig. 8.22) as the current value when the transfer curve crosses $v_{GS} = 0$ V (the ordinate axis).

For the JFET an important bias relationship between bias drain current (i_D) and gate-source voltage (v_{GS}) is given by the following equation:

$$i_D = I_{DSS}\left(1 - \frac{v_{GS}}{V_P}\right)^2 \tag{8.16}$$

where I_{DSS} and V_p are fixed values for a particular JFET and are specified for the device or can be obtained from the device characteristic.

An example of FET biasing can be obtained from the circuit of Fig. 8.23. Using the JFET device whose transfer characteristic is shown in Fig. 8.22, it is possible to determine the dc bias condition for any value of v_{GS} set by the bias battery, V_{GG}.

For the value of $V_{GG} = 1$ V, the resulting value of v_{GS} is -1 V (assuming that no gate current flows—an excellent assumption). We can then solve for the value of drain bias current:

$$I_D = 10(1 - \tfrac{1}{4})^2 = 5.6 \text{ mA}$$

Summing the voltage drops from the supply voltage V_{DD} to ground we can solve for V_{DS}:

$$V_{DS} = V_{DD} - I_D R_D = 15 - (5.6 \times 10^{-3})(1.5 \times 10^3) = 6.6 \text{ V}$$

Thus, the JFET is biased to operate at

$$V_{DS_Q} = 6.6 \text{ V}, \qquad I_{D_Q} = 5.6 \text{ mA}$$

which can be seen to be in the operating region of the device characteristic in Fig. 8.21.

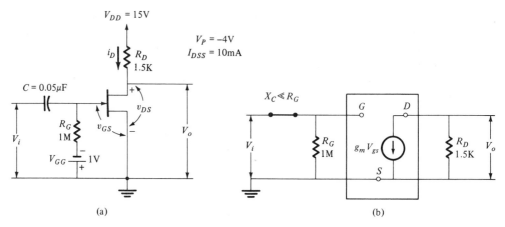

Figure 8.23 JFET amplifier (a) circuit; (b) ac equivalent circuit.

Ac operation. Having established the dc bias of the JFET it is now possible to determine some of the ac circuit features such as input and output impedance and voltage gain. An ac equivalent circuit is drawn in Fig. 8.23(b). The dc batteries are replaced by their near 0-Ω resistance, the capacitors are replaced by a short circuit,* and the circuit is redrawn to the neater form shown in Fig. 8.23(b).

The input impedance is seen to be that of the gate bias resistor, $R_G = 1$ M in this example. The gate appears as an open circuit (typically about 10^8–10^{10} Ω, for the JFET). We therefore have

$$R_i = R_G = 1 \text{ M}$$

The circuit output impedance is seen directly to be the drain resistance, R_D, so we can also write

$$R_o = R_D = 1.5 \text{ K}$$

We see that the circuit has good amplifier impedance characteristics with high input impedance and low output impedance.

The output ac voltage is easily calculated as the voltage across R_D due to the current source $g_m v_{gs}$, or

$$v_o = -g_m v_{gs} R_D$$

The value of g_m can be obtained from either the device specifications at the operating point or from the following relation between device parameters and bias point:†

$$g_m = \frac{2I_{DSS}}{V_P}\left(1 - \frac{v_{GS_Q}}{V_P}\right) \tag{8.17}$$

---

*It should be noted that although X_C at a midfrequency of say 1000 Hz is about 3.2 K (for the present circuit), this impedance is considerably smaller than the 1-M impedance of R_G, and essentially all the ac signal v_i appears across R_G so that X_C can be considered a short circuit.

†Using $i_D = I_{DSS}[1 - (v_{GS}/V_P)]^2$ we can derive the formula for g_m:

$$g_m = \frac{\partial i_D}{\partial v_{GS}}\bigg|_{Q\text{-point}} = \frac{2I_{DSS}}{V_P}\left(1 - \frac{v_{GS_Q}}{V_P}\right)$$

which in the present example is

$$g_m = \frac{2(10 \times 10^{-3})}{4}\left(1 - \frac{1}{4}\right) = 3.75 \times 10^{-3} = 3.75 \text{ mmho}$$

Using the calculated value of g_m, the circuit voltage gain is then

$$A_v = g_m R_D = -(3.75 \times 10^{-3})(1.5 \times 10^3) = -5.6$$

which shows a magnitude gain of 5.6 between output voltage and input.

If, for example, the input were a 10-mV, rms signal, then the output signal magnitude would be 56 mV, rms, which is still a small signal (compared to the supply voltage of 15 V). Thus, the FET small-signal amplifier circuit of Fig. 8.23 provides a voltage gain of 5.6 while having an input impedance of about 1 M and an output impedance of 1.5 K. Comparing this single FET stage with the BJT amplifier stages previously considered we note that the BJT stage has a larger voltage gain but much lower input impedance, which would result in loading an input source connected to it and therefore lower the overall voltage amplification.

Figure 8.24 shows a more practical version of an FET amplifier stage—this circuit developing the gate-source bias condition using only the single supply voltage V_{DD} and a bias resistor R_S. In analyzing the amplifier circuit we shall first determine the dc bias condition and then calculate the ac voltage gain and input and output impedances.

Dc bias. Using the device transfer characteristic [Fig. 8.24(b)] we can determine the gate-source bias voltage and quiescent drain current by drawing a self-bias line as shown in Fig. 8.24(b). For $R_S = 0.5$ K the self-bias line is drawn from the 0 axis point to intersect any point of $-V_{GS} = 0.5I_D$ (since $-V_{GS} = V_S = I_D R_R$), such as $I_D = 4$ mA and $V_{GS} = -2$ V. The intersection of the self-bias line and device transfer

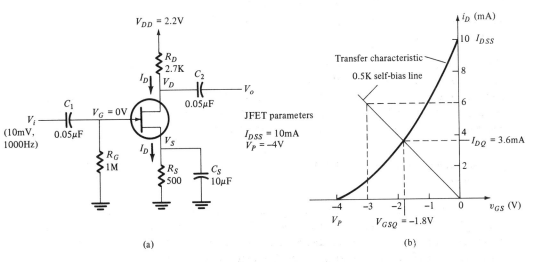

(a)   (b)

Figure 8.24 JFET amplifier circuit (a) circuit; (b) transfer characteristic.

characteristic is the quiescent or operating point for the FET in the present circuit and is seen to be

$$V_{GS_Q} = -1.8 \text{ V}, \qquad I_{D_Q} = 3.6 \text{ mA}$$

We can now calculate the drain quiescent voltage V_{D_Q} and drain-source quiescent voltage V_{DS_Q} as follows:

$$V_{D_Q} = V_{DD} - I_{D_Q}R_D = 22 - (3.6 \text{ mA})(2.7 \text{ K}) = 12.3 \text{ V}$$

and

$$V_{DS_Q} = V_{D_Q} - V_{S_Q} = 12.3 - 1.8 \text{ V} = 10.5 \text{ V}$$

Ac operation. The ac input impedance is again the gate bias resistor R_G (since the FET input impedance is typically 10^8–10^{10} and can be considered an open circuit) so that

$$R_i = R_G = 1 \text{ M}$$

The output impedance is the drain bias resistance:*

$$R_o = R_D = 2.7 \text{ K}$$

The ac voltage gain with bypass capacitor C_S connected is

$$A_v = -g_m R_D$$

(At 1000 Hz, for example, the impedance of capacitor C_S is approximately 16 Ω, which then makes the ac equivalent impedance of C_S and R_S in parallel negligible compared to the 2.7-K impedance of R_D.)

The FET transconductance, g_m, can be determined using Eq. (8.17)

$$g_m = \frac{2I_{DSS}}{V_P}\left(1 - \frac{v_{GS_Q}}{V_P}\right) = \frac{2(10 \text{ mA})}{4 \text{ V}}\left(1 - \frac{-1.8 \text{ V}}{-4 \text{ V}}\right)$$
$$= 2.75 \text{ mmho} = 2750 \text{ } \mu\text{mho}$$

which can be used to obtain the circuit voltage gain of

$$A_v = -g_m R_D = -(2750 \times 10^{-6})(2.7 \times 10^3) = -7.43$$

If the source resistor were not bypassed by a capacitor, the voltage gain would be reduced by the factor $1 + g_m R_S$, which, in the present example, would result in a voltage gain of -3.13. If the 500-Ω source resistance were partially bypassed (two series resistors, one bypassed and the other not), the voltage gain would be somewhere between the lower value of -3.13 and the upper value of -7.43.

dc Biasing Using Universal JFET Curves

We shall now cover a procedure for analyzing a practical JFET circuit, such as that of Fig. 8.24(a), using a set of universal JFET curves. Figure 8.25 is derived by normalizing the JFET transfer characteristic and making a simpler nomograph format for calculation of dc bias and ac gain for the standard JFET self-bias circuit.

---

*The FET does have a drain-source resistance r_d which is typically 50 K to 1 M and can be neglected compared to typical bias resistor values (R_D) less than 20 K.

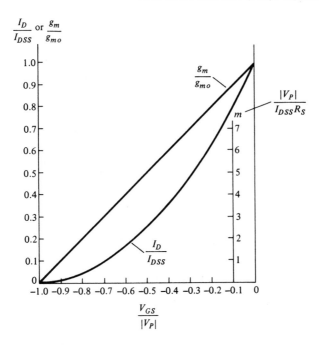

Figure 8.25 Universal JFET characteristics.

The procedure is carried out as follows (see Fig. 8.26)

1. Draw a line from the 0 point on the lower right axis through a point on the m axis at

$$m = \frac{|V_P|}{I_{DSS}R_S} = \frac{|4V|}{(10 \text{ mA})(0.5 \text{ K})} = 0.8 \qquad (8.18)$$

2. The intersection of this line with the I_D/I_{DSS} curve defines the operating point

$$\frac{V_{GS}}{|V_P|} = -0.42 \quad \text{and} \quad \frac{I_D}{I_{DSS}} = 0.33$$

in the present example. This then gives

$$V_{GS_Q} = -0.42|V_P| = -0.42(4 \text{ V}) = -1.68 \text{ V}$$
$$I_{D_Q} = 0.33I_{DSS} = 3.3 \text{ mA}$$

From these values of operating point we can calculate the drain voltage and drain-source voltage at the operating condition determined above. This results in

$$V_{D_Q} = V_{DD} - I_{D_Q}R_D = 22 - (3.3 \text{ mA})(2.7 \text{ K})$$
$$= 22 - 8.9 = 13.1 \text{ V}$$

and

$$V_{DS_Q} = V_{D_Q} - V_{S_Q} = 13.1 - 1.68 = 11.4 \text{ V}$$

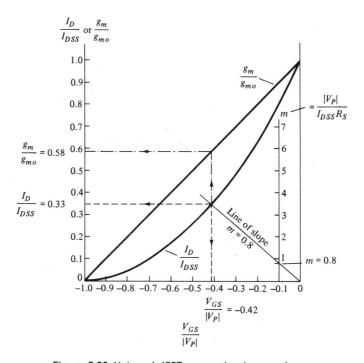

Figure 8.26 Universal JFET curves showing sample use.

Ac circuit calculations can now be carried out. The value of g_m is obtained from the g_m/g_{mo}* curve ($= 0.58$), and the JFET transconductance is

$$g_m = 0.58 g_{mo} = 0.58 \frac{2I_{DSS}}{V_P} = 0.58 \frac{2(10 \times 10^{-3})}{4}$$

$$= 2.9 \times 10^{-3} = 2900 \ \mu\text{mho}$$

The circuit ac voltage gain is then

$$A_v = g_m R_D = -(2.9 \times 10^{-3})(2.7 \times 10^3) = -7.83$$

The input ac resistance of the circuit is

$$R_i = R_G = 1 \text{ M}$$

and the ac output resistance is

$$R_o = R_D = 2.7 \text{ K}$$

Compare these answers with the previous calculations to see that the use of these universal JFET curves simplifies the determination of the dc operating point and the device transconductance at the operating point while giving good results.

$$*g_{mo} = \frac{2I_{DSS}}{|V_P|}.$$

MOSFET Device

The enhancement-type MOSFET device (covered in Chapter 6) has a characteristic such as that shown in Fig. 8.27. The voltage V_T is the threshold voltage at which the device channel is formed between drain and source (typically, $V_T = 2\text{--}4$ V). As the gate-source voltage is increased above this threshold value the drain current increases, as shown in Fig. 8.27(a). From Fig. 8.27(b) we also see that pinch-off

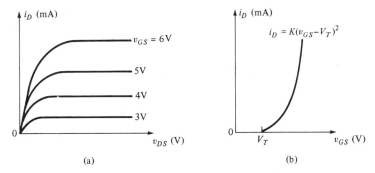

(a)   (b)

Figure 8.27 Enhancement-type *n*-channel MOSFET (a) Drain characteristic; (b) Transfer characteristic.

occurs for values of v_{DS} less than $v_{GS} - V_T$, and therefore values of v_{DS} larger than this value are necessary for current conduction to occur. It is possible to determine the drain current in the region where $v_{DS} > v_{GS} - V_T$ (in which a channel exists) using the following equation:

$$i_D = K(v_{GS} - V_T)^2 \tag{8.19}$$

where K, depending on the device physics and geometry, is typically 0.3. The quantity I_{DSS} used with JFET devices has no meaning with enhancement-type MOSFETs since $i_D = 0$ when $v_{GS} = 0$.

MOSFET operation. The operation of a MOSFET in an amplifier circuit can be examined using the basic circuit of Fig. 8.28. As usual, the dc bias and ac operation are considered separately.

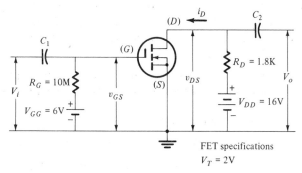

Figure 8.28 Enhancement MOSFET amplifier circuit (*n*-channel).

Dc bias. With the gate-source voltage v_{GS} set at 6 V by the supply voltage V_{GG} and the value of the threshold voltage $V_T = 2$ V for the chosen MOSFET, the dc bias drain current is

$$I_{D_Q} = K(V_{GS_Q} - V_T)^2 = 0.3(6 - 2)^2 = 4.8 \text{ mA}$$

At this quiescent drain current the voltage from the drain to source is determined by summing the voltage drops around the drain-source loop, resulting in

$$V_{DS_Q} = V_{DD} - I_{D_Q}R_D = 16 - (4.8 \text{ mA})(1.8 \text{ K}) = 7.36 \text{ V}$$

Ac operation. Having established an operating point we can next determine the ac voltage gain of the circuit and its input and output impedances. The voltage gain depends directly on the MOSFET transconductance g_m (or g_{fs}), which can be obtained from either the manufacturer's typical specifications, by direct measurement using appropriate instrumentation, or by approximate calculation using*

$$g_m = 0.6(V_{GS} - V_T) \quad \text{(mmhos)} \tag{8.20}$$

where V_T is the device threshold voltage and V_{GS} is the fixed gate-source bias voltage. Using Eq. (8.20) for the present example values,

$$g_m = 0.6(6 - 2) = 2.4 \text{ mmho} = 2400 \ \mu\text{mho}$$

As with the JFET, the circuit voltage gain is simply

$$A_v = \frac{V_o}{V_i} = -g_m R_D = -(2400 \times 10^{-6})(1.8 \times 10^3) = -4.32$$

Since the gate-source resistance of the FET is typically greater than $10^{14} \ \Omega$, the circuit input resistance is that of the gate bias resistor,

$$R_i = R_G = 10 \text{ M}$$

The circuit output resistance is the drain bias resistance R_D in parallel with the drain-source ac resistance. Manufacturer specifications usually list the reciprocal output conductance as g_{oss}.† A value of $g_{oss} = 10 \ \mu\text{mho}$, for example, is equivalent to a resistance of

$$r_o = \frac{1}{g_{oss}} = \frac{1}{10 \times 10^{-6}} = 100 \text{ K}$$

which is much larger than $R_D = 1.8$ K, and the loading effect on R_D (and therefore the voltage gain) is negligible. In usual practice the values of R_D and FET output conductance are selected so that very little loading occurs, and the circuit voltage gain is calculated using $A_v = -g_m R_D$.

Practical MOSFET circuit. A practical enhancement MOSFET amplifier circuit is shown in Fig. 8.29(a). The most noticeable difference between this circuit and the

---

*We can express the MOSFET transconductance g_m as

$$g_m = \frac{\partial i_d}{\partial v_{gs}}\bigg|_{V_{DS} \text{ constant}} = 2K(V_{GS} - V_T)$$

which is typically $2(0.3)(V_{GS} - V_T) = 0.6(V_{GS} - V_T)$ for the bias voltage V_{GS}.
†The subscripts of g_{oss} indicate output with gate source shorted (with $V_{GS} = 0$ V).

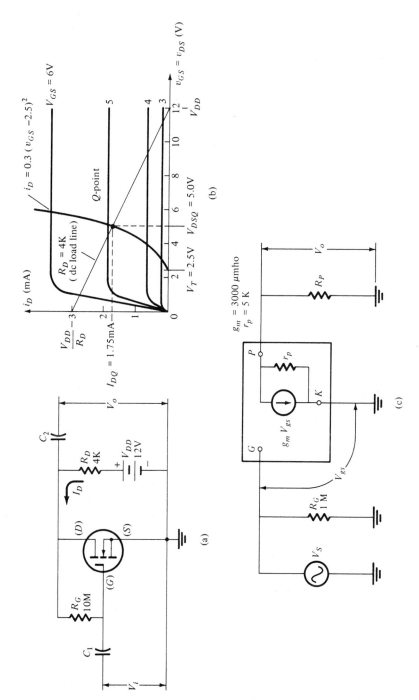

Figure 8.29 Practical *p*-channel enhancement MOSFET amplifier circuit.

circuit of Fig. 8.28 is the use of only one power supply in the present circuit. Connecting R_G from drain to gate accomplished the necessary setting of V_{GS} to bias the FET device. Since no gate current will flow, the value of V_{GS} equals the valus of V_{DS} at the quiescent condition, or

$$V_{GS_Q} = V_{DS_Q}$$

The only requirement for proper circuit biasing is that V_{GS} be greater than V_T, which is assured as long as the supply voltage V_{DD} is chosen to be greater than the FET threshold voltage V_T. A large value of R_G (> 10 M) can be chosen since there is practically no dc current flow through it due to the high input impedance between gate and source.

For the values of $V_{DD} = 12$ V and $R_D = 4$ K the drain-source dc bias voltage can be solved mathematically using the drain-source circuit equation

$$V_{DS} = V_{DD} - I_D R_D = 12 - 4 \times 10^3 (I_D)$$

and the device drain-current relation (for $V_T = 2.5$ V) is

$$I_D = K(V_{GS} - V_T)^2 = 0.3(V_{DS} - 2.5)^2$$

The solution of these two equations is necessary to solve for the value of V_{DS}. An easier procedure than mathematical solution is the graphical procedure used to determine the intersection of the lines representing these equations as shown in Fig. 8.29(b). Using the device drain characteristic a dc load line for $R_D = 4$ K can be drawn by intersecting a straight line from the horizontal at $V_{DS} = V_{DD} = 12$ V to the vertical axis at $I_D = V_{DD}/R_D = 3$ mA. This load line for $R_D = 4$ K represents the mathematical equation $V_{DS} = V_{DD} - I_D R_D$. The intersection of this load line with a transfer characteristic line* representing $I_D = K(V_{GS} - V_T)$, which for $V_{GS} = V_{DS}$ and $V_T = 2.5$ V is $I_D = 0.3(V_{DS} - 2.5)^2$, results in the quiescent operating point of

$$I_{D_Q} = 1.75 \text{ mA}, \qquad V_{DS_Q} = 5 \text{ V}$$

Since the drain and transfer characteristics are representative of the device operation, the only adjustment to the operating point is made by changing V_{DD} or, more usually, R_D. Larger values of R_D will cause the quiescent point to go to lower values of I_{D_Q} and V_{DS_Q}, while lower values of R_D will do opposite.

Ac calculations are the same as previously covered, so that

$$R_i = R_G = 10 \text{ M}$$
$$R_o = R_D = 4 \text{ K}$$
$$g_m|_{Q\text{-point}} = 2K(V_{GS_Q} - V_T) = 0.6(5 - 2.5) = 1.5 \times 10^{-3}$$

from which we calculate

$$A_v = -g_m R_D = -(1.5 \times 10^{-3})(4 \times 10^3) = -6$$

*The transfer curve of $I_D = K(V_{GS} - V_T)^2$ can be drawn onto the drain characteristic by plotting points, in this example for $I_D = 0.3(V_{DS} - 2.5)^2$. A few values of V_{DS}, with calculation of I_D, will give sufficient data to sketch the curve.

PROBLEMS

§ 8.2

1. Assuming an ideal diode in the circuit of Fig. 8.30, draw the output waveform for input signals of (a) 30 V, peak sinusoidal, (b) 20 V p-p, square wave.

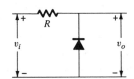

Figure 8.30

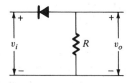

Figure 8.31

2. Draw the resulting output waveform for a sinusoidal signal of 20 V, peak applied to the circuit of Fig. 8.31.

3. Draw the resulting output waveform for a 20-V p-p square-wave signal applied to the circuit of Fig. 8.31.

4. Repeat Problem 2 for the circuit of Fig. 8.32.

5. Repeat Problem 3 for the circuit of Fig. 8.32.

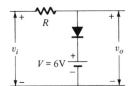

Figure 8.32

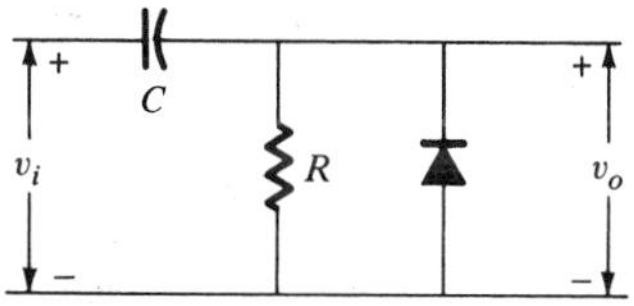

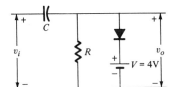

Figure 8.33

6. Draw the output waveform for a 15-V peak square wave applied to the circuit of Fig. 8.33.

7. For a 15-V peak square wave applied to the circuit of Fig. 8.34, draw the resulting output waveform.

Figure 8.34

§ 8.3

8. Compare the operation of ideal voltage and current amplifiers with regard to input and output resistance.

9. A practical voltage amplifier having a gain of 80, input resistance of 10 K, and output resistance of 5 K drives a load of 10 K. What is the voltage developed across the load for an input of $V_i = 15$ mV, rms?

10. If the load in Problem 9 is made 20 K, calculate the resulting output voltage.

11. A practical voltage amplifier having voltage gain of 65 with input resistance of 2 K is driven by a source of 10 mV, rms having source resistance of 500 Ω. What output voltage is developed if the amplifier drives a load much larger than the amplifier output resistance?

12. What current is driven into a 2-K load by a practical current amplifier having current gain of 120, input resistance of 5 K, and output resistance of 1 K when driven by a current of 100 μA, rms?

13. What current is driven into the load of Problem 12 if the load is changed to 3 K?

14. What is the output current from a current amplifier having current gain of 70 and an input resistance of 1 K when it is driven by an input current of 40 μA, rms from a source having 500-Ω resistance? Assume that the amplifier output resistance is very small compared to the load resistance.

§ 8.4

15. Calculate the dc bias currents I_E and I_C for the *CB* circuit of Fig. 8.35 for $R_E = 750$ Ω.

16. If R_E is selected so that $I_E = 2.5$ mA in the *CB* circuit of Fig. 8.35, calculate the voltage V_{CB}.

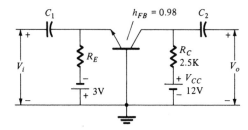

Figure 8.35

17. Calculate the ac output voltage of the *CB* circuit of Fig. 8.35 for an input voltage of 6 mV, rms for the dc bias condition established in Problem 16.

18. Determine the bias currents I_B and I_C for the *CE* circuit of Fig. 8.36 when $R_B = 120$ K.

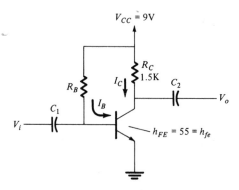

Figure 8.36

19. If R_B is selected to give $I_B = 60 \ \mu\text{A}$, calculate the resulting bias voltage V_{CE}.

20. Calculate the voltage gain of the circuit of Fig. 8.36 for $h_{ie} = 600$.

21. If R_B is selected so that $I_{C_Q} = 3.5 \ \text{mA}$, calculate the circuit voltage gain.

22. Calculate the bias voltage V_B and emitter current I_E at dc bias for the circuit of Fig. 8.37.

23. Calculate the voltage drop across the collector resistor ($I_C R_C$) and the dc bias voltage V_{CE} for the circuit of Fig. 8.37.

24. Determine the transistor bias points V_{CE_Q} and I_{C_Q} for the circuit of Fig. 8.37.

25. What value of resistance R_E will result in a collector bias current of 1.25 mA?

26. At a collector bias current of 1.25 mA, what is the voltage drop across R_C and the bias voltage V_{CE_Q}?

27. Does bypass capacitor C_E still do a good job of acting as an ac short across bias resistor R_E at a frequency of 1000 Hz? At what frequency does X_{C_E} equal R_E (and no longer acts as a good bypass capacitor)?

28. What value of bypass capacitor is needed to provide good bypass action ($X_{C_E} \leq 0.1 R_E$) down to $f = 20 \ \text{Hz}$?

29. Calculate the ac input resistance of the circuit in Fig. 8.37.

30. What is the ac output resistance for the circuit of Fig. 8.37?

31. Calculate the ac voltage gain of the circuit of Fig. 8.37.

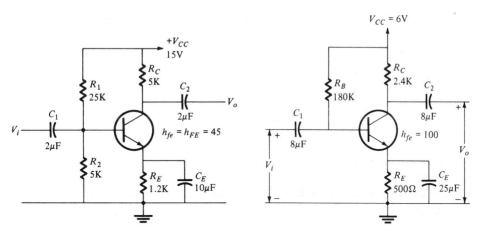

Figure 8.37   Figure 8.38

32. Calculate the bias currents I_{B_Q} and I_{C_Q} for the circuit of Fig. 8.38.

33. Calculate the bias voltage V_{CE_Q} for the circuit of Fig. 8.38.

34. What is the ac output voltage for an input of 6 mV, rms for the circuit of Fig. 8.38?

35. Calculate the input and output impedances (for $X_{C_E} \ll R_E$) for the circuit of Fig. 8.38.

§ 8.5

36. Calculate the bias point V_{DS_Q} and I_{D_Q} for the JFET amplifier circuit of Fig. 8.39 using the universal JFET curves of Fig. 8.26.

37. Calculate the value of g_m at this bias point (set in Problem 36) using the universal JFET curves of Fig. 8.26.

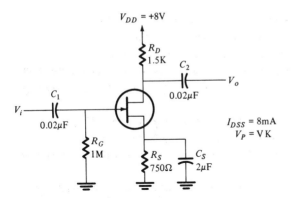

$V_{DD} = +8V$

R_D
1.5K

C_2
0.02μF

V_o

C_1
0.02μF

V_i

$I_{DSS} = 8mA$
$V_P = V K$

R_G
1M

R_S
750Ω

C_S
2μF

Figure 8.39

38. What values of R_S and R_D would result in a bias condition of $I_{D_Q} = 4$ mA in the circuit of Fig. 8.39?

39. Calculate the voltage gain of the circuit in Fig. 8.39 if R_D is kept at 1.8 K while R_S is adjusted so that g_m is 2200 μmho at this bias point.

40. For a bias drain current of 4 mA, calculate the bias voltage V_{DS_Q} for the MOSFET circuit of Fig. 8.40.

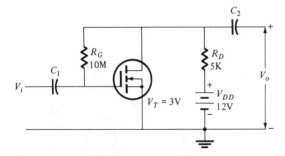

C_2

R_G
10M

R_D
5K

C_1

V_i

V_o

$V_T = 3V$

V_{DD}
12V

Figure 8.40

41. At the bias point set in Problem 40 for the circuit of Fig. 8.40, calculate the value of the MOSFET g_m.

42. If the value of g_m for the MOSFET in Fig. 8.40 is 2400 μmho, calculate the circuit voltage gain.

9

Multistage and Large-Signal Amplifiers

9.1 MULTISTAGE AMPLIFIERS

It should be apparent that a single amplifier stage is not sufficient for most purposes. When a number of stages are connected together consideration must be paid to the method of coupling these stages (e.g., capacitor coupling, transformer coupling, direct coupling), to the loading effect of each stage, to the frequency range of the signal being amplified, and to the amount of signal distortion resulting.

RC Coupling

The most common type of coupling for nonintegrated circuit amplifiers is capacitor coupling (also called *RC* coupling), as previously indicated in single-stage circuits. Figure 9.1 shows a two-stage FET amplifier. A few considerations must be made in this two (or more) stage connection which are not fully apparent when analyzing one stage at a time. First, the loading of the source by the first stage, of the first stage by the second stage, and of the second stage by the load must all be considered. In the present FET amplifier, however, these loadings are quite minimal and can be neglected, as will be shown. The voltage gain of the two amplifier stages is the *product* of the gains of each stage, as will also be shown.

For a practical signal source, such as that from a phono-cartridge, a signal is obtained of typically, say, 10 mV (in the frequency range 100–16,000 Hz—most of the audio range) from a source having an impedance of 50 K, typically. This signal is obtained from the cartridge when not connected to any circuit (or load). In the present circuit the input impedance of the first FET stage would act as a voltage divider of the unloaded signal (V_s). Figure 9.2(a) shows the resulting input voltage V_{i1} reduced from the unloaded source value V_s by the voltage divider of R_s and the first FET stage input impedance—in this case approximately $R_{G1} = 1$ M. The result-

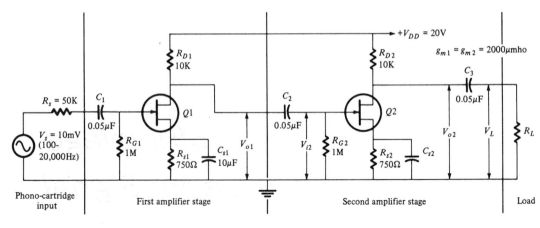

Figure 9.1 Two stage RC-coupled FET phono amplifier.

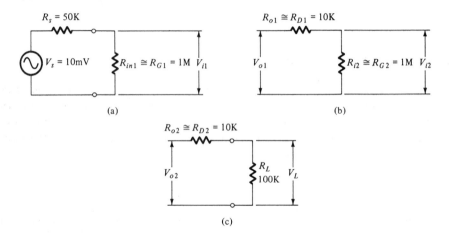

Figure 9.2 Loading effects in multistage amplifier (a) input stage (b) second stage (c) load.

ing input signal driving the first stage is thus

$$V_{i1} = \frac{R_{G1}}{R_{G1} + R_s} V_s = \frac{1\,M}{1\,M\Omega + 50\,K}(10\,mV) = 0.95(10\,mV) = 9.5\,mV$$

It seems apparent that little loading reduction of the input occurs for this FET amplifier stage, which may not be true in BJT circuits, as we shall see later.

The voltage gain of stage 1 is approximately

$$A_{v1} = -g_m R_{D1} = -(2000 \times 10^{-6})(10 \times 10^3) = -20$$

(neglecting the output impedance of FET $Q1$ and also the 1-M input impedance of stage 2), so that

$$V_{o1} = V_{i2} = A_{v1}V_{i1} = -20(9.5\,mV) = -190\,mV$$

Using identical stages for this example we can calculate the output of stage 2 to be

$$V_{o2} = A_{v2}V_{i2} = -(g_m R_{D2})V_{i2} = -(2000 \times 10^{-6})(10 \times 10^3)(-190 \text{ mV})$$
$$= 3800 \text{ mV} = 3.8 \text{ V}$$

The overall gain of the two stages is

$$A_v = \frac{V_{o2}}{V_{i1}} = \frac{3.8 \text{ V}}{9.5 \text{ mV}} = 400$$

which is the same as the product of the stage gains:

$$A_v = A_{v1}A_{v2} = 20 \times 20 = 400$$

The overall circuit gain, however, should include any loading effects from unloaded source voltage, V_s, to the loaded output voltage, V_L. The loading of stage 1 is seen in Fig. 9.2(b) to reduce V_{o1} to

$$V_{o1} = V_{i2} = \frac{R_{i2}}{R_{o1} + R_{i2}}V_{o1} = \frac{1 \text{ M}}{1 \text{ M} + 10 \text{ K}}(-190 \text{ mV}) = -188 \text{ mV}$$

Including this loading effect gives an output of

$$V_{o2} = A_{v2}V_{i2} = -20(-188 \text{ mV}) = 3.76 \text{ V}$$

If this output is connected to another amplifier stage or other circuit, then the effect of this load would be to further reduce the output voltage. With a load $R_L = 100 \text{ K}$ the load voltage delivered by the two-stage amplifier would be

$$V_L = \frac{R_L}{R_L + R_{o2}}V_{o2} = \frac{100 \text{ K}}{100 \text{ K} + 10\text{K}}(3.76 \text{ V}) = 3.42 \text{ V}$$

We see here that starting with two separate amplifier stages each having gains of 20 the resulting overall circuit gain is less than the "ideal" value of 400 due to the loading between stages and the loading of the source and load so that in this case an overall gain of only 342 is achieved. It should be noted that the effects of loading are real considerations that must be taken into account when connecting amplifier stages together or when connecting source or load circuits to an amplifier. Suitable results still are obtained, and high gains are easily achieved, if required, by adding additional amplifier stages. In fact, with *IC* circuits, multistage amplifier units are readily available for a great variety of uses.

Two BJT *RC*-coupled amplifier stages are shown in Fig. 9.3. A signal V_s from a source having a resistance R_s is capacitor-coupled as the ac input of stage 1 (V_{i1}). An amplified output V_{o2} results at the collector of transistor $Q1$, this signal being coupled through capacitor C_2 as input (V_{i2}) to stage 2. The signal is again amplified and coupled to any external load (V_{o2}) through capacitor C_3. Capacitors C_{E1} and C_{E2} act to bypass bias resistors R_{E1} and R_{E2}, respectively, so that they act as dc bias elements but are shorted out (bypassed) for ac signal operation. Resistor pairs R_1, R_2 and R_3, R_4 act as dc voltage-divider bias components.

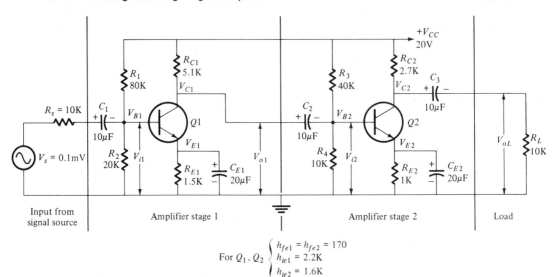

For Q_1, Q_2 $\begin{cases} h_{fe1} = h_{fe2} = 170 \\ h_{ie1} = 2.2\text{K} \\ h_{ie2} = 1.6\text{K} \end{cases}$

Figure 9.3 Two BJT RC-coupled amplifier stages.

dc Bias. The dc base voltage is approximately set by the voltage-divider action of R_1 and R_2 (and R_3, R_4) so that

$$V_{B1} = \frac{R_2}{R_1 + R_2} V_{CC} = \frac{20\text{ K}}{80\text{ K} + 20\text{ K}} (20\text{ V}) = 4\text{ V}$$

$$V_{B2} = \frac{R_4}{R_1 + R_2} V_{CC} = \frac{10\text{ K}}{40\text{ K} + 10\text{ K}} (20\text{ V}) = 4\text{ V}$$

For each transistor biased on with $V_{BE} = 0.5$ V the respective emitter voltages are

$$V_{E1} = V_{E2} = 4 - 0.5\text{ V} = 3.5\text{ V}$$

The emitter and collector currents are then calculated to be

$$I_{E1} = \frac{V_{E1}}{R_{E1}} = \frac{3.5\text{ V}}{1.5\text{ K}} = 2.3\text{ mA} \simeq I_{C1}$$

$$I_{E2} = \frac{V_{E2}}{R_{E2}} = \frac{3.5\text{ V}}{1\text{ K}} = 3.5\text{ mA} \simeq I_{C2}$$

The voltages measured from the collector to ground and also collector to emitter (transistor dc bias voltage) are then calculated to be

$$V_{C1} = V_{CC} - I_{C1}R_{C1} = 20 - (2.3\text{ mA})(4.7\text{ K}) \simeq 9.2\text{ V}$$

$$V_{CE1} = V_{C1} - V_{E1} = 9.2 - 3.5\text{ V} = 5.7\text{ V}$$

$$V_{C2} = V_{CC} - I_{C2}R_{C2} = 20 - (3.5\text{ mA})(2.7\text{ K}) = 10.55\text{ V}$$

$$V_{CE2} = V_{C2} - V_{E2} = 10.55 - 3.5\text{ V} = 7.05\text{ V}$$

Thus $Q1$ is biased to operate at $I_{C1} = 2.3$ mA, $V_{CE1} = 5.7$ V, while $Q2$ is biased to operate at $I_{C2} = 3.5$ mA, $V_{CE2} = 7.05$ V.

Notice that the voltage-divider bias did not require any transistor parameters (with the understood assumption that the value of $h_{FE}R_E$ is larger than R_2, typically by a factor of 10).

ac Operation. The ac source signal of $V_s = 2$ mV derived from a source impedance of 10 K is ac-coupled to amplifier stage 1 through capacitor C_1. The ac impedances of coupling capacitors $C_1 = C_2 = C_3 = 10$ μF are (at the lowest frequency of operation, say, 100 Hz for the present)

$$X_C = \frac{1}{2\pi f C} = \frac{1}{6.28(100)(10 \times 10^{-6})} \simeq 160 \; \Omega$$

which is sufficiently less than the kilohms of impedance of R_s and R_1, R_2 (or R_3, R_4). At higher frequencies the capacitor impedance becomes even smaller and continues to be neglected in ac calculations.

The input impedance seen looking into amplifier stage 1 [Fig. 9.4(a)] is (including the circuit impedance from base to ground)

$$R_{i1} = R_1 \| R_2 \| h_{ie1} = 80 \text{ K} \| 20 \text{ K} \| 2.2 \text{ K} = 1.9 \text{ K}$$

so that the input ac voltage to stage 1 is

$$V_{i1} = \frac{R_{i1}}{R_{i1} + R_s} V_s = \frac{1.9 \text{ K}}{1.9 \text{ K} + 1 \text{ K}} (0.1 \text{ mV}) \simeq 0.0655 \text{ mV}$$

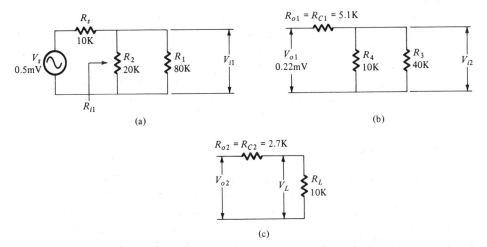

(a)

(b)

(c)

Figure 9.4 Calculating loading effects between amplifier stages.

The input impedance looking into stage 2 is similarly calculated to be

$$R_{i2} = R_3 \| R_4 \| h_{ie2} = 40 \text{ K} \| 10 \text{ K} \| 1.6 \text{ K} = 1.3 \text{ K}$$

The ac gain of stage 1 can be calculated as

$$A_{v1} \simeq \frac{-h_{fe1}R_{C1}}{h_{ie1}} = \frac{-170(5.1 \text{ K})}{2.2 \text{ K}} \simeq -394$$

so that

$$V_{o1} = A_{v1}V_{i1} = (-394)(0.0655 \text{ mV}) \simeq -25.8 \text{ mV}$$

The ac input to stage 2 is then reduced by loading [see Fig. 9.4(b)] to

$$V_{i2} = \frac{R_{i2}}{R_{i2} + R_{o1}} V_{o1} = \frac{1.3 \text{ K}}{1.3 \text{ K} + 5.1 \text{ K}}(-25.8 \text{ mV}) \simeq -5.2 \text{ mV}$$

The ac gain of stage 2 is

$$A_{v2} = \frac{-h_{fe2}R_2}{h_{ie2}} = \frac{-170(2.7 \text{ K})}{1.6 \text{ K}} \simeq -286.9$$

and

$$V_{o2} = A_{v2}V_{i2} = -286.9(-5.2 \text{ mV}) = 1.5 \text{ V}$$

The resulting ac voltage across a load of $R_L = 10$ K would then be [see Fig. 9.4(c)]

$$V_L = \frac{R_L}{R_{o2} + R_L} V_{o2} = \frac{10}{2.7 + 10}(1.5 \text{ V}) = 1.2 \text{ V}$$

The overall *circuit* gain (considering loading) is thus

$$A_v = \frac{V_L}{V_s} = \frac{1.2 \text{ V}}{0.1 \text{ mV}} = 12,000$$

The *RC* coupling just considered is probably the most common type used in discrete component circuits where dc signal amplification is not required. Other common types of coupling circuits are dc or coupling (as in differential and operational amplifiers and many IC circuits) and transformer coupling (for tuned AM or FM frequency operation and some large-signal amplifier circuits). The differential amplifier circuit covered next is one of the most common type of dc-coupled circuits. It provides far more than merely a means of directly coupling signals between multiple stages. It also provides dual input operation with the ability to amplify the desired input signal(s) while rejecting in-phase noise or unwanted signals.

Differential amplifier

An emitter-coupled differential amplifier is shown in Fig. 9.5 using BJTs. The two-transistor circuit comprises, essentially, one amplifier stage. Direct coupling of outputs V_{o1} and V_{o2} as input to another differential amplifier is usually made, and the frequency range of such a circuit is from dc signals up to some fairly high frequency (typically in the 100-MHz range). This circuit arrangement not only affords a very wide frequency operating range including dc operation but also provides a flexible use of input and output using opposite phase signals and most important of all a means of improving the signal to noise ratio by rejecting the in-phase (noise) components while amplifying the out-of-phase (signal) components. By amplifying one while reducing the other the circuit achieves a common-mode rejection ratio of considerable value—typically 10^5 (as will be demonstrated).

Figure 9.6 shows a single input (V_{i1}) and the resulting outputs at V_{o1} and V_{o2}. Referring to Fig. 9.6(a), the input to the base of Q1 (V_{i1}) is amplified and inverted, resulting in the signal shown from output V_{o1}. In addition the input signal appears,

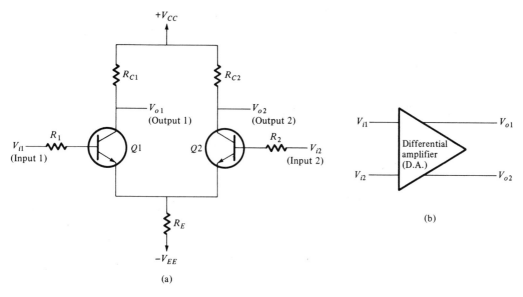

Figure 9.5 Differential amplifier circuit. (a) circuit (b) block symbol.

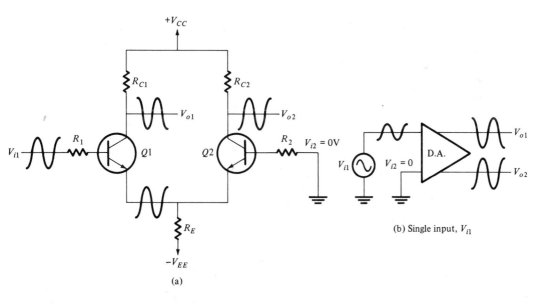

Figure 9.6 Operation of differential amplifier (a) circuit; (b) block symbol.

as shown, by emitter-follower action at the common-emitter terminal and is, of course, in phase with the signal (V_{i1}) applied to the base of $Q1$. Since the $Q2$ base voltage is 0 V (ground input at V_{i2}), the voltage developed from the emitter of $Q2$ to ground (or base of $Q2$) acts essentially the same as would an *opposite phase* input measured from the base of $Q2$ to the emitter of $Q2$. Thus, the input signal to $Q1$ not only results in an opposite phase output signal at the collector of $Q1$ but also operates transistor $Q2$ producing the signal from the collector (V_{o2}) shown in Fig. 9.6(a), which is *in phase* with input V_{i1}.

If a single-ended input is applied as V_{i2} [Fig. 9.7(a)] opposite phase, amplified outputs are also obtained. A very important application of differential amplifiers is the use of a differential (or floating) input signal [Fig. 9.7(b)] in which neither input terminal (nor output terminal) is the reference, or ground, terminal, and the resulting output can be obtained from either input to reference (ground) or as a floating (non-referenced) signal between output terminals.

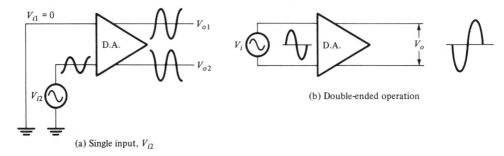

(a) Single input, V_{i2}
(b) Double-ended operation

Figure 9.7 Single- and double-ended circuit operation.

Common-Mode Rejection

A most important feature of differential amplifiers is their common-mode rejection: the ability to amplify out-of-phase signals at the inputs while greatly attenuating or rejecting in-phase inputs. Figure 9.8 shows the action of applying the same (in-phase) signal to both inputs and a difference signal between inputs. Referring back to Fig. 9.6(a), it should be apparent that the result of applying the same input signal to transistor stages $Q1$ and $Q2$ at each output is amplified in-phase and opposite phase signals which cancel out only if the gains of each stage are identical. In a practical circuit the gains of both sides of the difference amplifier are close enough so that only a small output signal results for in-phase inputs. For opposite phase

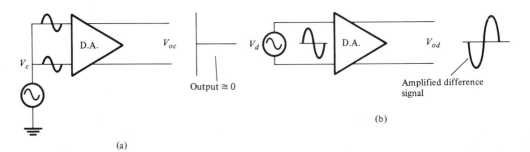

V_{oc}
Output $\cong 0$
V_{od}
Amplified difference signal

(a)
(b)

Figure 9-8 Common-mode rejection action of difference amplifier.

inputs the amplified signals due to both inputs add at each output to result in a large amplified output signal.

The common-mode rejection of a differential amplifier circuit is defined as the ratio of the differential gain to common gain, which is typically a very large number:

$$\text{CMRR (common-mode rejection ratio)} \equiv \frac{A_d}{A_C}$$

The common-mode rejection ratio is generally given in decibels.* For example, a value of 80 dB would be equivalent to

$$\text{CMRR} = 20 \log \frac{A_d}{A_c} = 80 \text{ dB}$$

$$\frac{A_d}{A_c} = \text{antilog } 4 = 10^4 = 10,000$$

If the amplifier differential gain is $A_d = 10$, then the common-mode gain is 10^{-3} ($A_c = A_d/\text{CMRR} = 10/10,000 = 1/1000$), or an attenuation of 1000.

A better version of differential amplifier uses a constant current source instead of a large resistor, R_E [Fig. 9.9]. The resulting value of CMRR for this circuit version is quite good, being typically 90–100 dB. The action of constant current stage $Q3$ is to provide a flexible adjustment of dc bias currents while acting as an open circuit (or high resistance) to an ac signal. The Zener diode $D1$ and resistor R_2 set the fixed value of emitter and collector current, which is held constant so that the ac signals effectively "see" an open circuit or high impedance looking into stage $Q3$.

A practical example of a differential amplifier is the CA3000 dc amplifier of Fig. 9.10. As shown in the block symbol of Fig. 9.10(a), the differential amplifier unit has two inputs with input 1 regarded as the inverting input (with reference to the output taken differentially from output 1 to output 2 or single-ended from output 1 to ground), while input 2 is noninverting. Two different voltages are generally used to provide $+V$ and $-V$ supply voltages. The numbers in circles are pin numbers for connecting the particular 10-pin IC socket. These circuits are usually packaged in a number of standard cases (e.g., 8- or 10-pin TO-5 case, 12-, 14-, or 16-pin DIP in-line).

The circuit details are provided in Fig. 9.10(b), which is manufactured as a single IC unit. There are a number of additional features of this circuit over the basic unit of Fig. 9.9. Stages $Q1$ and $Q4$ act as emitter-followers so that the input impedance of each input is quite large (195 K typically) compared with that of the circuit of Fig. 9.9 (typically 2 K). Stages $Q2$ and $Q3$ are then the differential amplifier stages with the addition of resistors R_4 and R_5 to aid in providing the high input impedance. Stage $Q5$ is the constant current source with a few internal points brought out for

---

*Decibel (dB) units are logarithmic and dB gain is defined as

$$A_{dB} = 20 \log_{10} \frac{V_o}{V_i}$$

For a voltage gain of $V_o/V_i = 1000$, the identical gain in dB units is

$$A_{dB} = 20 \log_{10} 1000 = 20(3) = 60 \text{ dB}$$

(Power of 10 numbers have simple logs; i.e., log 10 = 1, log 100 = 2, log 1000 = 3, log 10^6 = 6. Other values will be covered more fully in Section 9.2.)

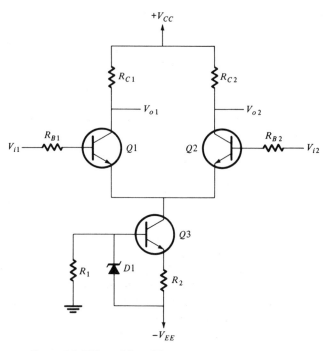

Figure 9.9 Differential amplifier circuit with constant current source.

flexible setting of the circuit operation. Various connections of terminals 4 and 5 to the minus supply or unconnected result in output dc quiescent voltages (at terminals 8 and 10) from -1.5 to $+4.2$ V. The circuit has two input terminals for out-of-phase or differential inputs and provides only a single output terminal. This circuit will provide an output V_o which is in phase with an input at V_{i2} and out of phase with an input at V_{i1}. The output is also out of phase for a differential input applied from terminal 1 to 2. The block symbol shown in Fig. 9.11(b) indicates that output is in phase with an input to terminal 2 by the plus $(+)$ sign and out of phase with an input at terminal 1 by the minus $(-)$ sign on the block symbol.

In the circuit of Fig. 9.11(a) transistor $Q3$ is used to provide a balanced circuit operation. The partial circuit of transistor $Q5$ provides the constant current source for dc biasing and good common-mode rejection. The output taken from transistor $Q4$ will be a linear signal in the single-ended voltage range from 0 V to $+V_{CC}$, whereas the full amplifier operates between $+V_{CC}$ and $-V_{EE}$ (usually the same magnitude).

Other forms of dc or direct coupling are available, although the differential amplifier is the most versatile and generally useful circuit. One example of a dc-coupled amplifier which is not in a differential amplifier form is the MOS transistor circuit shown in Fig. 9.12. Three stages are shown connected together. The first stage has previously been discussed with dc bias set so that the gate-source bias voltage equals the drain-source bias voltage and will properly operate as long as $V_{DD} > V_T$ for the FET used. The value of V_{DS_Q} for transistor $Q1$ also sets the same gate bias

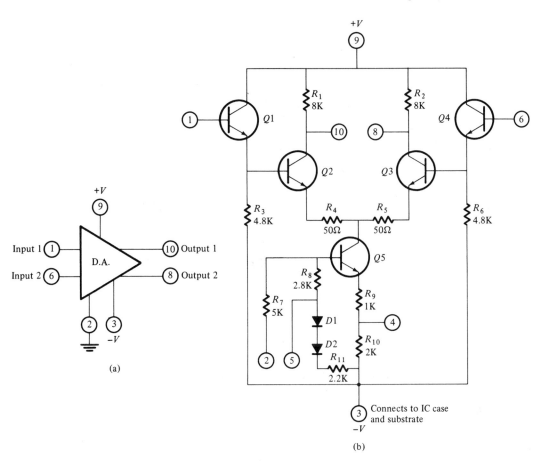

Figure 9.10 CA 3000 differential amplifier (a) symbol (b) schematic.

voltage for stage 2. The resulting dc bias voltage at the drain of $Q2$ then sets the gate bias voltage of $Q3$. Input and output are shown capacitively coupled here but could be direct-coupled if the input or output dc levels would not affect or be affected by the amplifier circuit.

A direct-coupled circuit using BJTs is shown in Fig. 9.13. In this circuit arrangement extra diodes are included to compensate for the voltage drop across forward-biased base-emitter junctions.

Transformer-Coupled Transistor Amplifier

Another method of coupling the ac signal between stages is transformer coupling. This type of circuit connection has been most widely used in tuned circuits as used in AM/FM radios and also in power amplifier stages to drive a load such as a speaker. Figure 9.14 shows a two-stage transformer-coupled BJT transistor amplifier as

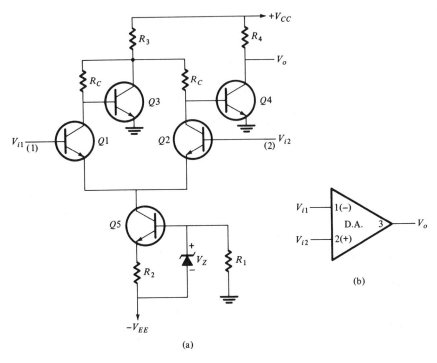

Figure 9.11 IC differential amplifier with differential inputs and single-ended output.

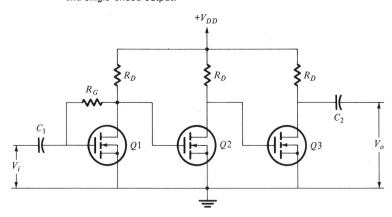

Figure 9.12 A multistage enhancement MOST microamplifier.

might be used as a low-frequency audio power amplifier. An ac input signal is coupled through transformer $T1$, causing the base current of transistor $Q1$ to vary around its bias setting. The input signal V_1 is stepped up by transformer $T1$, amplified by stage $Q1$, and then coupled to stage $Q2$ by step-down transformer $T2$. It would seem that stepping up the signal would be desired to increase the voltage. However, the trans-

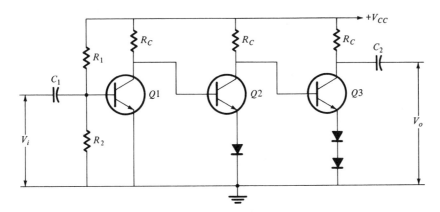

Figure 9.13 BJT direct coupled amplifier.

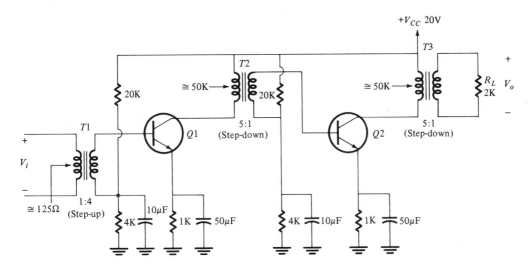

Figure 9.14 Two-stage transformer coupled BJT transistor amplifier.

former not only transforms or steps the voltage up or down; it also affects the current and thereby the ac impedance seen looking into the transformer. In this case it was appropriate to use a step-down transformer so that the impedance looking into stage $Q2$ was stepped up looking into transformer $T2$. Let's first consider how this impedance transformation is accomplished.

Transformer impedance transformation. Figure 9.15 shows a basic transformer configuration. An applied voltage V_1 results in a voltage V_2 dependent on the transformer turns ratio:

$$\frac{V_1}{V_2} = \frac{N_1}{N_2} = a \qquad \text{(transformation ratio)} \qquad (9.1)$$

In addition, the resulting current I_2 through load Z_L causes an input current which is

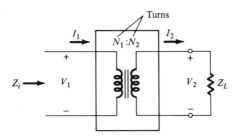

Figure 9.15 Basic transformer configuration.

also related by the transformer turns ratio:

$$\frac{I_1}{I_2} = \frac{N_2}{N_1} = \frac{1}{a} \tag{9.2}$$

Determining the input impedance looking into the transformer [Eqs. (9.1) and (9.2)],

$$Z_i = \frac{V_1}{I_1} = \frac{aV_2}{(1/a)I_2} = a^2\frac{V_2}{I_2} = a^2 Z_L \tag{9.3}$$

The transformer is seen to provide an input impedance which is the factor a^2 times the impedance connected across the secondary winding. Since a is the ratio of primary to secondary turns, a step-down transformer will act to provide an input impedance which is greater by the factor a^2 than the impedance across the transformer secondary.

In the circuit of Fig. 9.14 the step-down transformer reduces the signal across its primary by the transformer turns ratio, but this resulting signal across the transformer primary may be larger due to the higher ac load impedance so that the overall effect is a voltage gain between the input to stage $Q1$ and stage $Q2$.

For similar reasons the choice of a step-down transformer results in a higher ac impedance seen as load by stage 2. Input transformer $T1$ is the only one here used to step up the signal, and this is proper only if the lowered input impedance looking into the amplifier is not too low to load the source (which provides V_i).

To see how the dc and ac circuit operation can be separately considered, Fig. 9.15 shows part of stage $Q1$ with its dc and ac equivalent circuit. For dc considerations the capacitors act as open circuits and the transformer as a short circuit (or, typically, small dc resistance). From Fig. 9.16 we can see that the dc bias circuit is the popular beta-independent circuit connection. In the ac equivalent circuit the voltage supply and capacitors are replaced by short circuits, leaving only the transformer signal directly driving the base-emitter of transistor $Q1$. Extending this ac equivalent simplification to the complete circuit results in the ac equivalent circuit shown in Fig. 9.17.

A number of circuit connections are used in addition to the standard coupling techniques discussed above. Many of these are special component connections to take advantage of particular devices or circuit features. An example of a popular multistage (or multidevice) connection is the Darlington circuit of Fig. 9.18. This circuit is used to obtain very high input impedance and low output impedance when using BJTs. Stage $Q2$ operates as an emitter-follower circuit having an ac input impedance, looking into the base, of

$$Z_{i2} \simeq h_{fe2}R_{E2} = 100\,(1\text{ K}) = 100\text{ K}$$

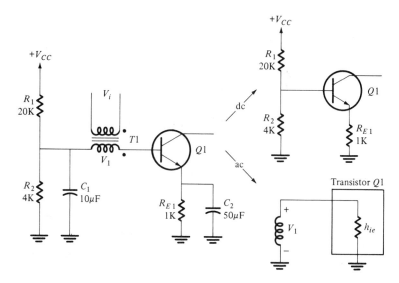

Figure 9.16 Partial circuits of Fig. 9.11 (a) First stage input.

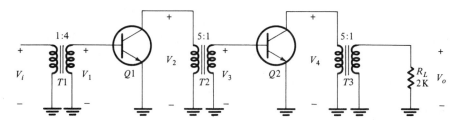

Figure 9.17 Small-signal ac equivalent of transformer-coupled amplifier.

Stage $Q1$ also acts as an emitter-follower with input impedance

$$Z_{i1} \simeq h_{fe1} R_{E1} = h_{fe1} Z_{i2} = 100\,(100\text{ K}) = 10\text{ M}$$

The ac impedance seen looking into the complete Darlington circuit* shown is then the base bias resistor R_B in parallel with the ac impedance Z_{i1}, resulting in

$$Z_i = \frac{R_B Z_{i1}}{R_B + Z_{i1}} = \frac{2 \times 10}{2 + 10} = 1.67\text{ M}$$

The value of R_B is limited by the dc base current required to bias the transistors. A resulting input impedance of nearly 2 MΩ is quite good for most applications. Note that a single FET circuit could easily provide an input impedance greater than the two-stage BJT Darlington circuit and is used when possible in present circuit design.

*For the complete Darlington circuit the ac input impedance, excluding R_B, is

$$Z_i = h_{fe1} h_{fe2} R_E = h_{fe}^2 R_E \qquad (\text{if } h_{fe1} = h_{fe2})$$

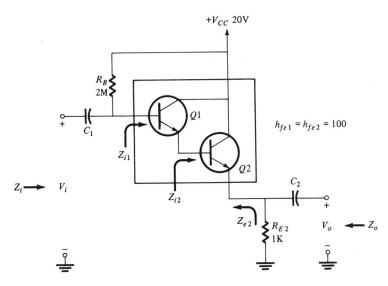

Figure 9.18 Darlington configuration.

The ac output impedance of the circuit Z_o is not the value of R_{E2} but that value in parallel with the ac impedance seen looking into the emitter of $Q2$. This impedance is quite low, usually much lower than R_{E2}. For example, if the circuit is driven by a source having 10-K resistance, the ac impedance looking into the emitter of $Q2$ is

$$Z_{e2} \simeq \frac{R_s}{h_{fe1}h_{fe2}} = \frac{10 \text{ K}}{100(100)} = 1\Omega$$

so that

$$Z_o \simeq Z_{e2} = 1 \text{ } \Omega$$

Thus, the circuit acts as an excellent impedance matcher—having very high input impedance so as not to load down the source and very low output impedance to act as a nearly perfect source for any following circuit. As with a standard emitter-follower circuit the Darlington circuit has a voltage gain of nearly unity with no phase inversion.

9.2 FREQUENCY CONSIDERATIONS

The frequency of the signal applied to an amplifier circuit can have considerable effect on the resulting gain of the circuit. Up to now the signal frequency has not been included in the circuit operation or the frequency has had no detrimental effect—the condition called midfrequency. The gain of a single amplifier stage will decrease at high frequency—the value at which the gain decrease being different for different transistors and circuits. When coupling stages by capacitor or transformer elements the overall circuit gain is greatly affected at both low and high frequencies. An example of how gain varies with frequency for a few types of coupling methods is shown in Fig. 9.19. Note that the horizontal axis is a logarithmic scale to permit a

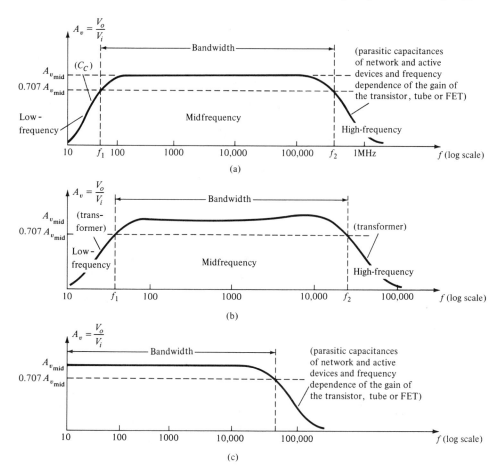

Figure 9.19 Gain vs. frequency for: (a) RC-coupled amplifiers; (b) transformer-coupled amplifiers; (c) direct-coupled amplifiers.

plot extending from the low- to the high-frequency regions. For each plot, a low-, high-, and mid-frequency region has been defined. In addition, the primary reasons for the decrease in gain at low and high frequencies have also been indicated within the parentheses. For the *RC*-coupled amplifier the decrease at low frequencies is due to the increasing reactance of the coupling capacitor, while its upper frequency limit is determined by either parasitic capacitive elements of the network and the active device (transistor) or the frequency dependence of the gain of the active device. Notice that while a dc-coupled amplifier such as a differential amplifier does not have decreased gain at low frequencies, it does show gain reduction at high frequencies due to stray wiring capacitance and the frequency dependence of the transistor devices themselves.

The approximately constant gain for most of the indicated frequency range is referred to as *midfrequency*. Accepted practice is to regard the condition at which the gain is 0.707 of the midfrequency gain (A_{mid}) as the point to describe either low-

or high-frequency cutoff. In decibel units the gain decreases 3 dB when it is reduced to $0.707A_{mid}$ and the upper and lower 3-dB frequencies are those at which the gain has dropped 3 dB (or to 0.707 of A_{mid}). The lower 3-dB frequency (f_1) for RC-coupled amplifier [Fig. 9.19(a)] is determined mainly by coupling capacitor C_C. In Fig. 9.19(a) the gain drops from its midfrequency value to $0.707A_{mid}$ (or by 3 dB) around 70 Hz. The upper 3-dB frequency (f_2) is shown to be about 500 kHz in Fig. 9.19(a) and is dependent mainly on stray capacitances and device frequency limitations. The bandwidth (BW) is then defined as the difference between the lower and upper frequency:

$$\text{BW} = f_2 - f_1 \qquad (9.4)$$

(or just f_2 for a dc amplifier).

General Low-Frequency Response
(*RC* Coupling)

For the common type of capacitor coupling of amplifier stages used with FET, BJT, or tube amplifiers the lower cutoff frequency $(f_1$ or $f_2)$ is dependent mainly on the value of the coupling capacitor (C) and the input resistance of the amplifier stage (R_i). In the frequency range called midfrequency the nonresistive components have slight effect so that the circuit gain remains essentially constant. A signal V_s is then amplified to V_o as shown in Fig. 9.20(a), with

$$A_{v\text{mid}} = A_{mid} = \frac{V_o}{V_s}$$

To consider the effect of the capacitor coupling a signal V_s into an amplifier stage, refer to Fig. 9.20(b), which shows the stage input impedance as an effective resistance R_i. The overall gain is reduced by the voltage division of impedances X_C and R_i, which is a function of the signal frequency. (At midfrequency the impedance X_C is so small compared to R_i that $V_i \approx V_s$ and the overall gain. To calculate the overall gain we first consider the voltage gain (or attenuation in this case) between V_s and V_i. This attenuation can be obtained using the voltage-divider rule, so that

$$\frac{V_i}{V_s} = \frac{R_i}{R_i - jX_C} = \frac{R_i}{\sqrt{R_i^2 + X_C^2}} \bigg/ -\tan^{-1}\frac{X_C}{R_i}$$

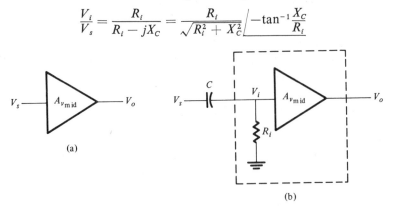

(a)

(b)

Figure 9.20 Amplifier for general low frequency response. (a) midfrequency; (b) low frequency.

The magnitude of the attenuation factor is thus

$$\text{Magnitude of } \frac{V_i}{V_s} = \frac{R_i}{\sqrt{R_i^2 + X_C^2}} = \frac{1}{\sqrt{1 + (X_C/R_i)^2}}$$

In particular, $V_i/V_s = 0.707^*$ when $X_C/R_i = 1$. This is a notable point to use since the gain has decreased by 3 dB† from its midfrequency value when it is $0.707A_{mid}$.

We see then that at the frequency where $X_C = R_i$ the overall gain has dropped by 3 dB from its midfrequency value (or to $0.707A_{mid}$). Since the 3-dB point is a common one to establish where the gain decrease begins to be considerable, we can identify the lower frequency of the overall amplifier circuit, or the lower 3-dB frequency, as

$$f_1 = f_L = \frac{1}{2\pi R_i C} \tag{9.5}$$

For example, a BJT amplifier having $R_i = 2$ K and a coupling capacitor $C = 0.5$ μF would have a lower cutoff frequency of

$$f_L = \frac{1}{2\pi R_i C} = \frac{1}{6.28(2 \times 10^3)(0.5 \times 10^{-6})} = 159 \simeq 160 \text{ Hz}$$

To reduce the lower cutoff frequency to, say, 20 Hz would require using a capacitor of

$$C = \frac{1}{2\pi f R_i} = \frac{1}{6.28(20)(2 \times 10^3)} = 3.98 \times 10^{-6} \simeq 4 \text{ } \mu\text{F}$$

General High-Frequency Response (*RC* Coupling)

A number of different capacitive effects result in the gain decreasing at frequencies above midfrequency. There are parasitic capacitances occurring between input, output terminals and ground and between input and output terminals. These capacitances are due to construction and connection of the circuit elements. In addition there is a stray wiring capacitance between amplifier stages due to connecting wires. There is also a decrease in active device gain as frequency increases, which will be considered later.

The effect of parasitic and stray wiring capacitance on the amplifier frequency response is to decrease the overall gain as the frequency increases. Figure 9.21(a) shows the parasitic capacitances of interest at high frequencies—these being a capacitance from input to ground, C_{ii}; a capacitance from output to ground, C_{oo}; and a capacitance between input and output terminals, C_{io}. There is also stray wiring capacitance, C_W—the total capacitive components resulting in an effective high-frequency capacitance, C_{high}. Figure 9.21(b) shows the output of one amplifier driving a second stage, with the midband output voltage source $A_{mid}V_i$ and output resistance R_{high} driving the effective capacitance, C_{high}. The voltage applied to the second amplifier stage is reduced from what it would be at midfrequency by the voltage-divider action

---

$^*1/\sqrt{1 + (1)^2} = 1/\sqrt{2} = 0.707$.
†$20 \log_{10}(V_i/V_s) = 20 \log(1/\sqrt{2}) = -20 \log \sqrt{2} = -20(0.151) = -3.01 \simeq -3$ dB.

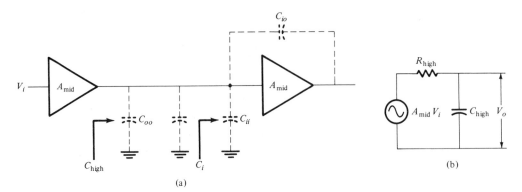

Figure 9.21 Amplifier circuit for general high frequency response. (a) distributed parasitic and stray capacitance; (b) effective high-frequency capacitance.

of capacitor C_{high} and resistor R_{high} as follows:

$$\frac{V_o}{A_{\text{mid}}V_i} = \frac{-jX_{C\text{high}}}{R_{\text{high}} - jX_{C\text{high}}} = \frac{X_{C\text{high}}}{\sqrt{X_{C\text{high}}^2 + R_{\text{high}}^2}} \bigg/ -90 - \tan^{-1}\frac{X_{C\text{high}}}{R_{\text{high}}}$$

In particular the magnitude of output to input voltage $V_o/V_i = 0.707A_{\text{mid}}$ when $R_{\text{high}} = X_{C\text{high}}$ at which frequency the gain is down by 3 dB from the midfrequency gain.

That is, the upper 3-dB cutoff frequency due to stray wiring and parasitic capacitance effects is

$$f_2 = f_H = \frac{1}{2\pi R_{\text{high}} C_{\text{high}}} \tag{9.6}$$

Determination of C_{high}. To calculate the effective value of capacitance which reduces the gain at higher frequencies we first shall consider how the parasitic capacitance between input and output, C_{io}, can be reflected to the input terminals. This reflection to the input is due to what is called the Miller effect, and the resulting capacitance value is often called the Miller capacitance.

Miller effect. Consider the simplified circuit of Fig. 9.22, which shows the parasitic capacitance from input to ground, C_{ii}, and that between input and output, C_{io}. The input voltage is V_i and the output, AV_i (where $A = A_{\text{mid}}$). These two capacitances can be equated to one equivalent capacitance C_i as follows:

$$I_i = I_{ii} + I_{io}$$
$$j\omega C_i V_i = j\omega C_{ii} V_i + (V_i - AV_i)j\omega C_{io}$$
$$C_i = C_{ii} + \underbrace{(1 - A_v)C_{io}}_{\text{Miller Capacitance}} \tag{9.7a}$$

Since $A_v = -g_m R_L$,

$$C_i = C_{ii} + (1 + g_m R_L)C_{io} \tag{9.7b}$$

Then, including stray wiring between stages and parasitic capacitance of output terminal to ground, we calculate a total high-frequency equivalent capacitance:

$$C_{\text{high}} = C_W + C_{oo} + C_i \tag{9.8}$$

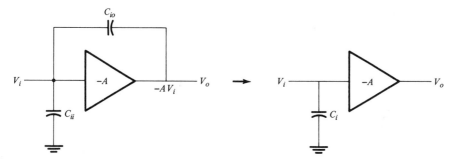

Figure 9.22 Circuit to calculate effective input capacitance, C_i.

[using C_i from Eq. (9.7a) or (9.7b)]. For example, in an *RC*-coupled amplifier having midfrequency gain of -55 and an output resistance of 5 K, the parasitic capacitances are determined to be $C_{ii} = 5\,\text{pF}$, $C_{io} = 3\,\text{pF}$, and $C_{oo} = 6\,\text{pF}$ with 15 pF due to stray wiring. Calculate the upper 3-dB frequency f_H.

Solution: Calculating C_i first,

$$C_i = C_{ii} + (1 - A_v)C_{io} = 5 + (1 + 55)3 = 173\ \text{pF}$$

The equivalent value of C_{high} is then

$$C_{high} = C_W + C_{oo} + C_i = 15 + 6 + 173 = 194\ \text{pF}$$

(note that the Miller capacitance value of 168 pF is most of the total equivalent value). Since $R_{high} = 5\ \text{K}$, the upper cutoff frequency is

$$f_H = \frac{1}{2\pi R_{high}C_{high}} = \frac{1}{6.28(5 \times 10^3)(194 \times 10^{-12})} = 164.2\ \text{kHz}$$

Asymptotic Gain-Frequency Plot

A simplified approach to plotting a gain vs. log of frequency curve uses straight-line asymptotes rather than exact curve values, as we shall see. The magnitude of voltage gain as a function of frequency can be expressed as

$$|A_v| = \frac{K_v}{\sqrt{1 + \left(\frac{f}{f_2}\right)^2}}$$

where f_2 is the upper cutoff frequency (upper 3-dB frequency). In decibel units

$$A_{dB}(f) = 20 \log |A_v(f)| = 20 \log \frac{K_v}{\sqrt{1 + \left(\frac{f}{f_2}\right)^2}}$$

$$= 20 \log K_v - 20 \log \sqrt{1 + \left(\frac{f}{f_2}\right)^2}$$

$$= K_{dB} - 20 \log \sqrt{1 + \left(\frac{f}{f_2}\right)^2} \qquad (9.9)$$

To plot this curve of gain vs. log of frequency consider the value of A_{dB} for very small and then very large frequency values. For frequencies much less than f_2 ($f \ll f_2$), Eq. (9.9) reduces to

$$A_{dB}(f) \simeq K_{dB} - 20 \log 1 = K_{dB} \tag{9.10a}$$

Thus at low frequencies the gain is constant, of value K_{dB}. At high frequencies ($f \gg f_2$) Eq. (9.9) can be expressed as

$$A_{dB}(f) = K_{dB} - 20 \log\left(\frac{f}{f_2}\right) \tag{9.10b}$$

Consider, for example, the gain decrease from K_{dB} when $f = 10f_2$ (a decade larger frequency than the cutoff value, f_2). The gain is then

$$A_{dB}(10f_2) = K_{dB} - 20 \log \frac{10f_2}{f_2} = K_{dB} - 20$$

or 20 dB down from the low-frequency gain value.* The curve of Eq. (9.9) is plotted in Fig. 9.23(a), showing the asymptotic straight-line gain of K dB at low frequencies up to a high-frequency break point and then decreasing at a rate of -20 dB/decade from the breakpoint (f_2) frequency.

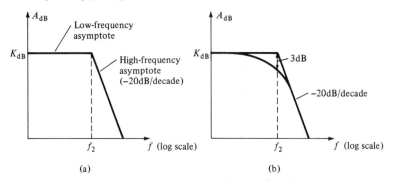

Figure 9.23 Gain-log frequency plots. (a) asymptotic; (b) actual.

These asymptotic straight lines are then easily drawn to approximate the actual gain-log frequency curve of Eq. (9.9). A more exact curve is then easily obtained from the asymptotic plot by noting that the actual gain is 3 dB less than the asymptotic value at frequency f_2,† as shown in the plot of Fig. 9.23(b).

While the gain of the amplifier decreases at frequencies above f_2, the output signal, originally 180° out of phase with the input, is shifted in phase from that value as given by the phase angle expression

$$\theta = -\tan^{-1}\left(\frac{f}{f_2}\right)$$

---

*The slope of the plotted curve for Eq. (9.9) on the log frequency scale is -20 dB/decade or -6 dB/octave, where decade is 10:1 frequency change and octave is 2:1 frequency change.

†At $f = f_2$ the gain is not K_{dB} as on the asymptotic plot of Fig. 9.23(a) but is reduced from K_{dB} by 3dB:

$$A_{dB} = K_{dB} - 20 \log \sqrt{1 + \left(\frac{f}{f_2}\right)^2} = K_{dB} - 20 \log \sqrt{2} \simeq K_{dB} - 3$$

At low frequencies $\theta \simeq 0$, and little phase shift (from the 180° midfrequency phase shift) occurs. At frequencies much larger than f_2 the phase shift approaches $-90°$, the phase shift being $-45°$ at f_2. A simple asymptotic plot of phase vs. log frequency is shown in Fig. 9.24 as a straight line of slope $-45°$/decade.

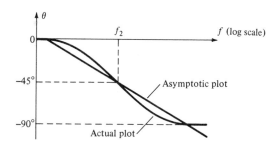

Figure 9.24 Phase shift versus log frequency.

Although not shown here, the *RC* amplifier response is 45° at the low cutoff frequency f_1, with the phase shift increasing to 90° near dc.

The gain-bandwidth relations of an *RC*-coupled amplifier can thus be specified as shown in Fig. 9.25 with the midfrequency gain constant between frequencies f_1 and f_2 and bandwidth equal to the difference $f_2 - f_1$. Recall that the gain decrease at low frequencies is essentially due to the coupling capacitor, while the gain reduction above f_2 is due to parasitic and stray wiring capacitances. The gain in the frequency band from f_1 to f_2 is the midfrequency gain set by the circuit resistive components and the active device parameters.

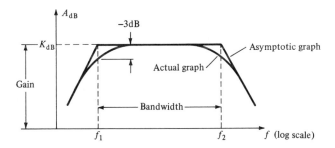

Figure 9.25 Gain-bandwidth relations for RC-coupled amplifiers.

It is interesting to note that the gain and bandwidth for a particular amplifier circuit cannot both be increased by changing resistive component values and that the gain-bandwidth product is essentially constant for the circuit. Since f_2 is usually much larger than f_1, we can approximately express the gain-bandwidth product as

$$\text{Gain-bandwidth} \simeq K_{\text{dB}} f_2 = (g_m R_{\text{high}}) \frac{1}{2\pi R_{\text{high}} C_{\text{high}}} = \frac{g_m}{2\pi C_{\text{high}}} = f_o$$

where $R_L = R_{\text{high}}$, and the resulting gain-bandwidth product is seen to be determined by the device ideal gain, g_m, and the circuit high-frequency capacitance, C_{high} (which is essentially fixed).

This can be demonstrated graphically by extending the gain-log frequency plot until the gain goes to unity (0 dB) as shown in Fig. 9.26. Since the gain decreases at the rate of 20 dB/decade until the frequency f_o (frequency at 0 dB), the asymptote is completely specified knowing f_o. If the gain is K_{dB} and upper cutoff frequency f_2 for a circuit high resistance of R_{high} changing the value of R_{high} to increase f_2, for example, will result in reducing the gain to K_{dB} as shown in Fig. 9.26. We see that an increase

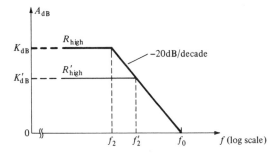

Figure 9.26 Gain-bandwidth plot showing upper frequency, f_o.

in bandwidth (f_2 greater) was accompanied by a decreased gain—the gain-bandwidth product remaining the same value. For all amplifiers having a high-frequency asymptote of -20 dB/decade the gain-bandwidth product can be shown to be the value of f_o, which is then an important figure of merit of the circuit. In practice the designer may trade gain for bandwidth (or vice versa) depending on the particular circuit requirements with the limit of trade being set by the value f_o.

Practical *RC*-Coupled Amplifier Circuits

Having covered the general action of *RC*-coupled circuits as a function of frequency it would now be helpful to consider specific *RC*-coupled amplifiers using FET, BJT, or vacuum-tube devices. The general concepts apply to all these circuits, but specific nomenclature will now be used for each type of device.

JFET Amplifier. A practical JFET amplifier is shown in Fig. 9.27. The low-frequency cutoff can be easily obtained using Eq. (9.4) since $R_i = R_g = 1$ M,

$$f_L = \frac{1}{2\pi R_i C} = \frac{1}{6.28(1 \times 10^6)(0.01 \times 10^{-6})} \simeq 16 \text{ Hz}$$

The low-frequency gain is also affected by the bypass action of capacitor C_S on bias resistor R_S, this action being omitted here as it is usually not the primary effect.

At midfrequency the gain is calculated as

$$A_{\text{mid}} = -g_m R_L = -(2500 \times 10^{-6})(5 \times 10^3) = -12.5$$

where $R_L = R_D = 5$ K.

To determine the upper frequency cutoff point we first calculate C_{high} using Eq. (9.7a) and then (9.8).

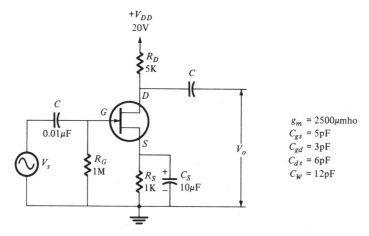

Figure 9.27 Practical JFET amplifier with capacitor coupling.

Using the JFET device nomenclature,

$$C_i = C_{gs} + (1 - A_{\text{mid}})C_{gd} = 5 + (1 + 12.5)3 = 45.5 \text{ pF}$$

and

$$C_{\text{high}} = C_i + C_{ds} + C_W = 45.5 + 6 + 12 = 63.5 \text{ pF}$$

Since the value of $R_{\text{high}} = R_D = 5$ K, we can calculate the high-frequency cutoff using Eq. (9.6):

$$f_H = \frac{1}{2\pi R_{\text{high}} C_{\text{high}}} = \frac{1}{6.28(5 \times 10^3)(6.35 \times 10^{-12})} \simeq 0.5 \text{ MHz}$$

In dB units the gain at midfrequency is

$$A_{\text{dB}} = 20 \log 12.5 = 22 \text{ dB}$$

and is less by 3 dB than that value at the lower 3-dB frequency of 16 Hz and also the upper 3-dB frequency of 0.5 MHz. The circuit bandwidth is approximately 0.5 MHz since $f_H \gg f_2$. The value of R_D could be changed to increase the bandwidth with resulting lower gain or the gain could be increased with resulting lower bandwidth as desired.

BJT Amplifier. A practical BJT amplifier circuit is shown in Fig. 9.28. The low-frequency cutoff point can be calculated using Eq. (9.4) after determining the value of R_i:

$$R_i \simeq \frac{R_1 R_2}{R_1 + R_2} = \frac{10(40)}{10 + 40} = 8\text{K}$$

($h_{fe} R_E = 80(1 \text{ K}) = 80 \text{ K} \gg R_i = 8$ K and can be ignored). The lower 3-dB cutoff frequency is then

$$f_L = \frac{1}{2\pi R_i C} = \frac{1}{6.28(8 \times 10^3)(2 \times 10^{-6})} \simeq 10 \text{ Hz}$$

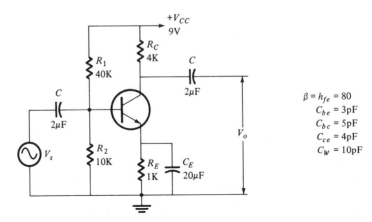

Figure 9.28 Practical BJT amplifier with capacitor coupling.

To calculate the midfrequency gain we first determine the value r_i for a dc emitter current of 1.3 mA (using dc analysis) to be

$$r_i \simeq 1.6 \text{ K}$$

and the midfrequency gain is then

$$A_{\text{mid}} = -\beta \frac{R_C}{r_i} = -80\left(\frac{4 \text{ K}}{1.6 \text{ K}}\right) = -200$$

The upper 3-dB cutoff frequency can now be calculated as follows:

$$C_i = C_{be} + (1 - A_{\text{mid}})C_{bc} = 3 + 201(5) = 1008 \text{ pF}$$
$$C_{\text{high}} = C_i + C_W + C_{ce} = 1008 + 10 + 4 = 1022 \text{ pF}$$

so that

$$f_H = \frac{1}{2\pi R_{\text{high}} C_{\text{high}}} = \frac{1}{6.28(4 \times 10^3)(1002 \times 10^{-12})} \simeq 39 \text{ kHz}$$

where $R_{\text{high}} \simeq R_C = 4 \text{ K}$.

The high-frequency cutoff is also limited by the variation of h_{fe} with frequency, but in this case (as in many) the circuit limits the cutoff frequency well below the device limitation. The relatively low bandwidth of about 39 kHz can of course be increased at the cost of reduced midfrequency gain from the value 200, which in dB units is

$$A_{\text{dB}} = 20 \log A_{\text{mid}} = 20 \log 200 = 46 \text{ dB}$$

Similar calculations can be made for a vacuum-tube amplifier circuit.

An increase in the number of cascaded stages can have a pronounced effect on the frequency response. For each additional stage the upper cutoff frequency is determined primarily by the stage having the lowest upper cutoff frequency. The low-frequency cutoff is primarily determined by that stage having the highest lower cutoff frequency. Obviously, therefore, one poorly designed stage can offset an otherwise well-designed cascaded system.

The effect of increasing the number of *identical* stages is demonstrated in Fig. 9.29. In each case the upper and lower cutoff frequencies of each of the cascaded stages is identical. For a single stage the cutoff frequencies are f_1 and f_2, as indicated. For two identical stages in cascade the drop-off occurs at a faster rate, and the bandwidth is considerably decreased.

It should be noted that a decrease in bandwidth is not always associated with an increase in the number of stages of the midband gain and can remain fixed, independent of the number of stages.

Two *RC*-coupled amplifiers are provided in Fig. 9.30 to show how both BJT and FET amplifiers are connected.

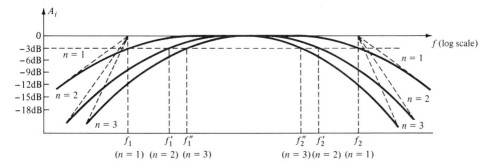

Figure 9.29 Effect of an increased number of stages on the cutoff frequencies and the bandwidth.

9.3 LARGE-SIGNAL AMPLIFIERS

After a signal has been sufficiently amplified by a number of small-signal amplifier stages to the level of a few volts it can be applied to a large-signal amplifier to drive an output device such as a speaker. A large-signal amplifier must operate efficiently and be capable of handling large amounts of power—typically, a few watts to hundreds of watts. Amplifier factors of greatest concern, then, are the power efficiency of the circuit, the maximum amount of power that the circuit is capable of handling, and the impedance matching of the circuit to the output device or transducer.

Amplifiers may be classified by their frequency range of operation, by type of application, or, in this case, by how they are biased to operate and thereby handle power efficiently. Up to now the small-signal amplifiers considered have all been class A. In class A operation the device (BJT or FET) is biased to operate so that the output signal is controlled by the input signal for a full input cycle (or 360° of operation). This is the least efficient operating condition since the device is biased so as to dissipate power (in a BJT circuit the collector current flows with a voltage across collector-emitter) even with no input ac signal applied. This condition is acceptable in small-signal amplifier circuits where the power dissipation may be only milliwatts but cannot be tolerated when many watts of power are to be handled by the circuit.

For power (large-signal) amplifiers class B operation is often used in a push-pull circuit configuration. A two-unit push-pull circuit operates in class B (in which

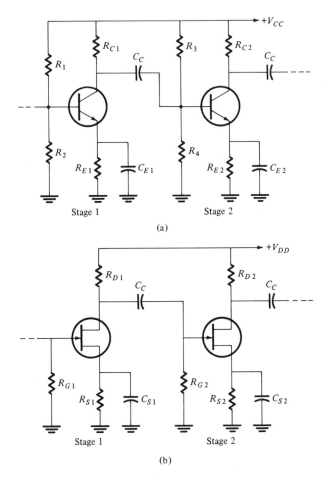

Figure 9.30 RC-coupled cascaded amplifier stages. (a) BJT; (b) FET

each device conducts for only one-half cycle—180° conduction) yet provides an overall output signal for the full input signal cycle. With no input signal applied neither class B stage conducts, giving good circuit efficiency.

In tuned circuits, such as in radio amplifiers, it is possible to bias circuit operation class C (less than 180° operation). This operating condition will be covered in Chapter 10. Where some distortion is acceptable, class AB (between 180° and 360° conduction) is used.

Transformer-Coupled Audio Power Amplifier

The most common class A amplifier connection uses transformer coupling to the load, as in Fig. 9.31. The load may typically be a speaker of 8- or 16-Ω impedance.

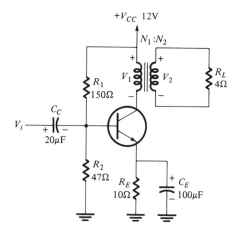

Figure 9.31 Transformer-coupled audio power amplifier.

Since this would be a very low impedance to drive directly, the transformer acts as an impedance matcher to result in a reasonable ac load for a BJT stage.

For transformer coupling the effective ac load impedance (R'_L) is a^2 ($a = N_1/N_2$) times the load resistance R_L. If, as usual, the transformer is connected as step-down, then R'_L is much larger than the typically low speaker impedance. For example, a turns ratio of $3:1$ would make a 4-Ω load appear as an effective ac load of

$$R'_L = a^2 R_L = (3)^2(4) = 36 \ \Omega$$

A graphical analysis of the operation of the circuit of Fig. 9.31 is shown in Fig. 9.32. The transistor is biased to operate in class A, as shown by the dc load line. Since the transformer dc resistance is generally quite low, the dc load line is shown ideally as a straight vertical line. For the dc bias component values shown, the collector quiescent current is about 330 mA. (V_B set by R_1, R_2 is about 3.2 V; V_E is then 2.5 V, and $I_C = I_E = 0.33$ A.) The operating point is thus $V_{CE_Q} = 12$ V, $I_{C_Q} = 0.33$ A.

For the ac load impedance of $36 \ \Omega$, $[a^2 R_L = (3)^2(4)]$, the ac load line of $36 \ \Omega$ is shown passing through the operating point. From the signal variations shown in Fig. 9.32 the value of the peak-to-peak signal swings are obtained as

$$V_{CE_{p-p}} = V_{CE_{max}} - V_{CE_{min}} = 22 - 2 = 20 \text{ V}$$

$$I_{C_{p-p}} = I_{C_{max}} - I_{C_{min}} = 0.61 - 0.05 \text{ A} = 0.56 \text{ A}$$

The ac power developed across the transformer primary can be calculated to be

$$P_o(ac) = V_{CE}(\text{rms})I_C(\text{rms})$$

$$= \frac{V_{CE}(\text{peak})}{\sqrt{2}} \cdot \frac{I_C(\text{peak})}{\sqrt{2}}$$

$$= \frac{V_{CE_{p-p}}/2}{\sqrt{2}} \cdot \frac{I_{C_{p-p}}/2}{\sqrt{2}} \tag{9.11a}$$

$$= \frac{(V_{CE_{max}} - V_{CE_{min}})(I_{C_{max}} - I_{C_{min}})}{8} \tag{9.11b}$$

$$= \frac{20(0.56)}{8} = 1.4 \text{ W}$$

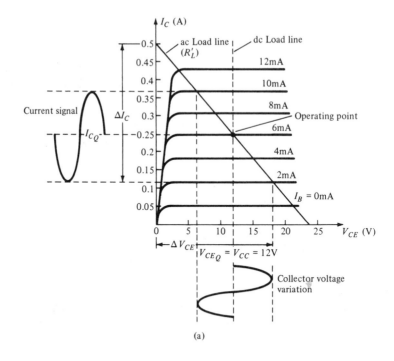

(a)

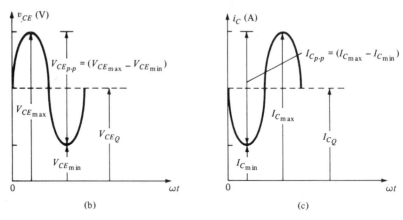

(b)   (c)

Figure 9.32 Graphical operation of transformer-coupled audio power amplifier.

The ac power delivered to the transformer and thereby to the load is then 1.4 W, while the power supplied by the dc voltage supply is

$$P_i(\text{dc}) = V_{CC}I_{C_Q} = 12\text{ V}(0.33\text{ A}) = 4\text{ W}$$

The power dissipated by the transistor is $P_d = P_i - P_o = 4 - 1.4 = 2.6\text{ W}$, and the efficiency is

$$\text{Efficiency} = \eta = \frac{P_o}{P_i} \times 100 = \frac{1.4\,\text{W}}{4\,\text{W}} \times 100 = 35\%$$

The efficiency of a transformer-coupled class A amplifier can at most be 50% when the largest signal swing delivers maximum power to the load, and practically is around 30–40%.

Distortion

If a full signal cycle of input signal does not result in a full output cycle, then distortion has occurred. Ideally, the output signal should be an amplified version of the input with no other change (except the 180° phase shift of the amplifier). While the circuit may provide faithful reproduction of a signal in the midfrequency band, it will have less gain for signals outside this range. Since Fourier analysis allows us to consider any complex signal to be made up of components (harmonics) of the basic or fundamental signal frequency, we can expect distortion to occur if the input signal contains components outside the midfrequency range of the particular amplifier. In addition, the amplifier itself may be poorly biased so that other than class A operation results and even a purely sinusoidal signal in the midfrequency band is distorted. The amount of this amplitude distortion is an important parameter of any power amplifier.

Any signal (music, voice, etc.) can be considered as being made of a fundamental component and harmonic components. To describe the distortion of a power amplifier one can apply a purely sinusoidal signal at a fundamental frequency (having no harmonics) and determine the amount of any harmonics in the resulting amplified output signal. If no distortion occurs, only the fundamental signal results. With distortion, varying amounts of harmonics are present. A wave analyzer (Fig. 9.33) can be used to measure the amount of distortion at each harmonic frequency and thus determine the total harmonic distortion. The wave analyzer can be set at the fundamental frequency and the signal magnitude measured. This measurement is repeated at each harmonic frequency. The fraction of each harmonic component to the fundamental component expressed in percent is a harmonic distortion component —there being second, third, fourth, etc., harmonic distortion components. Usually the first few harmonic components contain most of the distortion. The total harmonic distortion can then be calculated as follows:

$$D_T = \sqrt{D_2^2 + D_3^2 + D_4^2 + \cdots} \tag{9.12}$$

where D_2 = second harmonic distortion, D_3 = third harmonic distortion, D_4 = fourth harmonic distortion, and D_T = total distortion.

Push-Pull Amplifier Circuit

The push-pull circuit connection allows class B bias for efficient operation while obtaining low distortion. A typical transistor push-pull circuit connection is shown

Figure 9.33 Wave analyzer.

in Fig. 9.34. The circuit requires an input transformer to produce opposite polarity signals to drive the two transistor inputs and an output transformer to drive the load in the push-pull mode of operation to be described.

The input transformer with center-tapped ground results in opposite polarity signals applied to the inputs to the two common-emitter stages (Fig. 9.34). During

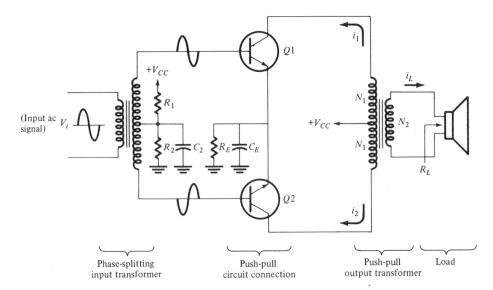

Figure 9.34 Push-pull circuit.

the positive half-cycle of the input signal transistor $Q1$ operates, resulting in a current i_1 through the upper half of the output transformer and a resulting half-cycle of signal through the secondary winding (N_2). The polarity of the signal applied to $Q2$ during the first half-cycle causes $Q2$ to remain in cutoff so it doesn't result in any signal through the output transformer. In fact, the dc bias of $Q2$ is in cutoff (class B operation).

During the negative half-cycle of the input stage $Q1$ remains biased in cutoff, and the ac signal now drives stage $Q2$, resulting in ac current i_2 through the lower half of the output transformer. The net result of each half-cycle through the primary of the output transformer is a complete cycle of operation across the load, R_L. Thus, the push-pull circuit is operated with each stage biased in efficient class B conduction, while each stage provides one-half of the output signal *which is combined in the output transformer into a full 360° conduction cycle*. The signal V_i must be large enough so that the required amplitude signal results as V_L providing the desired output power to load R_L.

Stages $Q1$ and $Q2$ could be operated even in class AB (for less efficient circuit operation) since the dc bias currents flow in opposite directions through the output transformer primary and cancel out. In fact, all even cycle harmonics (or in-phase components of the primary signal) tend to cancel out, further reducing any distortion in the output signal.

The push-pull circuit connection is very popular for use as a power amplifier. A number of various forms are used which provide the full output conduction while operating separate stages in class B for good efficiency. Since input and output transformers take the greatest amount of space and limit the frequency range of circuit operation, transformerless push-pull circuits are generally used to drive large-signal loads. These circuits may use complementary transistors connections, that is, *npn* and *pnp*. The input signal causes each type transistor to operate on opposite half-cycles [Fig. 9.35(a)], providing push-pull operation. Since the transistors are of opposite type they will conduct on opposite half-cycles of the input signal. During the positive half-cycle of the input signal, for example, the *pnp* transistor will be reverse-biased and will not conduct. The *npn* transistor, on the other hand, will be biased into conduction by the positive half-cycle signal with a resulting half-cycle of output across the load (R_L). During the negative half-cycle of input signal the *npn* transistor is biased off, and the output half-cycle developed across the load is due to the operation of the *pnp* transistor at this time.

Figure 9.35(b) shows another version of a transformerless push-pull amplifier, requiring only a single power supply. In this circuit complementary transistors $Q1$ and $Q2$ act as phase splitters. During the positive half-cycle of input *pnp* transistor $Q1$ is biased off, while $Q2$ conducts, providing an inverted, amplified output at the collector. During the negative half-cycle of input *npn* transistor $Q2$ is biased off, while transistor $Q1$ conducts and by emitter-follower action provides a negative half-cycle signal to $Q3$, as shown. During one half-cycle transistor $Q4$ conducts and provides a positive half-cycle of output, while transistor $Q3$ conducts on the other half-cycle, resulting in the negative swing of the output signal.

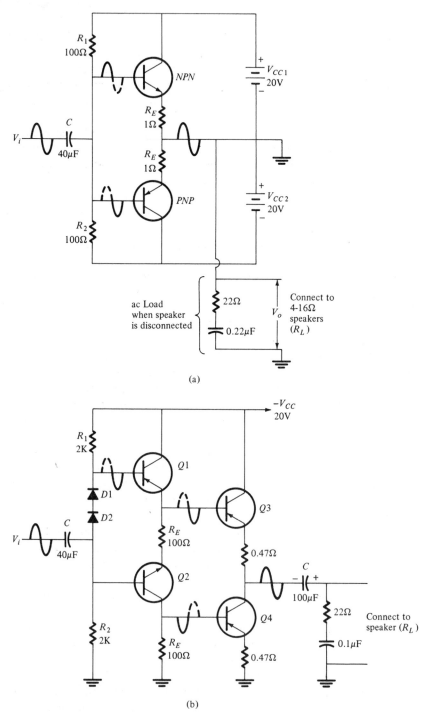

Figure 9.35 Transformerless push-pull circuits (a) Two supply version; (b) Single supply version.

PROBLEMS

§ 9.1

1. For the two-stage FET circuit of Fig. 9.1, calculate the voltage gain of stage 1 if R_{D1} is changed to 8.2 K.

2. Using the FET circuit of Fig. 9.1, calculate the overall voltage gain for $R_{D1} = R_{D2} = 8.2$ K (all other values remaining as shown in Fig. 9.1) for $R_L \gg R_{o2}$.

3. If the two-stage FET amplifier of Fig. 9.1 drives a 50-K load, calculate the output voltage for an input of $V_s = 10$ mV, rms using the circuit values shown in Fig. 9.1 except for $R_{D1} = R_{D2} = 8.2$ K.

4. Calculate the dc base and emitter bias voltages for the circuit of Fig. 9.3 if R_{E1} is changed to 1.8 K and R_{E2} is changed to 1.2 K.

5. Using new values of $R_{E1} = 1.8$ K and $R_{E2} = 1.2$ K in Fig. 9.3, calculate the dc bias collector currents.

6. Using new values of $R_{E1} = 1.8$ K and $R_{E2} = 1.2$ K in Fig. 9.3, calculate the dc bias collector voltages and collector-emitter voltages.

7. Calculate the values of R_{i1} and R_{i2} for the circuit of Fig. 9.3 using new values of $R_{E1} = 1.8$ K and $R_{E2} = 1.2$ K ($h_{ie1} = 2.5$ K, $h_{ie2} = 1.8$ K).

8. Calculate the ac voltage gains of stages 1 and 2 using the circuit of Fig. 9.3 and new values of $R_{E1} = 1.8$ K and $R_{E2} = 1.2$ K.

9. Calculate the resulting ac voltage across a 12-K load in the circuit of Fig. 9.3 for new values of $R_{E1} = 1.8$ K and $R_{E2} = 1.2$ K. What is the value of the overall circuit voltage gain?

10. Draw the circuit diagram of a differential amplifier.

11. What is common-mode rejection?

12. If a differential amplifier has a common-mode rejection ratio of 10^5 and a differential gain of 100, what is the value of common-mode gain?

13. What is the value of common-mode gain for a differential amplifier having CMRR $= 100$ dB and $A_c = 10^{-3}$?

14. What value of CMRR in dB units results for a differential amplifier with $A_d = 150$ and $A_c = 1/2000$?

15. Draw the circuit diagram of a differential amplifier with constant current source.

16. Calculate the input impedance seen looking into a 20:1 step-down transformer connected to a 16-Ω load.

17. What transformer turns ratio will match an 8-K output impedance to a 10-Ω load?

18. Calculate the ac input impedance of a Darlington circuit, as in Fig. 9.18, for new values of $h_{fe1} = h_{fe2} = 60$ and $R_{E2} = 1.5$ K ($h_{ie1} = 3.5$ K, $h_{ie2} = 675\Omega$).

19. Calculate the ac output impedance of a Darlington circuit as in Fig. 9.18 for new values of $h_{fe1} = h_{fe2} = 60$ and $R_{E2} = 1.5$ K (R_s is still 10 K, $h_{ie1} = 3.5$ K and $h_{ie2} = 675\Omega$).

§ 9.2

20. Calculate the low cutoff frequency for an amplifier having input resistance of $R_i = 1.5$ K and coupling capacitor $C = 0.2$ μF.

21. What value of coupling capacitor is required to have a cutoff (lower 3-dB) frequency of 20 Hz for an amplifier having an input resistance of $R_i = 5$ K?

22. Calculate the upper cutoff frequency for an amplifier having values of $R_{\text{high}} = 2$ K and $C_{\text{high}} = 200$ pF.

23. Calculate the value of C_{high} for an amplifier having a midfrequency gain of $A_{\text{mid}} = 85$, stray wiring capacitance of 12 pF, and amplifier terminal capacitances of $C_{ii} = 6$ pF, $C_{oo} = 9$ pF, and $C_{io} = 4$ pF.

24. Calculate the value of low-frequency cutoff for the circuit of Fig. 9.27 with C changed to 0.005 μF.

25. For R_D changed to 3.6 K and $g_m = 3000$ μmho in the circuit of Fig. 9.27, calculate the value of A_{mid}.

26. Using the capacitance values of Fig. 9.27 with new values of $R_D = 3.6$ K and $g_m = 3000$ μmho, calculate the upper cutoff frequency for the amplifier of Fig. 9.27.

27. Calculate the midfrequency gain of the BJT amplifier circuit in Fig. 9.28 for new values of $R_C = 5.1$ K and $h_{fe} = 55$.

28. Calculate the lower 3-dB frequency for the circuit of Fig. 9.28 with new values of $C = 0.5$ μF and $h_{fe} = 55$.

29. Calculate the value of the upper 3-dB frequency for the circuit of Fig. 9.28 with new values of $R_C = 5.1$ K and $h_{fe} = 55$ (all other circuit values remaining as listed in Fig. 9.28).

30. Calculate the dB voltage gain of the amplifier in Fig. 9.28 for new values of $R_C = 5.1$ K and $h_{fe} = 55$.

§ 9.3

31. Calculate the resulting output power delivered to a load if the resulting voltage swing is 15 V p-p with a current swing of 120 mA p-p.

32. Calculate the ac power delivered to an 8-Ω load by a voltage swing of 6 V peak.

33. If an average dc current of 25 mA is drawn from a 9-V supply by an amplifier circuit, calculate the input dc power delivered to the amplifier.

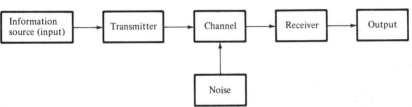

10

Communications

10.1 GENERAL COMMUNICATIONS CONCEPTS

The methods commonly used for information transmission and reception are generally well known. Voice (audio) information is transmitted through space by AM (amplitude modulation) or FM (frequency modulation) techniques, through wire as audio over phone lines, or by a number of other less common methods of signal transmission. Visual (and audio) information is, of course, sent as TV signals and transmitted over cables or through the air. Digital data are transmitted by a wide variety of methods—over cables, through the air, by pulse coding, frequency coding, and amplitude coding, as examples.

Any communication network can be rather simply described by similar basic parts with some generally common characteristics. Figure 10.1 shows the block diagram of a general communications system. The *information source* originates the message or information that is to be sent via a transmitter into the *channel*. The transmitter processes the information and converts it to a form suitable for transmission. This signal processing by the transmitter may be as simple as direct correspondence between electrical current over telephone wires and air pressure variations arising from the human voice; it may be an encoding process such as Morse code or more complex encoding such as amplitude modulation, frequency modulation, and phase or pulse modulation, as examples.

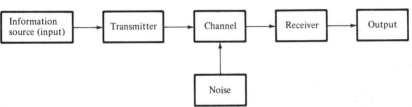

Figure 10.1 Block diagram of a communication system.

This information (encoded or modulated) is then sent to another point or place via a channel, which may be simply a pair of wires, the air, or the vacuum of free space. The channel characteristics are most important in setting the limits of the transmitter (and receiver) and are of greatest concern in designing the system. Transmission through the channel will result in signal attenuation, distortion, and, worst of all, the introduction of *noise* into (or onto) the signal. If the channel is a pair of wires, then the wire-distributed impedances cause attenuation and distortion to various frequency components of the transmitted signal. Noise can come from signals picked up from other nearby lines or through electromagnetic signals around it. These electromagnetic signals can cause problems in transmission through air or space as well and originate from solar radiation, lightning, motors or other electrical disturbances. The channel's signal to noise ratio is quite important in specifying how well the channel operates. The signal and noise picked up by the receiver must then be converted back as close to the original transmitted information as possible with as much of the noise eliminated as possible. The processes of decoding and demodulating are the inverse of those used in transmission.

Some of the major types of circuits used in communications systems can be seen using signal processing in an AM transmitter, for example. The acoustic energy from the person speaking is converted into electrical energy by the microphone. This electrical signal source is generally too weak or low in amplitude for transmission over an air channel or long wire or cable channel. Amplifiers operating in the audio-frequency range signal (about 30–18,000 Hz, typically) increase the signal strength so that it can be transmitted. Audio frequencies are, however, not suitable for long-distance transmission, as the air channel tends to attenuate sound in a relatively short distance. Higher-frequency transmission is capable of much longer distances; in fact, it is possible for electromagnetic radiation transmission to go around the globe or out into space.

An oscillator circuit, radio frequency (or rf) in this case, is used to generate these high-frequency signals used to carry the audio information. The process of placing the audio signal on the carrier signal is called *modulation*, and in this discussion is AM or amplitude modulation. The rf signal, now amplitude-modulated, is amplified so that a suitable large signal can be applied to the transmitting antenna for conversion to electromagnetic radiation into the air (or free space).

After an introduction to AM and FM transmission and reception techniques some of the basic circuits—tuned amplifiers, oscillators, modulation, and demodulating techniques—will be covered and applied to AM and FM transmission and reception.

10.2 AM TRANSMISSION/RECEPTION CONCEPTS

AM Transmission

Figure 10.2 of an AM transmitter indicates the basic parts for obtaining an amplitude-modulated signal. The radio frequency (rf) oscillator generates an rf carrier signal in the range 550–1600 kHz for the AM broadcast band on a precise frequency

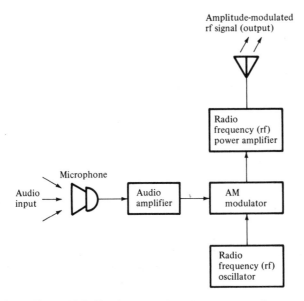

Figure 10.2 Signal passage through an *am* transmitter.

assigned in this broadcast range. The precise oscillator carrier frequency is then a sinusoidal signal as shown in Fig. 10.3(a). An audio frequency signal (30–18,000 Hz) is produced by sound waves striking the microphone, and the resulting electrical signal is raised to the level of a few volts by the audio amplifier [Fig. 10.3(b)]. The modulator circuit then uses the audio signal to modulate the rf carrier signal by varying the amplitude of the carrier with the audio signal [Fig. 10.3(c)]. The modulated rf signal is then amplified by amplifier circuits tuned to the rf frequency band and sent out as electromagnetic radiation from the transmitting antenna.

A voltage which varies sinusoidally at the carrier frequency (from the rf oscillator) can be expressed as

$$v = V \cos \omega_c t$$

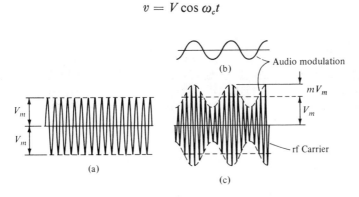

Figure 10.3 AM signals (a) carrier (rf) signal (b) audio-frequency signal; (c) amplitude-modulated carrier signal.

where V is the rms voltage amplitude and ω_c the carrier radian frequency. This is the signal shown in Fig. 10.3(a). The audio signal [Fig. 10.3(b)] can be expressed as

$$v_m = V_m(m \cos \omega_m t)$$

where m is a modulating factor (with $m = 1$ being 100% modulation) and ω_m is the audio or modulating radian frequency. The modulated carrier signal [Fig. 10.3(c)] can then be expressed as

$$v = V(1 + m \cos \omega_m t) \cos \omega_c t \tag{10.1}$$

which is just $V \cos \omega_c t$ for $m = 0$ (no modulation). Using a trigonometric relationship* it is possible to rework Eq. (10.1) into the form

$$v = V \cos \omega_c t + 0.5\, mV \cos(\omega_c + \omega_m)t + 0.5\, mV \cos(\omega_c - \omega_m)t \tag{10.2}$$

Equation (10.2) more clearly shows that the resulting modulated signal contains three frequency components: the carrier frequency, a frequency ω_m (rad/sec) *above* the carrier, and another frequency ω_m (rad/sec) *below* the carrier frequency. These upper and lower components around the carrier frequency are referred to as signal *side bands*. For the pure tone assumed here a pair of side bands is generated. If the actual modulating signal contains many frequency components, as in voice or music, then the resulting modulated signal contains side bands extending above and below the carrier frequency by a frequency range set by the highest frequency of the applied modulating signal. Thus, the transmitted signal has a bandwidth of twice the highest modulating frequency. In the AM commercial broadcast band, for example, the station carrier frequencies would have to be separated by 40 kHz if the full 20-kHz audio band were desired. This wide separation does not usually occur, especially in an area with many stations, and the station's modulating bandwidth is reduced from this full audio range.

Modulations greater than 100% are not allowed as they cause objectionable side-band *splattering*. Figure 10.4 shows the modulated carrier signal for a number of modulating conditions. For the condition of less than 100% modulation [Fig. 10.4(a)] the carrier amplitude peak voltage remains greater than zero. It becomes

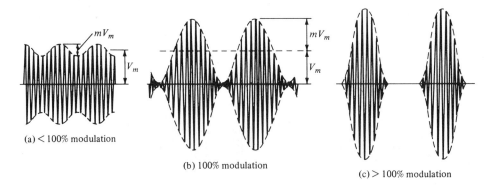

(a) < 100% modulation

(b) 100% modulation

(c) > 100% modulation

Figure 10.4 Various modulation condition.

*$\cos A \cos B = \frac{1}{2}[\cos(A + B) + \cos(A - B)]$.

zero [Fig. 10.4(b)] for 100% modulation at the negative peaks of the modulating signal. For up to 100% modulation the output signal is described by Eq. (10.2). Greater than 100% modulation, however, results in a discontinuous output signal [Fig. 10.4(c)], which is no longer a purely sinusoidal signal. The overmodulated signal results in more than two side bands, this side-band splattering causing interference in adjacent station channels. Obviously, transmission at greater than 100% modulation is not allowed.

It is often desireable to measure the amount of modulation of a signal. Using the waveform of Fig. 10.5, the amount of modulation can be obtained from the oscilloscope display by measuring the peak-peak voltages at maximum and minimum points of the signal, as shown, and calculating m using*

$$m = \frac{A - B}{A + B} \tag{10.3}$$

If $B = 0$, $m = 1$ (100% modulation) as expected, while $B = A$ results in $m = 0$ (0% modulation) and no modulation is present.

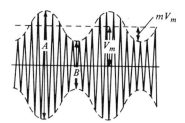

Figure 10.5 Waveform for measuring modulation (m).

Amplitude Modulation Circuit

We have seen that carrier and modulating signals are applied to a circuit called a modulator resulting in an amplitude-modulated signal which is transmitted at the carrier frequency with a bandwidth of twice the highest modulating signal frequency. To achieve this modulating action on a carrier signal requires a nonlinear circuit. Mixing the carrier and modulating signal with a diode or other nonlinear element or using an amplifying device in its nonlinear operating range will result in the desired modulated carrier signal.

Figure 10.6 shows a circuit which can be used as a small-signal modulator. The rf carrier signal is applied to the base, and the af signal to the emitter with a resulting modulated signal in the collector circuit. A tuned circuit at the collector is tuned to the rf carrier frequency and has a bandwidth in the range of twice the highest signal

*From the waveform in Fig. 10.5,

$$mV = \frac{1}{2}\left(\frac{A}{2} - \frac{B}{2}\right) = \frac{1}{4}(A - B)$$

$$V_m = \frac{B}{2} + \frac{1}{2}\left(\frac{A}{2} - \frac{B}{2}\right) = \frac{1}{4}(A + B)$$

$$\frac{mV_m}{V_m} = m = \frac{\frac{1}{4}(A - B)}{\frac{1}{4}(A + B)} = \frac{A - B}{A + B}$$

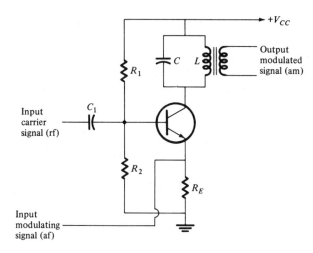

Figure 10.6 Circuit for small-signal amplitude modulation.

frequency. The modulated signal resulting from this circuit is a small signal and has to be coupled through an rf power amplifier to drive the transmitting antenna.

AM detection (demodulation). The received AM signal must be demodulated or detected to extract the af signal from the rf carrier. A combination detector and filter circuit acts to obtain the audio signal from the modulated carrier and then remove the carrier leaving the desired signal.

Figure 10.7 shows the modulated rf signal applied first to a diode circuit which clips the signal, resulting in one having an average value. This rectified signal is then

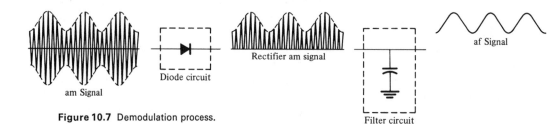

Figure 10.7 Demodulation process.

applied to a filter circuit which follows the signal peaks or signal *envelope*, reproducing the audio signal. A simple circuit for obtaining demodulation is shown in Fig. 10.8. The AM signal received by an antenna is amplified to the level of a few volts and then coupled, through a circuit tuned (L_1, C_1) to the rf carrier frequency, to the detection diode (D_1). The rectified signal is then filtered by capacitor C_2 (and resistor R_2) with the resulting audio-frequency signal, coupled through capacitor C_3, the desired audio signal.

AM receiver. An overall block diagram of the AM receiver is shown in Fig. 10.9. The signal received by the antenna is quite small (microvolts) and is passed through

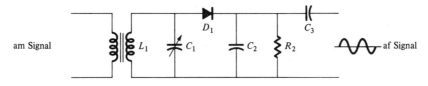

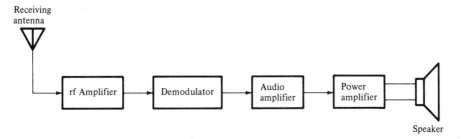

Figure 10.8 AM demodulation circuit.

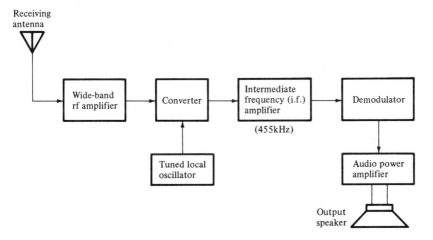

Figure 10.9 AM receiver, block diagram.

a number of rf amplifier stages bringing it to the levels of a few volts. The AM signal is then demodulated, and the resulting audio signal is again amplified to drive a speaker.

Superheterodyne operation. Most receiver circuits use *heterodyne* operation to provide better station tuning, selection, and sensitivity. Figure 10.10 shows the block diagram of a superheterodyne receiver. The signal received by the antenna is first amplified by a wide-band rf amplifier, which amplifies any signals received in the full AM frequency band. The rf signal is then heterodyned or beat with an oscillator frequency obtained from an adjustable oscillator controlled by the tuning knob of the receiver. The resulting signal is still a modulated rf signal but at beat frequencies

Figure 10.10 Superheterodyne receiver, block diagram.

at the sum and difference of the rf and oscillator frequencies. Using the difference frequency from the convertor acts to shift the center frequency of the carrier down to an intermediate frequency, which retains the original modulated information. Using a local oscillator tunable from 1 to 2 MHz results in an intermediate frequency of 455 kHz for rf carrier signals in the range of 550–1600 kHz. By converting the received signal into a fixed intermediate frequency it is possible to then use very selective high-gain tuned intermediate-frequency (i.f.) stages designed to operate at one fixed frequency. Various settings of the local oscillator beat various carrier frequencies down to the i.f. value, so that tuning the oscillator acts to select the signal to be further amplified and detected. Carrier signals received at other than that set to be beat by the oscillator to the i.f. value are also converted but do not pass through the i.f. stages which are highly selective to the intermediate frequency.

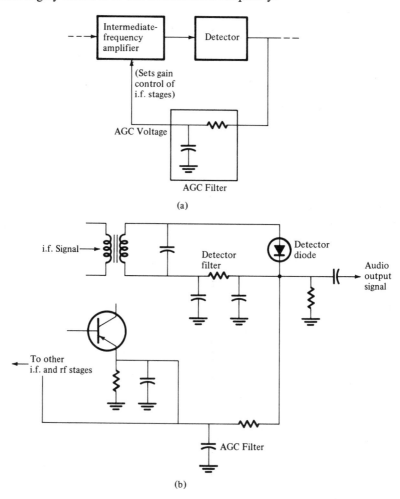

Figure 10.11 Automatic gain control (AGC).

Automatic gain control (AGC). The strength of the signal received from different stations or even from the same station at various times can change considerably. Without the automatic gain control function, to be described, the output signal volume would be almost constantly changing and would be, obviously, quite disturbing to listen to. Receivers are therefore provided with automatic volume control (AVC) or automatic gain control (AGC) circuitry to eliminate this problem. In an AM radio receiver the audio voltage obtained at the detector output is directly related to the strength of the signal received. An AGC control voltage can be obtained from this audio signal by further filtering to develop a voltage which is related to the amplitude of the detected signal. Figure 10.11(a) shows, in block form, the action of the AGC filter. The received signal is converted to i.f. and amplified, before detection. The larger the amplified i.f. signal, the larger the amplitude of the detected signal and the larger the AGC voltage. This AGC voltage is therefore small for small audio signal amplitudes and larger for larger audio signal amplitudes. Then AGC voltage is used to control the bias of the rf and i.f. amplifier stages, thereby controlling the stage gains. This AGC voltage is used so that larger AGC voltage results in reduced amplifier stage gain. In most actual radio circuits the AGC voltage is developed as delayed AGC, which results only when the audio signal increases above a predetermined signal level to provide a smoother-acting gain control.

10.3 FREQUENCY MODULATION (FM)
CONCEPTS

Instead of varying the amplitude of a carrier signal with an audio frequency signal, it is possible to vary the *frequency* of the carrier signal. This concept of *frequency modulation* was conceived about 1933 by Edward Armstrong, one of the early geniuses of the radio field. Armstrong conceived the most important principles in both AM and FM radio, including the early regenerative receiver, the superheterodyne receiver, and later the concept of FM transmission. The story of Armstrong's life is a most fascinating story of the history of radio electronics up to the early 1950s.* Realizing that the source of most noise in AM transmission and reception is amplitude-sensitive (e.g., lightning, radiation from flourescent devices, electronic machinery), Armstrong conceived the idea of keeping the carrier amplitude constant and varying the carrier's frequency at the audio signal rate. Not only does this result in the virtual elimination of much unwanted noise but it also provides operation around a higher carrier with greater bandwidth for much higher fidelity than was possible in AM operation. Figure 10.12 shows frequency modulation of a carrier signal. Notice that the amplitude of the frequency-modulated signal does not vary, whereas the frequency of the signal is higher for higher audio signal amplitude. If the audio signal amplitude is held constant, then the FM signal remains at its center carrier frequency (f_c). The resulting frequency derivation (f_d) from the carrier follows the amplitude

---

*A most absorbing account of Armstrong's life and work is given in *Man of High Fidelity: Edwin Howard Armstrong*, by Lawrence Lessing, Bantam Books, New York, 1969.

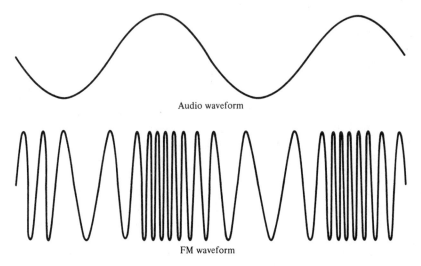

Audio waveform

FM waveform

Figure 10.12 Frequency modulation.

variation of the audio signal (f_s). Not only does the carrier frequency deviate according to the amplitude of the audio signal, but the rate of change of the FM signal also depends on the audio frequency.

Frequency modulation is usually specified by an index of modulation:

$$m_f \equiv \frac{f_d}{f_s} \qquad (10.4)$$

where f_s is the modulating or signal frequency and f_d is the resulting frequency deviation of the carrier. In FM broadcasting the maximum allowable frequency deviation is 75 kHz with a maximum audio frequency of 15 kHz (so that $m_f = \frac{75}{15} = 5$). Indices of modulation greater than unity are considered wide-band frequency modulation.

One important difference between amplitude and frequency modulation is in the amount of transmitted energy. In AM signals 100% modulation occurs when the carrier amplitude varies between zero and twice its (unmodulated) value. The power transmitted has to increase by 50% from 0 to 100% modulation, making the circuit design more complicated for efficient operation over this power range. FM transmission, on the other hand, is always at the same amplitude or power level so that the number of side bands produced in FM transmission and the energy at the carrier and side-band frequencies depend on the modulation index, m_f. Although the relationship is relatively complex, the number of side bands increases with larger values of modulating index, and the energy transmitted is distributed among the side bands and carrier center frequency so that for some values of modulating index the side-band energy is greater than the carrier energy. Although an FM wave theoretically has an infinite number of side bands even for a single tone modulating signal, only a limited number of side bands contain sufficient energy to be important. It is usually assumed that the side bands extend around the carrier frequency by an amount equal to the sum of the modulating frequency and carrier frequency deviation. Thus, the bandwidth requirement of an FM system is *greater* than twice the modulating frequency. To allow this larger bandwidth commercial FM transmission is made on carrier frequencies ranging from 88 to 108 MHz.

FM transmission is provided by a number of circuit arrangements including the reactance-tube transmitter [Fig. 10.13(a)] and the standard Armstrong transmitter [Fig. 10.13(b)]. In the reactance tube transmitter the audio signal varies the reactance

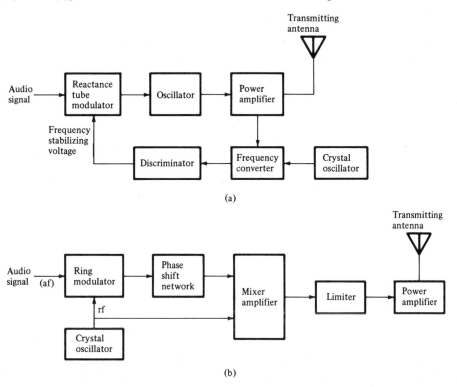

Figure 10.13 (a) Reactance tube FM transmitter; (b) Armstrong FM transmitter.

of the tube, which acts as a trimming capacitor in the oscillator circuit it is connected to. Modern transmitter units use varactor diodes as voltage-controlled capacitive reactances. The main effect is the variation of oscillator carrier frequency by the audio signal. Where practical considerations require a lower oscillator frequency than the carrier value, frequency multiplier stages may be used before the power amplifier stage to set the desired carrier frequency. The modulated carrier is transmitted by the antenna. A part of this signal is generally connected back to be used to stabilize the carrier frequency. A crystal oscillator provides the frequency standard which the frequency converter uses to compare to the signal to be transmitted. The frequency difference applied to the discriminator results in a dc voltage to the reactance tube (or varactor diode) to stabilize the oscillator carrier frequency. This crystal control is quite important in keeping transmission at the allowed carrier frequency within the frequency tolerance permitted. Figure 10.13(b) shows an Armstrong FM transmitter. A crystal controlled oscillator signal is combined with the input audio signal in a ring modulator whose output signal is phase-shifted and combined (mixed) with the crystal oscillator signal for stabilizing purposes. The resulting FM signal is amplitude-limited to provide a purely FM modulated signal to the power amplifier.

FM Reception

The FM radio wave received by an antenna is generally small enough to require some amplification before it can be processed (Fig. 10.14). The FM signal is then

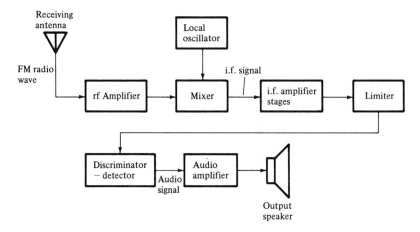

Figure 10.14 FM Receiver, block diagram.

combined with a local oscillator signal in a mixer to obtain an i.f. signal (as in AM reception), with the audio information still carried as frequency variations. A series of i.f. amplifier stages are used to amplify the selected FM signal to the useful level of a few volts. A limiter circuit is used to ensure that no amplitude variation occurs. This FM modulated signal is then applied to discriminator-FM detection circuits to convert the FM frequency variations into their original af variations. The detector essentially provides a voltage output varied by the frequency variation of the FM signal. The audio signal is then amplified and used to drive output speakers.

Two basic circuits for converting an FM signal to audio are the discriminator [Fig. 10.15(a)] and the ratio detector [Fig. 10.15(b)]. Both provide essentially the same circuit action of developing an output voltage dependent on the frequency of the input signal. This voltage-frequency relationship is shown in Fig. 10.15(c) as a frequency response curve. With the demodulator circuit aligned so that no voltage results at the i.f. frequency, frequency variations above and below the carrier produce directly corresponding voltages. For a range of frequencies around circuit resonance the resulting voltage is linear with change in frequency.

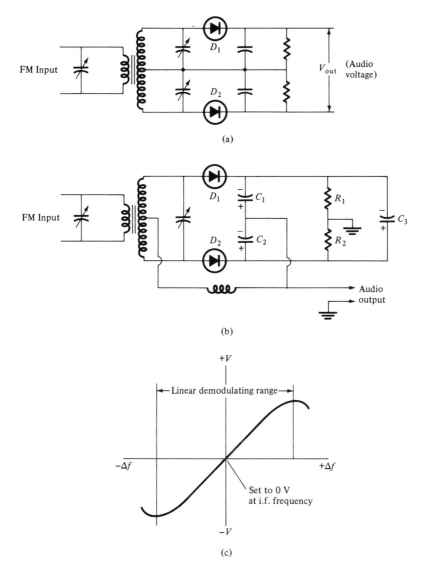

Figure 10.15 (a) FM discriminator; (b) ratio detector; (c) frequency response curve.

10.4 OSCILLATOR CIRCUITS

An important class of electronic circuits are those which provide a sinusoidally varying signal at a frequency which can be adjusted or set to a desired value. An oscillator is a circuit in which part of the output signal is fed back or connected back to the input to add to the input signal (positive feedback) so that the circuit amplifica-

tion increases until it goes into the desired state of oscillation. Various types of oscillators are built to provide either a very stable fixed frequency, a widely adjustable
frequency range, a low-frequency signal, a high-frequency signal, or a nonsinusoidal-
type signal.

Feedback Concepts

Part of the output signal from an amplifier circuit can be fed back to the input,
this feedback resulting in modified circuit operation. This feedback can be either
voltage or current, and the feedback connection may be made either in series or shunt
with the input. Each of these combinations results in somewhat different effects. If
the feedback is connected so as to *oppose* the input, then negative feedback results,
generally providing lowered amplifier gain and higher input impedance. This type of
feedback is often used to provide a more stabilized gain setting. When the output is
connected back to the input to aid the input, then the positive feedback occurs, and
under proper conditions the circuit will go into oscillation, usually varying sinusoidally, at a specific frequency set by particular circuit components.

A basic feedback circuit is shown in Fig. 10.16. The amplifier has a gain A
($= V_o/V_i$), where V_i is not the direct input signal V_s but that value added to a fraction of the output signal (V_o), where β is the fraction of the output fed back to the
input. Expressing the overall circuit gain (V_o/V_s) as A_f (gain with feedback), we can
obtain*

$$A_f = \frac{A}{1 - \beta A} \tag{10.5}$$

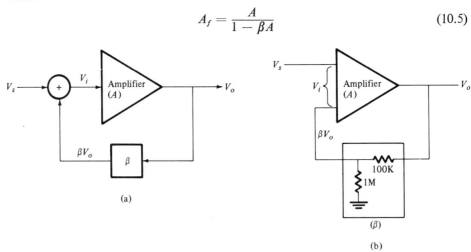

(a)

(β)

(b)

Figure 10.16 Feedback circuit (a) basic block diagram;
(b) circuit schematic.

$$*V_o = AV_i = A(V_s + \beta V_o)$$
$$(1 - \beta A)V_o = AV_s$$
$$A_f \equiv \frac{V_o}{V_s} = \frac{A}{1 - \beta A}$$

If the amplifier is a single-stage (or an inverting) amplifier, then the value of A_f is less than A (negative feedback). For example, $A = -100$, $\beta = \frac{1}{10}$:

$$A_f = \frac{-100}{1 - (\frac{1}{10})(-100)} = \frac{-100}{1 + 10} = -9.1$$

The gain, with feedback, is thus reduced from -100 (without feedback) to -9.1 with improvements in gain stability, input, and output impedance.

Feedback Conditions

Considering all the possible conditions for the feedback circuit using Eq. (10.5), we can specify

1. If $\beta A < 0$ (negative), then A_f is less than A, and the circuit is a negative feedback amplifier.
2. If $0 < \beta A < 1$, then $A_f > A$ and the circuit acts as a positive feedback amplifier.
3. If $\beta A \geq 1$, then $A_f \to \infty$, and the circuit goes into oscillation ($\beta A = 1$ is the Barkhausen criterion for oscillation).

As the values of β and A are adjusted so that βA approaches 1 the feedback circuit goes into oscillation. Since β and A are usually frequency-dependent, this condition for oscillation may occur over a limited frequency range and can be set to a specific value by specific circuit component values. Of the many types of sinusoidal oscillators a number of the more popular ones to be described are the LC oscillators (Hartley and Colpitts), crystal RC phase shift, and Wien bridge oscillator circuits.

LC Oscillator Circuits

A basic LC oscillator circuit is that shown in Fig. 10.17. For impedances Z_1 and Z_2 chosen as inductors and impedance Z_3 as a capacitor the circuit is called a Hartley oscillator [Fig. 10.17(b)], while use of capacitors for Z_1 and Z_2 with Z_3 an inductor results in a Colpitts oscillator circuit [Fig. 10.17(c)].

The frequency of oscillation of the Hartley circuit can be determined from

$$f_o \simeq \frac{1}{2\pi\sqrt{LC}} \tag{10.6}$$

where $L = L_1 + L_2 + 2M$, the total inductance of the coil (M is the coil mutual inductance). A coil of impedance $L = 800 \ \mu H$ and a 120-pF capacitor used in a Hartley oscillator circuit would provide a sinusoidal signal at a frequency calculated to be

$$f_o = \frac{1}{6.28\sqrt{(800 \times 10^{-6})(120 \times 10^{-12})}} \simeq 514 \ \text{kHz}$$

For a Colpitts oscillator the resonant (oscillation) frequency is also given by

$$f_o \simeq \frac{1}{2\pi\sqrt{LC}}$$

but in this case $C = C_1 C_2 / (C_1 + C_2)$, the series equivalent of the two capacitors.

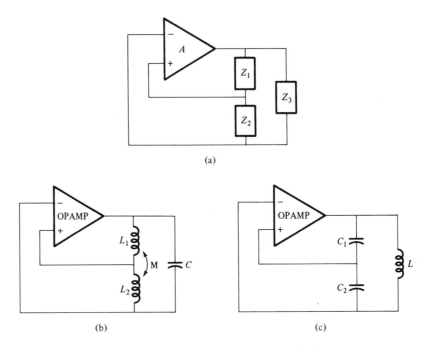

Figure 10.17 (a) Basic LC oscillator circuit; (b) Hartley oscillator; (c) Colpitts oscillator.

Using a 400-μH coil with two 100-pF capacitors would result in an oscillator frequency of

$$f_o = \frac{1}{2\pi\sqrt{(400 \times 10^{-6})(50 \times 10^{-12})}} \simeq 1.126 \text{ MHz}$$

where C was calculated to be

$$C = \frac{C_1 C_2}{C_1 + C_2} = \frac{100(100)}{100 + 100} = 50 \text{ pF}$$

Single-stage circuit versions of Hartley and Colpitts oscillator circuits using FET and bipolar transistor circuits are shown in Fig. 10.18. An RFC (radio-frequency coil) coil of low dc impedance and very high ac impedance is used as the collector load impedance for the BJT and drain load impedance for the FET circuit. Both Hartley oscillator circuits used a tapped coil for the inductor (L), and an adjustable capacitor (C) is used to adjust or vary the oscillator frequency. In the Colpitts oscillator circuit form the two capacitors C_1 and C_2 are usually ganged and adjusted as a pair to vary the oscillator frequency. Note the slightly different appearance of the feedback connection in Fig. 10.18(c) and Fig. 10.18(d), which both connect output back to input and are electronically the same.

In all these circuits the frequency is essentially set as given in Eq. (10.5). It also is important that the gain A be large enough so that the condition $\beta A = 1$ is met. For a fixed value of β it is necessary that the circuit amplification be large enough so that oscillation occurs.

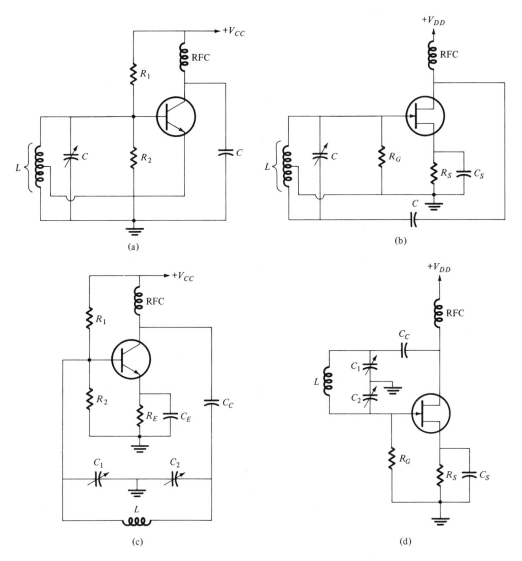

Figure 10.18 (a) BJT Hartley; (b) JFET Hartley; (c) BJT Colpitts;
(d) JFET Colpitts.

Crystal Oscillator Circuit

Where very precise frequency is desired a crystal element is used to set the oscillator circuit frequency. A crystal is cut so that it has electromechanical resonance at
a specific frequency. Actually, crystals exhibit series resonance at which frequency
the crystal acts as a very low-impedance element or parallel resonance at which
frequency the crystal acts as a very high-impedance element. There are many ways of
connecting the crystal in an oscillator circuit. For example, the Colpitts circuit form

can be modified to act as a crystal controlled oscillator circuit, as shown in Fig. 10.19(a). The crystal operates here in its parallel impedance mode, providing the highest impedance and greatest feedback at the crystal's parallel resonance frequency. Figure 10.19(b) shows a Pierce crystal oscillator circuit with the crystal acting in its series resonance mode providing least impedance and greatest feedback at the crystal's series resonance frequency.

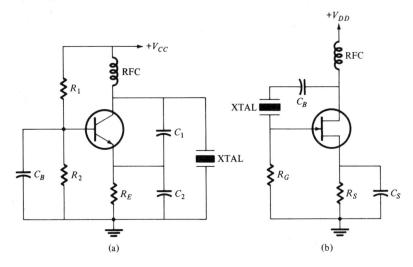

Figure 10.19 Crystal oscillator circuits. (a) modified Colpitts; (b) Pierce circuit.

Crystal oscillators develop a very stable, precise frequency which can be altered only by changing crystals. Transmission over AM, FM, or other frequency bands must be made at very precise frequencies with great stability maintained over a wide range of supply voltage and temperature requiring use of crystal controlled oscillators. Crystals range in frequency from a few kilohertz to tens of megahertz for operation in the crystal's fundamental mode of oscillation (and in the hundreds of megahertz operating on overtone mode). Crystals also are manufactured with various sized and shaped cuts for operation in the various frequency ranges, different temperature ranges, and filter or oscillator applications.

RC Phase Shift Oscillation

A popular oscillator circuit using a resistor-capacitor network to set the oscillator frequency is the phase shift circuit shown in Fig. 10.20. The OPAMP circuit provides the gain of $-A$ required so that the signal fed back through the *RC* phase shift network (β) provides the required $\beta A = 1$ for circuit oscillation. At the conditional $\beta A = 1$ the circuit oscillates at a frequency of

$$f_o = \frac{1}{2\pi\sqrt{6}\,RC} \tag{10.7}$$

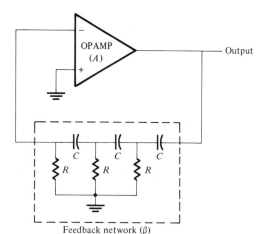

Figure 10.20 *RC* phase shift oscillator using OP amp.

Feedback network (β)

provided the amplifier gain A is greater than 29 (because $\beta = \frac{1}{29}$ at this frequency). For example, values of $R = 1\ K$ and $C = 1000$ pF would set the circuit oscillating at a frequency of

$$f_o = \frac{1}{2\pi(2.45)(1 \times 10^3)(1000 \times 10^{-12})} \simeq 65\ \text{kHz}$$

Adjustment of the oscillator frequency can be made by varying either R or C.

The phase shift network provides a phase shift of 180° to develop the required positive feedback, and the circuit oscillates at this frequency. Other networks may be used to give the required 180° phase shift, although the RC circuit of Fig. 10.20 is quite popular.

Practical FET and BJT transistor circuit versions are shown in Fig. 10.21. For the FET circuit [Fig. 10.21(a)] the amplifier will have a gain greater than 29 if the values of g_m and R_D are chosen so that

$$|A_v| = g_m R_D > 29$$

For example, a JFET having $g_m = 2000\ \mu$mho and a drain resistance of 18 K will result in an amplifier gain of -36, enough to provide for oscillation at the frequency set by the feedback network. If the feedback components are $R = 10$ K and $C = 0.01\ \mu$F, then the oscillator will operate at

$$f_o = \frac{1}{2\pi\sqrt{6}\ RC} = \frac{1}{6.28(2.45)(10 \times 10^3)(0.01 \times 10^{-6})} \simeq 650\ \text{Hz}$$

Changing C to 1000 pF would change the operating frequency by a factor of 10, to 6.5 kHz. Changing C to 100 pF and R to 1 K would result in an operating frequency of 65 kHz. It should be noted here that reducing R any lower might result in loading the amplifier, thereby reducing the gain, and no oscillation would occur. The capacitor value could still be lowered to, say, 10 pF, but the stray circuit capacitance would affect the circuit operation. At $C = 10$ pF the frequency of operation would be 650 kHz, showing that upper frequencies about 1 MHz begin to bring practical circuit effects, limiting the upper range to this low-megahertz limit. At the lower fre-

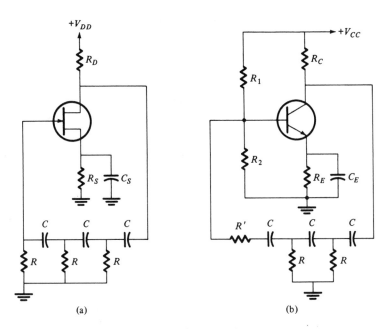

Figure 10.21 Transistor RC phase-shift oscillator circuits.

quency the values of R up to about 10 M and C to, say, 1000 μF would result in cutoff frequencies below 1 Hz, by which value the effect of C_E would result in lowered gain, possibly preventing the circuit from oscillating. Although the calculation of frequency can be made for any given set of values for R and C, there are practical circuit limits that must always be taken into consideration.

The BJT oscillator circuit [Fig. 10.21(b)] is seen to be slightly modified. The third RC section does not have a direct connection to ground. Since the ac impedance of R_1 in parallel with R_2 provides an impedance to ground, the resistor R' is chosen so that in series with $R_1 \| R_2$ it provides an effective impedance of R, as desired. The amplifier gain must be greater than 29, which requires a transistor with gain h_{fe} such that

$$h_{fe} > 23 + 29\frac{R}{R_C} + 4\frac{R_C}{R} \tag{10.8}$$

For $R = 10$ K and $R_C = 10$ K we would need

$$h_{fe} > 23 + 29\left(\frac{10\text{ K}}{10\text{ K}}\right) + 4\left(\frac{10\text{ K}}{10\text{ K}}\right) = 23 + 29 + 4 = 56$$

so that a transistor having h_{fe} of, say, 60 would just cause oscillation to occur.

Note here the practical effect of choosing R greater than R_C, requiring very high transistor gain. Typically $R_C \simeq R$ for practical values of h_{fe} around 50.

The oscillator frequency for this circuit is also affected by the R/R_C ratio, where

$$f_o = \frac{1}{2\pi RC\sqrt{6 + 4\frac{R_C}{R}}} \tag{10.9}$$

which for $C = 1000$ pF, with $R = R_c = 10$ K, would result in

$$f_o = \frac{1}{2\pi(10 \times 10^3)(1000 \times 10^{-12})\sqrt{6 + 4\left(\dfrac{10\,\text{K}}{10\,\text{K}}\right)}} \simeq 5\,\text{kHz}$$

Wien Bridge Oscillator

Another circuit using resistors and capacitors to set the oscillator frequency is the Wien bridge circuit shown using an OPAMP in Fig. 10.22. The circuit form is seen to be a bridge connection. Resistors R_1 and R_2 and capacitors C_1 and C_2 form the frequency adjustment elements, while resistors R_3 and R_4 form part of the feedback path. The OPAMP output is connected as the bridge input at points a and c. The bridge circuit output at points b and d is the input to the OPAMP.

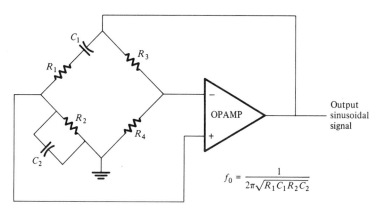

Figure 10.22 Wien bridge oscillator circuit using OPAMP amplifier.

The frequency of oscillation can be calculated to be

$$f_o = \frac{1}{2\pi\sqrt{R_1 C_1 R_2 C_2}} \tag{10.10a}$$

If the values are chosen identically so that $R_1 = R_2 = R$ and $C_1 = C_2 = C$, then

$$f_o = \frac{1}{2\pi RC} \tag{10.10b}$$

For sufficient loop gain it is necessary that the ratio of R_3 to R_4 be

$$\frac{R_3}{R_4} = \frac{R_1}{R_2} + \frac{C_2}{C_1} \tag{10.11a}$$

which for identical resistors and capacitors is

$$\frac{R_3}{R_4} = 2 \tag{10.11b}$$

For example, $R = 10$ K and $C = 0.001$ μF would result in an oscillator frequency of

$$f_o = \frac{1}{6.28(10 \times 10^3)(0.001 \times 10^{-6})} = 15.9 \text{ kHz}$$

A practical Wien bridge oscillator circuit is shown in Fig. 10.23. This is exactly the same bridge connection as in Fig. 10.22 but is redrawn in a more common form.

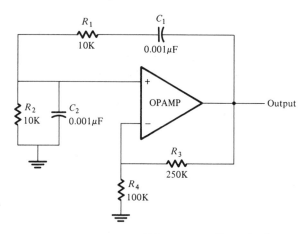

Figure 10.23 Practical Wien bridge oscillator circuit.

PROBLEMS

§ 10.2

1. Show how the trigonometric relationship

$$\cos A \cos B = \tfrac{1}{2}[\cos (A + B) + \cos (A - B)]$$

is used to express Eq. (10.1) in the form of Eq. (10.2).

2. What are the side bands expressed in Eq. (10.2)?

3. If $f_c = 900$ kHz and $f_m = 10$ kHz, what are the side-band frequencies?

4. If an unmodulated carrier is 25 V and the modulating carrier signal has 40 V peak, what is the modulation index?

5. What is the difference between the received rf signal and the i.f. signal?

6. For an AM radio intermediate frequency of 455 kHz, what local oscillator frequency will tune in a 1.6-MHz radio station signal?

7. For an AM-modulated wave described by

$$v(t) = [60 + 30 \cos 2\pi(6 \times 10^3)t] \cos 2\pi(880 \times 10^3)t$$

(a) What are the carrier frequency and the modulating frequency of the AM wave?
(b) What is the value of the modulation index, m?
(c) What is the amplitude of the unmodulated carrier?

8. What is a superheterodyne receiver? What are its advantages?

9. Draw the block diagram of a superheterodyne AM receiver and list typical values.

§ 10.4

10. What does the value of β represent in a feedback circuit?

11. How do positive and negative feedback differ?

12. How does the gain with feedback differ from the gain without feedback?

13. How do Colpitts and Hartley oscillators differ?

14. What is the resonant frequency of a Hartley oscillator having total impedance $L = 650\ \mu H$ and capacitance 240 pF?

15. What is the resonant frequency of a Colpitts oscillator having $C_1 = 150$ pF, $C_2 = 250$ pF, and $L = 300\ \mu H$?

16. Calculate the resonant frequency of an RC oscillator circuit having $R_1 = R_2 = R = 2.4$ K and $C_1 = C_2 = C = 2500$ pF.

17. What value equal capacitors are needed to provide an oscillator frequency of $f_o = 80$ kHz using $R_1 = R_2 = R = 4.4$ K?

18. Calculate the oscillator frequency of a phase shift oscillator circuit having $R = 7.2$ K and $C = 0.002\ \mu F$.

19. What value capacitor is needed in a phase shift oscillator to provide a resonant frequency of 12.5 kHz using $R = 5$ K?

20. Calculate the resonant frequency of a Wien bridge oscillator circuit having $R_1 = R_2 = R = 3.6$ K and $C_1 = C_2 = C = 1200$ pF.

21. What value equal capacitor is needed to provide a resonant frequency of 75 kHz using equal resistors $R = 4.7$ K in a Wien bridge oscillator circuit?

22. Draw the circuit diagram of a phase shift oscillator using FETs.

23. Draw the circuit diagram of an FET RC oscillator circuit.

24. Draw the circuit diagram of an FET Wien bridge oscillator.

11

Control Systems

11.1 GENERAL

Electrical circuits and mechanical devices must both be considered to understand the functioning of a large variety of control systems. Any group of components which operates some device is a control system. A system which drives a telescope to a desired position, for example, is a control system. This control system can operate either closed-loop or open-loop. In an *open-loop* system the telescope would be driven to the desired position, but no facility is included to check if the instrument has actually reached this position. A closed-loop system would additionally provide signal *feedback* from the telescope drive so that some comparison can be made between the desired position and the present position and an *error signal* developed to correct for this difference. An open-loop system can be symbolically represented by a block diagram such as that of Fig. 11.1(a), while a closed-loop system can be represented by the block connection of Fig. 11.1(b).

In Fig. 11.1(a) the actuating signal or input signal, S_i, is provided to operate the electromechanical driving components. This command signal may be simply obtained by someone moving a dial to a desired setting, thereby generating a command signal, or the input signal could result from some other system (such as a computer). The driver unit then positions the telescope to an output position, indicated by S_o. There is no way for the system to check that S_o is the desired value of S_i after the driver moves the telescope. The more common arrangement [Fig. 11.1(b)] derives a feedback signal, S_f, from the output signal. A feedback circuit develops a signal from the output which is then compared to the input signal. In the usual arrangement a comparator circuit subtracts the feedback from the input to provide a signal to actuate the telescope position driver. When the feedback signal is quite different from the input signal a large actuating signal is present. As the output signal approaches the desired position the actuating signal decreases.

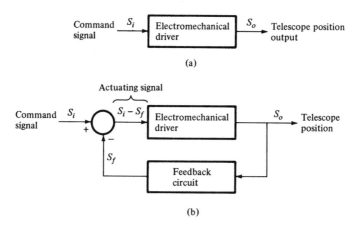

(a)

(b)

Figure 11.1 Open and closed loop control systems.

11.2 BLOCK DIAGRAMS

One popular way of describing the interaction of various parts of a control system is the representation by a block diagram. Each part of the control system may be represented by a transfer function (Fig. 11.2). The simple resistor divider of Fig. 11.2(a) may be represented by the block diagram of Fig. 11.2(b), where the transfer function of the block is

$$T = \frac{R_2}{R_1 + R_2}$$

and the input is then

$$V_o = TV_i$$

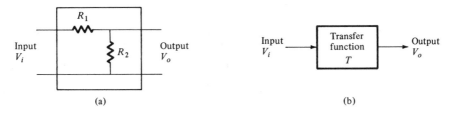

(a)   (b)

Figure 11.2 Block transfer function.

A circuit may be broken up into more than one block as shown in Fig. 11.3(a). If we assume an ideal amplifier of voltage gain A and infinite input impedance then the divider network and amplifier can be represented by separate blocks as shown in Fig. 11.3(b). The overall gain is such that

$$V_o = TAV_i$$

and the overall network can be simplified to a single block as shown in Fig. 11.3(c), if this simplified form is useful.

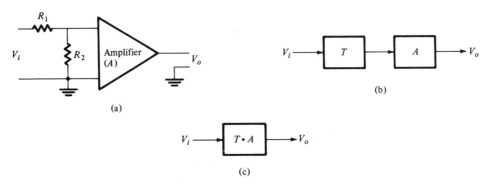

(a)

(b)

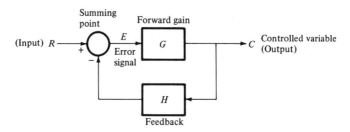

(c)

Figure 11.3 Two block stages in series.

Feedback

When part or all of the output signal is fed back and subtracted from the input the usual form of feedback circuit results. In most control systems various parts or blocks are electrical or mechanical elements or components. A more generalized nomenclature for representing a feedback control circuit is shown in Fig. 11.4.

Figure 11.4 Typical feedback control loop.

The overall transfer function of the feedback control loop can be obtained by solving for the ratio of output to input, C/R. This ratio is calculated as follows:

$$C = GE \tag{11.1}$$

$$E = R - HC \tag{11.2}$$

Inserting Eq. (11.1) in (11.2),

$$C = G(R - HC)$$

Solving for C,

$$C = \frac{GR}{1 + GH} \tag{11.3}$$

so that

$$\frac{C}{R} = \frac{G}{1 + GH} \tag{11.4}$$

Equation (11.4) shows that the effect of the feedback is a transfer function which reduces the gain from the forward gain G by the factor $1 + GH$ or 1 *plus* the forward gain (G) times the feedback (H). Consider a network having a forward gain $G = 1000$ and feedback of $H = 0.1$. We find that the overall gain is then

$$\frac{C}{R} = \frac{G}{1 + GH} = \frac{1000}{1 + 1000(0.1)} = 9.9$$

If the product of forward gain times feedback is much larger than 1,

$$\frac{G}{1 + GH} \longrightarrow \frac{G}{HG} = \frac{1}{H} \qquad \text{(for } HG \gg 1) \tag{11.5}$$

The gain for such a circuit is approximately $1/H$ ($1/0.1 = 10$ in the above example). This type of circuit is used in analog computers to obtain precise amplifier gain using precision resistors to set the feedback factor H. In many control systems where it may not be possible to set $GH \gg 1$, the factor 1 is still important. If the loop gain GH approaches unity and if the phase shift around the loop approaches 180°, then the denominator $1 + HG$ approaches zero and the ratio C/R goes to infinity, which in practical terms means that the system becomes unstable and goes into oscillation. Since this is undesirable in a control system, it is necessary to investigate under what conditions this can occur for the particular system and adjust various parts so that it is avoided. System stability is covered more fully later in this chapter.

Breaking down a complex system into its various components and representing them by a feedback block diagram allows for analysis and possible modification or adjustments in a well-defined manner. The techniques of block diagram manipulation and simplification have been well developed. While a number of block diagram simplification rules can be defined and used for relatively simple systems, the transfer function of larger systems can easily be obtained using signal flow graph analysis.*

Block Diagram Simplification

A block diagram representing the interconnections of various parts of a system including any existing or added feedback is a means of describing systems in a standardized manner. Table 11.1 shows a number of block diagram simplifications which may be applied to help simplify more complex systems so that the overall system transfer function expression can be obtained. Figure 11.5 shows the steps in reducing a given system block diagram so that the transfer function expression for the overall system can be obtained. Starting with the loop G_1, H_1 we recall that such a feedback loop can be expressed by

$$\frac{G_1}{1 + H_1 G_1} \tag{11.6}$$

Figure 11.5(b) shows the loop replaced by a single block whose transfer function is given by Eq. (11.6). Combining the two blocks in series results in a single block [Fig. 11.5(c)] with the transfer function

$$\frac{G_1 G_2}{1 + H_1 G_1} \tag{11.7}$$

*A short simple description of signal flow graph analysis is provided in *Automatic Control Systems*, by Ben Zeines, Prentice-Hall, Englewood Cliffs, N.J., 1972, pp. 64–74.

Transformation	Block diagram	Equivalent block diagram	Equation
Cascaded blocks	$R \to G_1 \to G_2 \to C$	$R \to G_1 G_2 \to C$	$\dfrac{C}{R} = G_1 G_2$
Eliminating a forward loop	$R \to G_1,\ G_2 \to C$	$R \to G_1 \pm G_2 \to C$	$\dfrac{C}{R} = G_1 \pm G_2$
Eliminating a feedback loop	$R \to G_1,\ H_1 \to C$	$R \to \dfrac{G_1}{1 \mp G_1 H_1} \to C$	$\dfrac{C}{R} = \dfrac{G_1}{1 \mp G_1 H_1}$
Moving a pickoff point beyond a block	$R \to G \to C$	$R \to G \to C$; $R \leftarrow \dfrac{1}{G}$	$\dfrac{C}{R} = G$
Moving a pickoff point a block ahead	$R \to G \to C$; C	$R \to G \to C$; $C \leftarrow G$	$\dfrac{C}{R} = G$
Moving a summing point beyond a block	$R_1,\ R_2 \to G \to C$	$R_1 \to G \to C$; $R_1 \to G$	$\dfrac{C}{R_1 \pm R_2} = G$
Moving a summing point ahead of a block	$R_1 \to G \to C$; R_2	$R_1 \to G \to C$; $\dfrac{1}{G} \leftarrow R$	$C = (R_1 G \pm R_2)$

Table 11.1

Figure 11.5(c) again forms a simple feedback loop, and the overall system transfer function can be obtained:

$$\frac{C}{R} = \frac{G_1 G_2/(1 + H_1 G_1)}{1 + H_2[G_1 G_2/(1 + H_1 G_1)]} = \frac{G_1 G_2}{1 + H_1 G_1 + H_2 G_1 G_2} \qquad (11.8)$$

Using the equivalent block diagrams shown in Table 11.1 is also sometimes helpful in rearranging the block diagram so that a modified circuit arrangement still results in the same overall transfer function. For example, the circuit of Fig. 11.6(a) can be modified to that of Fig. 11.6(c), the latter system having the same transfer function as the original one. Obviously considerable manipulation is possible with practical systems and component factors dictating choice. Where more complex systems are involved the signal flow graph technique is more systematic in simplifying and manipulating the system block diagram.

11.3 BODE PLOT

Having obtained a transfer function of a system it is possible to investigate the frequency response of the system. In particular it is important to determine whether

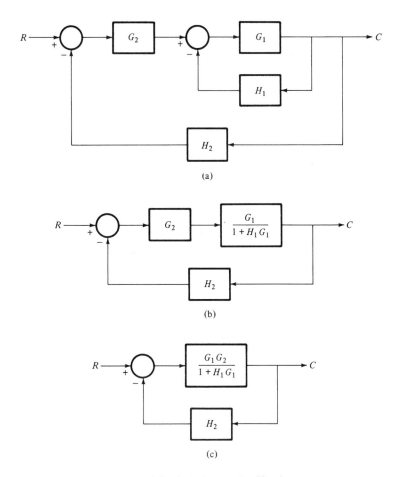

Figure 11.5 Block diagram simplification.

the loop gain approaches unity with a 180° phase shift to examine the system's stability. One way of graphically describing the system transfer function is the use of Bode plots—two graphs, one of gain (or magnitude) vs. frequency and the other of phase angle vs. frequency. To allow for a large frequency scale the $\log_{10}$ of frequency is used as the abscissa axis. For example, consider the simple RC circuit of Fig. 11.7. The transfer function can be expressed using LaPlace notation* as

$$G(s) = \frac{1}{1 + (s/RC)} = \frac{1}{1 + sT} \tag{11.9}$$

where

$$T = \frac{1}{RC}$$

*See the coverage of LaPlace techniques in *Automatic Control Theory*, by B. DeRoy, Wiley, New York, 1966, Chap. 2.

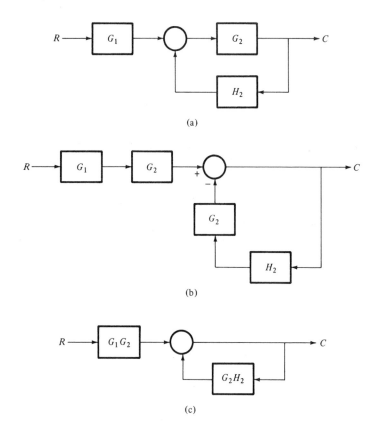

(a)

(b)

(c)

Figure 11.6 Manipulating a system block diagram.

Setting $s = j\omega$ (ω is radian frequency) results in

$$G(j\omega) = \frac{1}{1 + j\omega T} \tag{11.10}$$

a complex transfer function having magnitude and phase angle as a function of frequency:

$$|G(j\omega)| = \frac{1}{\sqrt{1 + (\omega T)^2}} \tag{11.11}$$

$$\angle G(j\omega) = \theta = \tan^{-1} \omega T$$

Thus, the transfer function of Eq. (11.10) can be expressed by the two factors in Eq. (11.11). A Bode plot is a plot of each of the expressions of Eq. (11.11) on a $\log_{10}$ frequency scale. Choosing values of R and C in Fig. 11.7 of $R = 100\,K$ and $C = 100$ μF would result in a Bode plot as shown in Fig. 11.8. An exact curve can be obtained by calculating and plotting the corresponding values for a number of frequency points. Simplified techniques are available to draw the Bode plots using straight-line and asymptotic approximations to the actual curves.

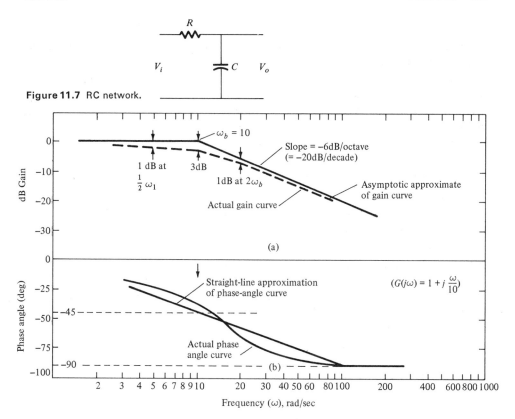

Figure 11.7 RC network.

Figure 11.8 Graphs of Bode magnitude and phase angle plots.

Gain Asymptotic Approximations

A critical point for the magnitude plot of Fig. 11.8(a), called a *breakpoint* (or corner frequency), is determined for Eq. (11.11) as the point where $\omega T = 1$. For frequencies much less than $\omega T = 1$ the expression $\sqrt{1 + (\omega T)^2}$ approaches 1 and the $\log_{10} 1$ is 0 dB. For frequencies so that ωT is much greater than $1 (\omega T \gg 1)$ the magnitude expression reduces to ωT, and $\log_{10} \omega T$ is plotted as a sloped line on the log gain, log frequency Bode plot. The critical point to determine is where this sloped line starts, this being the breakpoint $\omega_b T = 1$ or $\omega_b = 1/T$. For the present example

$$\frac{1}{T} = RC = (100 \times 10^3)(100 \times 10^{-6}) = 10$$

so that $\omega_b = 10$ rad/sec. The Bode plot of Fig. 11.8(a) shows this breakpoint, with log gain 0 dB before the breakpoint and dropping at a constant slope of either 6 dB/octave or 20 dB/decade.* In the present example the curve drops off by 6 dB at 20 rad/sec and 20 dB at 100 rad/sec.

*An octave is a doubling of the frequency, while a decade is a 10-fold increase in frequency.

The transfer function of Eq. (11.10) is then easily plotted on a Bode plot by determining ω_b and drawing a fixed slope line (slope of 6 dB/octave or 20 dB/decade) from this frequency point.

This straight-line approximation plot is a quick, easy way of drawing the Bode magnitude plot. The exact magnitude plot can also be easily sketched using the approximate lines plotted as asymptotes. In particular, Fig. 11.8(a) shows that at $\frac{1}{2}\omega_b$ the actual gain plot should be 1 dB less than the asymptotic line, at ω_b the actual gain is 3 dB less, and at $2\omega_b$ it is again 1 dB less than the asymptotic curve.

Phase Shift Asymptotic Curve

The asymptotic approximation Bode plot of the phase angle or phase shift for Eq. (11.11) is shown in Fig. 11.8(b). For frequencies ωT much less than 1 the value $\tan^{-1} \omega T$ is very small and asymptotically shown as a 0° phase shift line. The first critical point to note is at $0.1\omega_b$ at which a straight line at slope 45°/decade is started. Notice, then, that at ω_b the phase shift is 45° (when $\omega_b T = 1$, $\tan^{-1} \omega_b T = \tan^{-1} 1 = 45°$) and that at $10\omega_b$ the phase shift has reached 90° at which point the asymptotic curve becomes a constant phase angle line. The actual phase angle curve can then be plotted by noting the phase angle corrections in Fig. 11.8(b) and listed in Table 11.2.

The rules above apply to the Bode plot only for a transfer function such as that of Eq. (11.10), which happens to be very common. Other basic equation forms can also be examined and a set of Bode diagram building blocks described for use in plotting a very wide variety of transfer function equations.*

Table 11.2

Corrections to Asymptotic Phase Angle Curve

ω	PHASE ANGLE CORRECTION
$0.05\omega_b$	$-3°$
$0.1\omega_b$	$-6°$
$0.3\omega_b$	$+5°$
$0.5\omega_b$	$+5°$
$1.0\omega_b$	$0°$
$2.0\omega_b$	$-5°$
$3.0\omega_b$	$-5°$
$10\omega_b$	$+6°$
$20\omega_b$	$+3°$

Stability Information from Bode Diagrams

The Bode diagram provides a good graphical description of the control system transfer functions. In particular, information on system stability can be easily read off the curves. Two stability factors are commonly obtained from the Bode diagrams.

*A description of the rules for plotting a variety of Bode plots is given in *Automatic Control Theory, op. cit.*, pp. 55–68.

An example of these stability factors is given in Fig. 11.9, which combines both magnitude and phase angle curves on a single graph for the transfer function

$$\frac{4}{(j\omega)(1 + j0.125\omega)(1 + j0.5\omega)} = \frac{4}{(j\omega)(1 + j\omega/8)(1 + j\omega/2)}$$

A *gain margin*, defined as the amount the gain differs from 0 dB occurring at a frequency point when the phase angle is −180°, is shown in Fig. 11.9 to be about 8 dB. The frequency at which this occurs is called the *phase crossover* frequency. Since a gain of 1 at a phase angle of 180° is indicative of system stability, the gain margin reflects how much the gain differs from a gain of 1 (0 dB) at a phase shift of 180°.

A *phase margin*, defined as the amount of phase shift from 180° occurring when the gain curve crosses 0 dB, is shown in Fig. 11.9 to be about 30°. The frequency at which this occurs is called the *gain crossover* frequency. The phase margin indicates how much the phase shift may vary before reaching the critical value of 180° at unity gain. The phase margin gives designers a good description of system stability with acceptable phase margins usually around 40°–60°, much less being too close to instability or too jittery and much more having too much stability or being too sluggish.

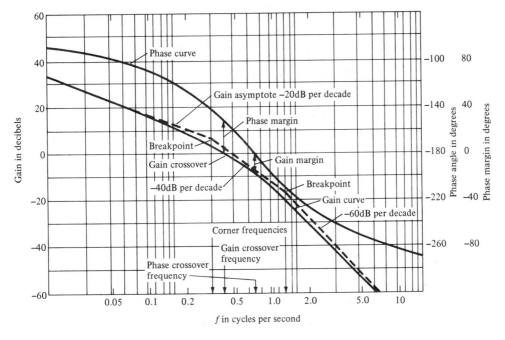

Figure 11.9 Graph of stability factors.

11.4 CONTROL SYSTEM COMPONENTS

Control systems are built using many different parts, some electrical, some mechanical, some electromechanical. There are a number of components that are used more frequently, and they are covered below. They include potentiometers, gear trains, synchros, and servos. Our concern here will be with those components more common in instrument servomechanisms—systems which control position, speed, or acceleration. In addition to the basic control system components mentioned above, the complete system usually contains amplifiers, feedback elements, power activators, and modulator-demodulator networks in some systems.

Potentiometers

Potentiometers used in control systems are usually precision wire-wound circular components, often with gearing. As a slider-arm pickoff is rotated from a starting point the amount of resistance between the slider arm and either extreme of the resistor varies. Using dc voltage excitation it is possible to pick off a dc voltage proportional to the angular shaft rotation. Figure 11.10(a) shows a potentiometer assembly, Fig. 11.10(b) the potentiometer symbol with dc excitation, and Fig. 11.10(c) a block dia-

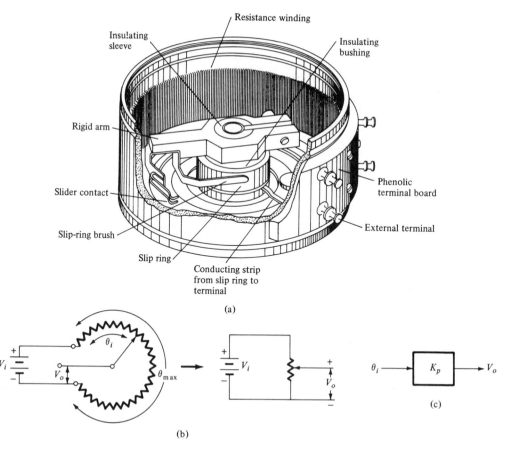

(a)

(b)

(c)

Figure 11.10 Potentiometer (a) assembly. (b) schematic. (c) block symbol.

gram representation. Assuming linear operation the output voltage is proportional
to the amount of rotation and excitation voltage so that

$$V_o = V_i \frac{\theta_i}{\theta_{max}} = K_p \theta_i \qquad (11.12)$$

where K_p, as shown in Fig. 11.10(c), is the potentiometer transfer function in volts
per radian. If the potentiometer shaft is connected to some device, for example, a
valve which rotates open, the resulting potentiometer voltage would be an indication
of the amount of valve opening which can be transmitted as an electrical dc voltage—
the potentiometer acting as a transducer device to convert angular rotation into
electrical voltage.

A pair of potentiometers may be used as an error detector to sense when the
positions of two different shafts are not the same and provide an electrical signal
proportional to this shaft displacement. Figure 11.11(a) shows such a circuit setup
with the representative block symbol shown in Fig. 11.11(b). The circuit acts as a

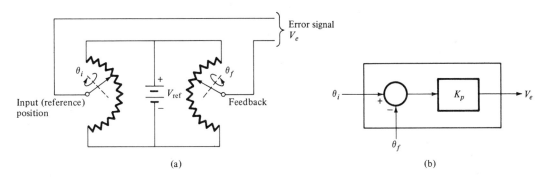

(a)   (b)

Figure 11.11 (a) Two-potentiometer error detector; (b) block diagram.

bridge circuit providing 0-V output when the bridge is balanced. The voltage due to
a difference between the input and feedback potentiometer settings is used as an
error signal to drive the mechanism connected to the feedback shaft. In a properly
adjusted system any change in reference position results in an error voltage, which
then drives some mechanical device until the feedback shaft position again aligns with
the reference shaft and no error voltage occurs. A diagram of a complete servomechanism system using potentiometers as error detectors is shown in Fig. 11.12.

An input shaft position setting, θ_i, resulting in a positional difference between the
wiper arm settings of input and feedback potentiometers, produces a dc voltage to a
dc amplifier circuit. The magnitude and polarity of this error signal drive the dc
motor to realign the feedback potentiometer wiper arm. The dc motor also reposi-
tions the device being controlled to produce the desired overall servomechanism
action.

The bridge reference or excitation voltage can be either a dc voltage as previously
shown or ac voltage excitation. With ac excitation the error signal magnitude is still
an indicator of the amount of shaft position difference between input and feedback

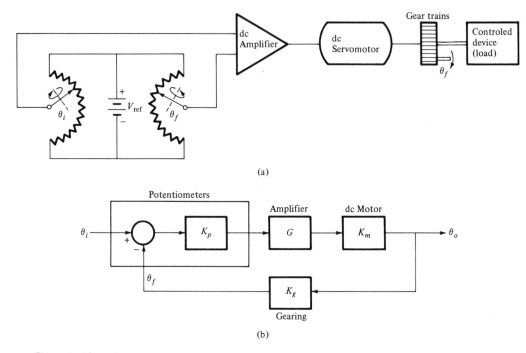

Figure 11.12 Position servomechanism (a) system diagram. (b) block diagram.

signals. The direction of the error signal is indicated, however, by the phase reversal of the ac signal instead of a voltage polarity reversal as with dc operation. The amplifier can then be an ac circuit and the motor an ac motor unit. When it is desired to operate partly with dc devices demodulator-modulator circuits can be used to go from ac to dc and back to ac again.

Synchros

Another very popular unit used as an error detector in control systems is the synchro (Fig. 11.13). Control or instrument synchros (as opposed to torque or driving synchros) provide for transmission of shaft position information for input and output with the ability to provide an error signal if the two shafts are not aligned. The usual connection of a transmitter (CX)-receiver (CR) synchro pair, shown in Fig. 11.14, can be used to remotely transmit shaft position information. The synchro consists of a rotor, shown in Fig. 11.14 to be excited by the ac line (typically 400 Hz in aircraft systems). The ac voltage is connected to the single salient pole rotor windings by means of slip rings and brushes. The magnetic field set up by the rotor and ac excitation is coupled to three separate stator windings positioned 120° apart. One terminal of each winding is connected to a common (central) point, while each of the other winding terminals forms a three-phase voltage for connection to the receiver synchro. The transmitter rotor is held fixed at the desired (or command) setting. Using stator

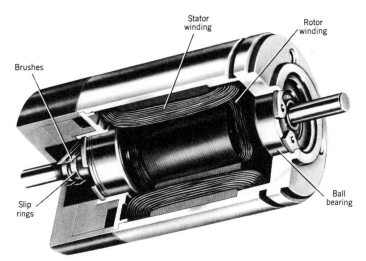

Figure 11.13 Cutaway view of a typical synchro. (Courtesy Weston Instruments, Inc; Transicoil Division).

winding S_2 as reference, the zero rotor position is defined as the rotor aligned with winding S_2. The three stator induced voltages can be mathematically expressed with reference to the central point as*

$$V_{S_1} = KV \cos \theta$$
$$V_{S_2} = KV \cos(\theta + 120°) \qquad (11.13)$$
$$V_{S_3} = KV \cos(\theta - 120°)$$

where θ is the rotor angle with respect to winding S_2.

The excited rotor winding of the receiver synchros similarly induces voltages in the three stator windings of the receiver. If the receiver and transmitter rotors are aligned, these induced voltages are equal, and no net current flows in the stator windings. If, however, the receiver rotor is at an angle different from the transmitter rotor, the transmitter-induced voltages and net currents circulate in the stator windings. Since the transmitter rotor is held fixed and the receiver rotor is, at most, lightly loaded, the resulting receiver stator and rotor magnetic fields interact to pull the receiver rotor into line with the transmitter rotor position. If the transmitter rotor is later moved again, the receiver rotor will be driven to line up with the transmitter rotor. Since the only connection between the transmitter and the rotor is the electrical wires, the transmitter could be mounted to the shaft under observation, and the receiver synchro could be used to drive a light load (indicator dial) at a distant display console.

*The stator voltages can also be expressed across pairs of terminals S_1, S_2, S_3 as
$$V_{S_1 S_2} = \sqrt{3} KV \cos(30° - \theta)$$
$$V_{S_2 S_3} = -\sqrt{3} KV \cos(30° + \theta)$$
$$V_{S_1 S_3} = KV \cos \theta$$

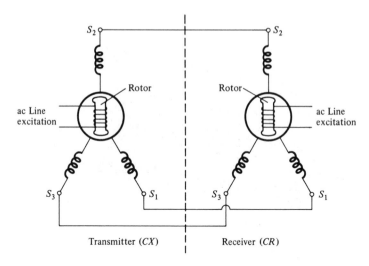

Figure 11.14 Synchro transmitter-receiver pair.

If the load to be driven is not light, then a synchro transformer can be used in place of the synchro receiver, as shown in Fig. 11.15. The synchro transmitter rotor may be positioned as desired with a set of voltages developed in the three windings as expressed in Eq. (11.13). These voltages are connected through a set of information lines to the stator windings of a control transformer. The control transformer rotor is, however, not the same as that of a receiver (the CT rotor is cylindrical) nor is any ac excitation applied to it. The control transformer rotor is set to a fixed zero angle position, typically perpendicular to the S_2 winding. A net voltage is induced in the CT rotor windings, which is a function of the CX stator voltages and thereby a function

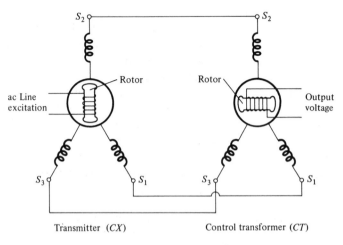

Figure 11.15 Synchro transmitter-control transformer pair.

of the *CX* rotor position. The output voltage of the control transformer can be expressed as

$$V_o = K_s \sin \theta_i \qquad (11.14)$$

where K_s is the control transformer sensitivity measured in volts per radian or volts per degree and θ_i is the angle or position of the transmitter shaft position rather than a corresponding shaft position as with a *CX-CR* pair. The resulting control transformer output voltage [Eq. (11.14)] can be used to drive some power device to position some other shaft or display.

Servomotors

A servomotor may be either dc or ac, with the latter type, most common in low-power applications, to be covered here. An ac servomotor is essentially a two-phase induction motor, of typically small size, for use in an instrument servomechanism system. Figure 11.16 shows the schematic diagram of an ac servomotor. A reference

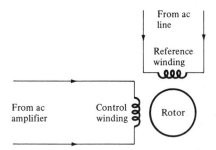

Figure 11.16 Schematic diagram of ac servomotor.

stator receives excitation from the fixed ac line voltage. A control stator winding is excited by the ac drive signal from an amplifier. An error signal obtained by, say, a synchro transmitter-control transformer pair is applied to an ac amplifier. The amplifier signal driving the servomotor control winding will vary in both amplitude and signal voltage phase. The servomotor speed will be proportional to the control winding signal amplitude, while the direction of rotation will depend on the phase of the drive signal. A typical control system using a servomotor is shown schematically in Fig. 11.17(a) and in block diagram form in Fig. 11.17(b). A command or input shaft position which will determine the load shaft position is connected to the rotor of the synchro transmitter. The *CX-CT* pair then provides an output voltage from the control whose amplitude and phase depend on the input shaft position (command) and the *CT* rotor shaft position (feedback). An error signal in the form of ac voltage V_t from the control transformer is applied to an ac amplifier, whose output signal drives the servomotor control winding. The servomotor then drives the load (through gearing in this example). As the load moves to the desired position the *CT* rotor is repositioned through the gearing and the error voltage driving the ac amplifier is reduced, until a null position is reached. This load position will be maintained unless the input (command) position is changed.

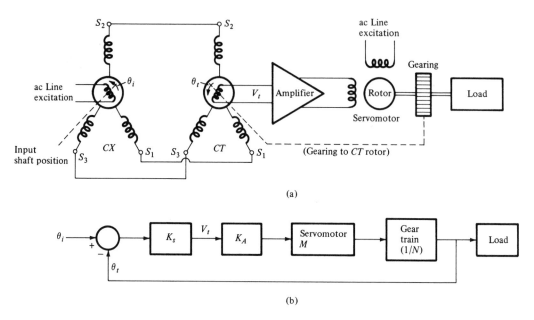

(a)

(b)

Figure 11.17 Position control system.

Gear Trains

Connection of various shafts is often made through gear trains rather than directly. If the number of teeth on each gear is different, a gearing ratio results. This gearing can be used in a servomotor system to increase the torque driving the load since the servomotor itself is a high-speed, low-torque device. The gear train transfer function [see Fig. 11.17(b)] is given by the teeth ratio as

$$\text{Transfer function} = \frac{\theta_o}{\theta_i} = \frac{N_o}{N_i} = \frac{1}{N} \tag{11.15}$$

where N is the ratio of teeth in output to input gears.

The control system of Fig. 11.17 will compare the command shaft position to the load shaft position, develop an error voltage, and drive the servomotor so that the load is moved to the command position. If the error is great, a large error signal results, and the servomotor is driven very fast, so fast that it may overshoot the desired position and then oscillate back and forth around the desired position until it finally settles down. One means of compensating for the drive at great position error without making the small error signal drive too sluggish is the use of a rate generator or tachometer in a second feedback path. An ac tachometer, as shown in Fig. 11.18, is a two-phase induction device. In the tachometer an ac reference voltage is applied to one stator winding, and the rotor is driven by the servomotor, resulting in an output voltage proportional to the servomotor speed from a second stator winding. The resulting output ac voltage can be expressed as

$$V_o = K_g \omega \tag{11.16}$$

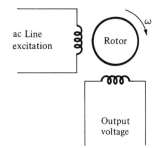

Figure 11.18 Ac tachometer schematic diagram.

where K_g is the rate generator sensitivity in volts per radian per minute (usually rated at 1000 rpm) and ω is the rotational speed of the rate generator rotor shaft in radians per minute. The rate generator transfer function can then be expressed as

$$\frac{V_o}{\theta_i} = K_g s \qquad (11.17)$$

where, θ_i is the input angular position (rotor position) and $\omega = s\theta$ in Laplace notation. Figure 11.19(a) shows a positional servo system schematic diagram. A command position input, θ_i, positions the synchro transmitter rotor, resulting in an error voltage from the synchro transformer dependent on the position of the CT rotor, θ_t, which is mechanically connected to the load. The error voltage is summed with the rate tachometer output voltage to give a resulting error signal which combines both position feedback via CT and rate feedback via the tachometer. The amplified

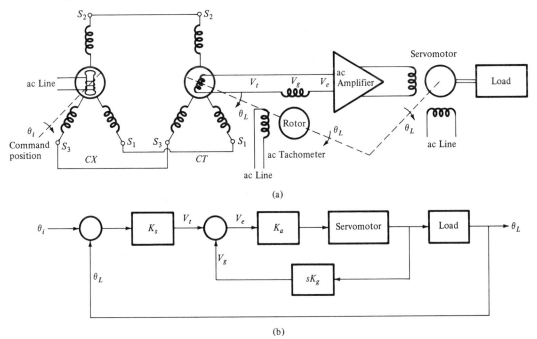

Figure 11.19 Positional servo system with rate feedback.

error voltage then drives the load through the servomotor to the desired position. Figure 11.19(b) shows the positional servo system in block diagram form, with two feedback loops, one positioned and one rate-controlled.

PROBLEMS

§ 11.2

1. Determine the transfer function of the network shown in Fig. 11.20 and calculate the value of the transfer function for $R_1 = 10$ K, $R_2 = 20$ K.

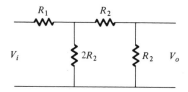

Figure 11.20 Network for problem 1.

Figure 11.21 Feedback network for problem 2.

2. Calculate the closed-loop gain of the feedback circuit of Fig. 11.21.
3. What is the difference between open-loop and closed-loop gain?
4. Calculate the gain of the feedback network of Fig. 11.22.

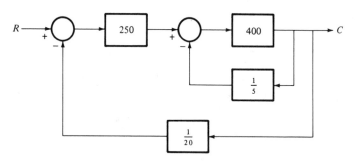

Figure 11.22 Feedback network for problem 4.

§ 11.3

5. What does a Bode plot represent?
6. Sketch the magnitude Bode plot for $100/[(j\omega_a)(1 + j\omega_b)]$ where $\omega_a = 10$ rad/sec and $\omega_b = 15$ rad/sec.
7. Define gain margin.
8. Define phase margin.
9. Sketch a Bode phase plot for the transfer function $100/[(j\omega_a)(1 + j\omega_b)]$.

12

Digital Computers

12.1 GENERAL

A digital computer means many different things to different people. To a business man it is a report generator and payroll calculator. To a scientist it is a lightning-fast complex calculator enabling solution of extremely long or complex problems. The computer is basically a very simple device which can carry out such operations as adding two numbers, subtracting, multiplying, and dividing. By cleverly using these operations, and a few other test-type instructions, the computer can be *programmed* to do almost anything. The word programmed is the clue to what can be done with a computer. Stated simply, the computer can do anything which *you* (or others) can program it to do. The computer knows nothing and cannot "think" for itself. It is the person using the machine who writes the program which directs the computer to act. The overall result of these simple steps may be the solution to some very important and involved problems, but each step itself is quite simple. If we describe the computer as a *general-purpose* digital computer, we mean a computer which operates on binary data (internally), handling any problem which is written for it. The general-purpose operation is important since the intended uses of the computer are unlimited. The binary (base 2) nature of the computer operation is important. The simplest way to operate with numbers using physical devices capable of the highest possible speeds is a binary or two-valued number. This is represented in a computer by, for example, a transistor conducting or not conducting, a switch ON or OFF, a light ON or OFF, a magnetic material magnetized clockwise or counterclockwise, etc. This simple nature of representing one of two possible conditions is why the binary operation is used. We shall study binary numbers in Section 12.2 and the ways of representing decimal numbers using groups of binary numbers in Section 12.3.

The final and also very important words to understand in describing the computer are the words *stored program*. The program used to operate the computer is stored in the memory of the computer, and each instruction is carried out as soon as the previous one is completed. In this way two factors are achieved. One is that the instructions can be carried out almost as fast as it is obtained from memory, which can presently be from millions to tens of millions per second; second, and more important, is that with the instruction stored in memory the instruction itself can be modified. Thus, to perform a program over and over again it is only necessary to modify that instruction, causing it to do the same basic series of operations but each time with some different information. There is actually another factor which makes today's computers so powerful, that is, the solid-state electronic advances which have occurred since modern computers were first built.

Having said a few words about the digital computer (stored-program, general-purpose, digital computer) we might quickly consider the differences between it and the analog computer to be covered in the next chapter. Basically, the analog computer is continuous-operating, must be rewired for each different problem solved, and is limited in accuracy by the physical tolerances of the components used. Also, the areas of use are restricted primarily to the scientific. The digital computer operates on discrete data, programs may be simply changed using the fast input devices available, the type of problem solved is unlimited, and the accuracy is limited only by the number size (word size) used. If greater accuracy is desired, it can be obtained, if programming effort is worth it. The main advantages of the analog computer over the digital are the simplicity of setting up the problem (programming is not needed) and of changing the problem, the fact that the solution obtained directly is a graph of the results in real time, and the comparatively lower cost of the analog computer. The words *real time* mean that the operation has an output when the input is present.

Figure 12.1 shows a general computer block diagram. Basically, there are five parts of a computer: the *input units*, providing both a means of entering the program and data into the computer; the *memory unit*, which stores this information; the *arithmetic unit*, which provides the functions of addition, subtraction, multiplication, and division; the *control unit*, which analyzes each instruction and then sends the signals to the different parts of the computer in order to carry out the instruction; and, finally, the *output units*, which provide the answers to the user. Considering first the input unit, there are a number of very useful and common forms of input. Punched card input is very common, as are paper tape, magnetic tape, the magnetic disk, and the cassette cartridge. These devices will be covered in Section 12.10.

The memory unit is probably either magnetic core or integrated circuit (IC) storage, although newer special devices are also used. The memory stores binary numbers, 0 or 1, and can provide these numbers at rates over 1 million/sec. The program is stored in the computer memory unit after being read in from punched card or magnetic tape or disk, for example. The control unit reads each instruction from memory, decodes or interprets the instruction, and then sends command signals to the appropriate computer units to carry out the instruction. When an instruction is completed the memory unit then calls for the next instruction from memory and continues in this manner. The control unit doesn't know what instruction is just performed or what is yet to come. It faithfully does what it is told by the program. The

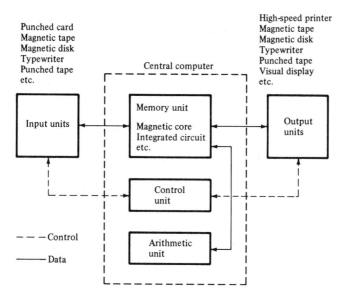

Figure 12.1 Digital Computer. Block diagram.

arithmetic unit is provided with the numbers to operate on by instructions which specify where in memory to get the numbers from. Once in the arithmetic unit an instruction to add, for example, will cause the arithmetic unit to perform that operation, and the answer is then available for use. Finally, the output unit provides such outputs as printed results on a high-speed printer or output on a punched card, punched paper tape, magnetic tape, or a magnetic disk. Details of these different computer parts are covered in later sections of this chapter.

12.2 NUMBER SYSTEMS

Recalling that binary (base 2) numbers are used in computers because of the simplicity of physical devices to represent the two possible states or digits in that system, we should now review what a number system is so that the simplest useful base of all (base 2) is understood. If decimal numbers were well understood, this section would not be necessary. Since this is generally not true, we shall go through the definitions and application of any base system, the base 2 being of direct interest as are bases 8 and 16.

Binary Number System

Just as the digits of the decimal (base 10) number system are 0, 1, ..., 9, with 10 a number made up of the digits 1 and 0, the binary (base 2) number system contains the digits 0 and 1, and the numbers of the system are made using these digits. The significance of a number system is that of position. Placing 0 in the rightmost position and 1 in the next leftward position makes the digits represent the number 10 in the decimal system. In the binary system these same digits in the same position would represent the number 2. A number can be expressed in general as

$$N = d_n \times R^n + \cdots + d_3 \times R^3 + d_2 \times R^2 + d_1 \times R^1 + d_0 \times R^0$$

For the decimal system R, which represents the *radix* or base, is used to give positional value to the digits represented by the subscripted d values. For example, the number 398 can be written as

$$N = 3 \times 10^2 + 9 \times 10^1 + 8 \times 10^0$$

The digits 3, 9, and 8 in the positions chosen for the example indicate how many of each positional values there are. In the present example there are 3 hundreds, 9 tens, and 8 units. Let's see how this concept carries over to the binary system. The number 101 (remember, only the digits 0 and 1 can be used) can be expressed as

$$N = 1 \times 2^2 + 0 \times 2^1 + 1 \times 2^0$$

or, 1 four, 0 twos, and 1 unit, which is 5 in decimal. Thus the number 101 in binary represents the same quantity as the number 5 in decimal.

As another example, consider the binary number 1011101 and find its decimal equivalent. By the scheme outlined above we get

$$N = 1 \times 2^6 + 0 \times 2^5 + 1 \times 2^4 + 1 \times 2^3 + 1 \times 2^2 + 0 \times 2^1 + 1 \times 2^0$$

or

$$64 + 0 + 16 + 8 + 4 + 0 + 1$$

which is 93 in decimal. The conversion of a given binary number into decimal is straightforward as shown.

Conversion of a number from decimal to binary is not so straightforward. What is the binary equivalent of the decimal number 98? One way to get the binary number would be to add the power of two parts needed for this number. That is, 128 (2^7) is too large; 64 (2^6) is needed, leaving 34, and then 32 (2^5) is needed, but 16 is not and also 8 and 4 are not needed; 2 (2^1) is needed but 1(2^0) is not, giving the final answer of 1100010. This is, of course, a poor way to go about the conversion. This is especially apparent if asked to convert a number such as 93,785.

A simple but effective method is outlined below. The number in decimal to be converted into binary is continuously divided by 2 (base of the system into which it is being converted) and the remainders marked down. *The converted answer is read as the last remainder back.* An example will show this more clearly.

EXAMPLE 12.1

Convert 98 into binary.

Solution:

$$\frac{98}{2} = 49 \qquad +0 \text{ remainder}$$

$$\frac{49}{2} = 24 \qquad +1 \text{ remainder}$$

$$\frac{24}{2} = 12 \qquad +0 \text{ remainder}$$

$$\frac{12}{2} = 6 \qquad +0 \text{ remainder}$$

$$\frac{6}{2} = 3 \qquad +0 \text{ remainder}$$

$$\frac{3}{2} = 1 \qquad +1 \text{ remainder}$$

$$\frac{1}{2} = 0 \qquad +1 \text{ remainder}$$

The answer is read as 1100010.

Octal Number System

Having laid the groundwork we can quickly learn the octal or base 8 number system. First, we must remember that the octal number system has only the digits $0, 1, \ldots, 7$. The number 8 is 10. Converting an octal number into decimal is straightforward. As an example, convert 375 octal into its decimal equivalent.

EXAMPLE 12.2

$(375)_8 = 3 \times 8^2 + 7 \times 8^1 + 5 \times 8^0 = 3 \times 64 + 7 \times 8 + 5 \times 1 = 253$ (decimal).

Another example is the conversion of 1127 (octal) into decimal.

EXAMPLE 12.3

$$(1127)_8 = 1 \times 8^3 + 1 \times 8^2 + 2 \times 8^1 + 7 \times 8^0 = 512 + 64 + 16 + 7$$
$$= 599 \text{ (decimal)}.$$

Using the previous scheme for converting decimal into another base system, let's see how division by 8 and use of the remainder provides us with the answer.

EXAMPLE 12.4

Convert 189 (decimal) into octal.

Solution:

$$\frac{189}{8} = 23 \qquad +5 \text{ remainder}$$

$$\frac{23}{8} = 2 \qquad +7 \text{ remainder}$$

$$\frac{2}{8} = 0 \qquad +2 \text{ remainder}$$

The answer is $(275)_8$. Check it to see that it's correct.

Hexadecimal Number System

Another common number system used with digital computers is the hexadecimal or base 16. The digits in this number system go from 0 to 15, with the digits for 10 through 15 represented usually by the letter symbols A through F. Thus, the single digit symbol B represents the digit for 11, while C represents 12, etc. The positional value is units, 16s, 256s, etc., so that a number such as $14A_{16}$ is the equivalent of decimal 330 as follows:

$$(14A)_{16} = 1 \times 16^2 + 4 \times 16^1 + A \times 16^0$$
$$= 1 \times 256 + 4 \times 16 + 10 \times 1 = 330_{10}$$

EXERCISE 12.5

Convert the given hexadecimal numbers to decimal: (1) 24, (2) C9, (3) 2D9.

Using the scheme previously presented to convert decimal into another base system, note the following example.

EXAMPLE 12.6

Convert 330_{10} into hexadecimal.

Solution:

$$\frac{330}{16} = 20 \quad +10 \text{ remainder } (10 = A)$$

$$\frac{20}{16} = 1 \quad + 4 \text{ remainder}$$

$$\frac{1}{16} = 0 \quad + 1 \text{ remainder}$$

The answer is $14A_{16}$.

Arithmetic Operations

Having become familiar with expressing numbers in other than the decimal system, let's consider using these other number systems in doing such arithmetic operations as addition and subtraction and a little multiplication. Adding binary, octal, or hexadecimal numbers is the same as adding decimal.

EXAMPLE 12.7

$$\left(\begin{array}{c} 10111 \\ + \ 1011 \\ \hline 100010 \end{array}\right)_2 ; \quad \left(\begin{array}{c} 73 \\ +67 \\ \hline 162 \end{array}\right)_8 ; \quad \left(\begin{array}{c} 2A \\ +73 \\ \hline 9D \end{array}\right)_{16}$$

An addition table for binary numbers would show that adding 0 and 1 results in 1 as expected and that adding 1 and 1 gives a sum digit of 0 and a carry of 1 to the next higher bit position. Thus, in the above binary addition, adding 1 and 1 gave a sum digit of 0 and a carry of 1 into the next left position. The addition of the three 1s gave a sum of 1 and again a carry of 1. The remainder of the addition follows in a similar manner. The main point to remember is that $1 + 1$ is 0 and a carry of 1. Adding octal numbers as in Example 12.7 also requires some extra thought at first. Adding 3 and 7 gave a sum of 2 and a carry of 1. Using either an addition table (not provided), using rote memory, or doing the addition in decimal for the two numbers and then converting into octal, such as $3 + 7$ is $(10)_{10}$ or $(12)_8$, are all possible methods. Choose your own.

Subtraction, also the same in other base systems as in decimal, is nonetheless more difficult because of lack of familiarity. The rules for subtracting are the same, in that a smaller number is subtracted from a larger, and if the number to be subtracted (subtrahend) is larger than that which it is subtracted from (minuend), the base value must be borrowed from a higher position and the value in that decreased by 1. A few examples in binary subtraction will show what is involved.

EXAMPLE 12.8

$$\begin{array}{ccc} 1011 & 1010 & 10000 \\ -\ 101; & -\ 101; & -\ \ 111 \\ \hline 110 & 101 & 1001 \end{array}$$

Subtracting $1 - 0$ is 1, $1 - 1$ is 0, and $0 - 1$ is 1 with a borrow of 1. In the first example in Example 12.8 the third term to the left has $0 - 1$. To subtract the larger digit from the smaller it is necessary to borrow 1 from the next higher position, reducing the top digit to 0. The subtraction of $0 - 1$ then results in 1 for that position and leaves $0 - 0$ for the highest position of the number. In the third example in Example 12.8 the first subtraction (units position) of $0 - 1$ cannot be done directly, and 1 must be borrowed from a higher position. However, since the first higher position with a 1 is the 2^4 position, the lower positions are reduced each by 1, as is the 2^4 position. In order, then, the subtraction of $0 - 1$ gives 1, subtracting the reduced $1 - 1$ gives 0, again $1 - 1$ gives 0, then $1 - 0$ gives 1, and finally $0 - 0$ gives 0.

12.3 CODES

In handling decimal numbers in a computer, which operates in binary, it is necessary to represent the decimal number using binary. There are numerous ways of representing a decimal number using groups of binary. These representations are called codes, the most commonly used of these being the binary-coded decimal(BCD)code. Decimal and BCD digits are shown in Table 12.1. In using BCD numbers *each* digit of the decimal number is coded as a BCD character. Thus, the decimal number 79 is represented as 0111 1001 in the BCD coding. Some examples should make this representation clear.

Table 12.1

Decimal/BCD Table

DECIMAL	BCD CODE
0	0 0 0 0
1	0 0 0 1
2	0 0 1 0
3	0 0 1 1
4	0 1 0 0
5	0 1 0 1
6	0 1 1 0
7	0 1 1 1
8	1 0 0 0
9	1 0 0 1

EXAMPLE 12.9

389 (decimal) is 0011 1000 1001 (BCD)
1982 (decimal) is 0001 1001 1000 0010 (BCD)
357 (decimal) is 0011 0101 0111 (BCD)

Computers operate in either BCD or straight binary as a rule. To facilitate handling input and output in straight binary, a simple coding scheme, binary-coded octal or hexadecimal, is employed. This is useful since the coding only requires the proper grouping of numbers to go from this coded form to straight binary. Although decimal numbers (or decimal coding of numbers) are desirable, the savings in computer circuitry and speed sometimes make the octal or hexadecimal coding preferable. Consider the proposed scheme. The use of three binary positions allows representing all the digits from 0 (000) to 7 (111). Using this grouping of three places gives the code form.

EXAMPLE 12.10

756 (octal) is 111 101 110 in binary-coded octal.
345 (octal) is 011 100 101 in binary-coded octal.
111011101 is 111 011 101 or 735 in octal.
01110110 is 001 110 110 or 166 in octal.

In the first example in Example 12.10 the grouping of 111 for 7, 101 for 5, and 110 for 6 is clear. As an important point it should now be clear that the resulting number is *also* the straight binary form of the octal number. Read as 111101110, this number is the same as octal 756. Thus, converting from octal to binary is just a matter of writing each octal digit using *three binary places*. When used with the computer, each octal digit is separately converted into binary-coded octal as it enters, but the binary numbers are then grouped to give the binary equivalent of the input octal number. It is this simple operation which makes using octal coding important since it is far easier for us to write and remember octal numbers than binary. If the binary number is not a multiple of three places, the digits must be grouped starting from the right. Thus, in Example 12.10, 01110110 was grouped on the right as 110, then going left a second group was 110, and for the remaining digits 01 a leading 0 (which doesn't change the number) is added so that 001 is the third grouping, giving the result of 166 (octal).

With hexadecimal coding *four binary digits* are grouped to provide a more compact means of representing binary. For example,

$$011101010011 = 0111\ 0101\ 0011 = 753 \qquad \text{(in hexadecimal)}$$

Similarly,

$$101101111111 = 1011\ 0111\ 1111 = B7F \qquad \text{(in hexadecimal)}$$

Most present large-scale computers (IBM 370, for example) use this hexadecimal coding to represent binary numbers.

EXAMPLE 12.11

$A25_{16}$ is 1010 0010 0101 in binary-coded hexadecimal.
$1B79_{16}$ is 0001 1011 0111 1001 in binary-coded hexadecimal.
11010110_2 is 1101 0110 or $D6_{16}$.
01011100_2 is 0101 1100 or $5C_{16}$.

Another code representation used adds an extra bit position to an existing code form to make the number of 1 bits an even number in one case or an odd number in another. The added bit is called a *parity* bit, and the scheme used can be either *even* or *odd* parity. With even parity the extra bit is chosen (either 1 or 0) to make the number of 1s even and with odd parity to make the total number of 1s odd. For example, adding a parity bit for even parity to BCD digits of decimal 8 would be 1000 $\underline{1}$; the underlined bit, the parity bit, is chosen to be 1 since that makes the total number of 1s even. With decimal 9 the BCD with parity would be 1001 $\underline{0}$ for even parity. For odd parity decimal 5 would be written as 0101 $\underline{1}$, and decimal 7 as 0111 $\underline{0}$. Parity is added because it allows finding or detecting that certain errors occurred.

12.4 BOOLEAN ALGEBRA AND LOGIC

The two-valued logic, called Boolean logic, allows manipulating and simplifying circuits and connections used in switching systems and in computer logic circuits. To deal with circuits whose inputs and outputs can by only 1 or 0, TRUE or FALSE, ON or OFF, etc., we shall use the algebra developed by the English mathematician George Boole. *Boolean algebra*, also called logical algebra, is a much simplified form of algebra, since the output or answer can only be 1 or 0. There are two basic logic operations which may appear new to us. These are the AND operation and the OR operation. Considering a pair of switches for explanation, the AND function would be demonstrated by the two switches in series. If both switch A AND switch B are closed, the light will be energized in Fig. 12.2. The OR operation, demonstrated by the two switches in Fig. 12.3, requires that either switch A OR switch B be closed for the light to be energized.

We shall now consider a more formal and useful way of describing the two basic functions. A *truth table* is a listing of *all* the possible variations for the number of inputs under consideration and the output of the specific function under considera-

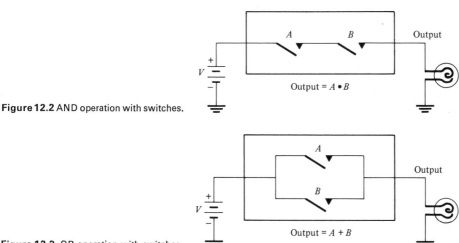

Figure 12.2 AND operation with switches.

Figure 12.3 OR operation with switches.

tion. The truth tables for the AND, OR, and NOT functions are shown in Table 12.2. We use the values 1 and 0 to represent the inputs and outputs.

Table 12.2

Truth Tables for AND, OR, and NOT Functions

A	B	$A \cdot B$	$A + B$	$\bar{A}$	$\bar{B}$
0	0	0	0	1	1
0	1	0	1	1	0
1	0	0	1	0	1
1	1	1	1	0	0

Logic Simplification

Once we obtain a logical expression we would want to simplify, if possible, for a more economic circuit. As a beginning consider the identities provided below:

$$A + A = A \qquad A + \bar{A} = 1 \qquad A + 1 = 1 \qquad A + 0 = A$$
$$A \cdot A = A \qquad A \cdot \bar{A} = 0 \qquad A \cdot 1 = A \qquad A \cdot 0 = 0$$

A few examples will show how logic simplifications are performed.

EXAMPLE 12.12

$$\begin{aligned}
\bar{A}\bar{B}C + \bar{A}B\bar{C} + A\bar{B}C + ABC &= \bar{A}(\bar{B}C + B\bar{C}) + A(\bar{B}C + BC) \\
&= \bar{A}(\bar{B}C + B\bar{C}) + AC(\bar{B} + B) \\
&= \bar{A}(\bar{B}C + B\bar{C}) + AC(1) \\
&= \bar{A}(\bar{B}C + B\bar{C}) + AC
\end{aligned}$$

EXAMPLE 12.13

$$\begin{aligned}
\bar{A}B + AB + \bar{A}\bar{B} &= (\bar{A} + A)B + \bar{A}\bar{B} \\
&= (1)B + \bar{A}\bar{B} \\
&= B + \bar{A}\bar{B}
\end{aligned}$$

While we were able to simplify the given logic expression, it is not obvious that the final expression could still be reduced to $B + \bar{A}$. Logic simplification techniques including use of Karnaugh maps are covered in more detail in texts on digital computer theory.*

Logic Block Diagrams

In practice, the AND circuits, OR circuits, and inverter circuits are made using electronic circuits of different types. Typical logic block symbols are shown in Fig. 12.4, which uses the ASA Standard symbols.

*_Introduction to Digital Computer Technology_, by L. Nashelsky, Wiley, New York, 1972.

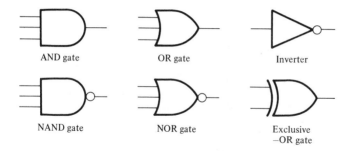

Figure 12.4 ASA standard logic symbols.

The logic diagram for implementing a given logic expression can be developed. These diagrams specify the interconnection of the AND gates, OR gates, inverters, NAND gates (AND-inverter combination), and NOR gates (OR-inverter).*

EXAMPLE 12.14

The logic diagram for the expression $\bar{A}B + AB + \bar{A}\bar{B}$ is shown in Fig. 12.5a.

EXAMPLE 12.15

With the logic circuit for $AB + A \cdot (\overline{B + C})$, we see an expression in which the partial term $B + C$ formed first is then inverted before being ANDed with the A term. Although the expression can be simplified, the circuit diagram for the given expression is shown in Fig. 12.5b.

EXAMPLE 12.16

The logic circuit diagram for $\bar{A}BC + A\bar{B}C + \bar{A}B\bar{C} + ABC$ is shown in Fig. 12.6 using NAND gates.

With this background we can now consider a practical example applying these techniques. To develop the logic circuit for adding two binary inputs we can first write the truth table and from it draw the logic circuit needed. With two inputs the logic circuit is called a half-adder, since in general when adding two numbers it is also necessary to add in the carry occurring in the lower binary position. If in the 2^0 position the two bits added are both 1, the sum is 0 and there is a carry to the next higher position of 1. The half-adder is only good for the 2^0 position since for any other position three inputs are required. However, two half-adder circuits can be

---

*A logical operation called De Morgan's rule can be used to alter the form of these inverted expressions. De Morgan's rule is

$$(\overline{A + B + C}) = \bar{A} \cdot \bar{B} \cdot \bar{C}$$

or

$$(\overline{A \cdot B \cdot C}) = \bar{A} + \bar{B} + \bar{C}$$

The rule states that an overall inversion to a logical expression in parentheses is equal to an expression with each term inverted and each AND and OR operation interchanged. De Morgan's rule is quite useful when handling the resulting output logical expression from NAND or NOR circuits.

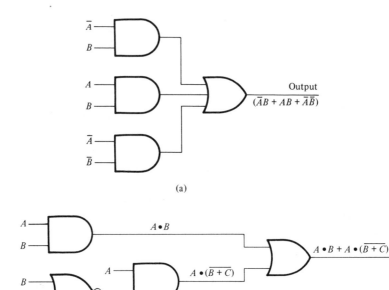

(a)

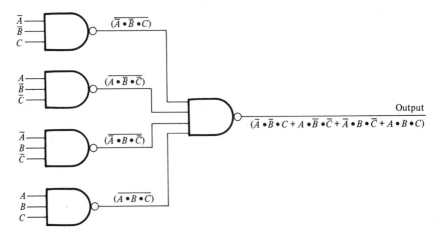

(b)

Figure 12.5 Logic diagram for (a) $\bar{A} B + AB + \bar{A} \bar{B}$; (b)
$AB + A\overline{(B+C)}$

Figure 12.6 Logic diagram for $\bar{A} B C + A\bar{B}C + \bar{A}B\bar{C} + ABC$

connected to provide what is called a full-adder, which has three inputs and provides
a SUM and CARRY output. Because the half-adder is simpler it will be left as an
exercise, and the full-adder circuit will now be developed. For a truth table we consider the circuit as having two outputs, each generated separately, which are operated
by three inputs. With three inputs the table must consider the eight possible combinations. Whenever any two of the inputs are 1 the SUM output is 0, or when all
inputs are 0 the SUM output is 0. When only one input is 1 or when all three inputs

are 1 the SUM output is 1. The CARRY output is 1 when any two inputs are 1 or
when all three inputs are 1. Otherwise the CARRY output is 0. The truth table for
the full-adder is shown in Fig. 12.7, and the logical expressions obtained from the
table are

$$\text{SUM} = \bar{A}\bar{B}C + \bar{A}B\bar{C} + A\bar{B}\bar{C} + ABC$$

$$\text{CARRY} = \bar{A}BC + A\bar{B}C + AB\bar{C} + ABC$$

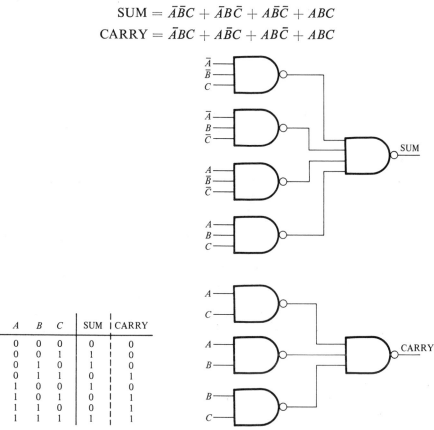

A	B	C	SUM	CARRY
0	0	0	0	0
0	0	1	1	0
0	1	0	1	0
0	1	1	0	1
1	0	0	1	0
1	0	1	0	1
1	1	0	0	1
1	1	1	1	1

Figure 12.7 Logic diagram of full-adder.

The SUM expression cannot be reduced much, but the CARRY can be reduced to
$\text{CARRY} = C(A + B) + AB$ (can you reduce the original expression to this form?).
The logic diagram for the complete full-adder is shown in Fig. 12.7 with a modified
SUM expression and the reduced CARRY expression implemented. The full adder
can be built using the gates shown and will provide the proper SUM and CARRY
values for any of the possible combinations of inputs that occur. Some variation of
the circuit is used in any computer to provide the addition operation.

12.5 COMPUTER MULTIVIBRATOR CIRCUITS

There are three basic multivibrator circuits, each providing two outputs which
are complements ($\bar{A}$ is the complement, or inverse, of A). The first circuit is an *astable*

multivibrator (Fig. 12.8), which, as the name implies, has no stable state. What this means is that the outputs cannot remain at either, say, $+5$ V or 0 V but must constantly oscillate back and forth between these two levels, at a rate determined by the circuit elements. The astable multivibrator is, then, a square-wave oscillator providing the basic timing signal or clock signal for a digital circuit.

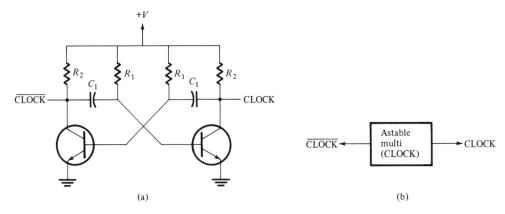

Figure 12.8 Astable multivibrator.

Second is the *monostable multivibrator circuit* (also called one-shot), which has only one stable condition. For the discrete circuits and block diagram shown in Fig. 12.9 the output market TRUE, for example, might be at 0 V as long as no signal is applied to the trigger input (the monostable circuit must be triggered or driven by a pulse in order to operate, unlike the astable, which operates without any external

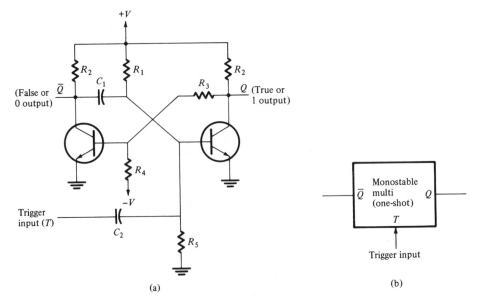

Figure 12.9 Monostable multivibrator.

inputs). When triggered the reference output will go to $+5$ V but remain at this level for only a fixed time interval, determined by the circuit elements. Some uses of the monostable circuit are for pulse reshaping (the pulse width is made independent of the input pulse shape) and for time delay purposes, where the time interval is needed to delay one computer operation from the next and the end of the monostable interval serves as an indication that a certain (fixed) amount of time has passed. This delay can be adjusted from microseconds to seconds.

The third, but most widely used, circuit is the *bistable multivibrator*, more commonly called a *flip-flop*. A discrete circuit and block diagram are shown in Fig. 12.10.

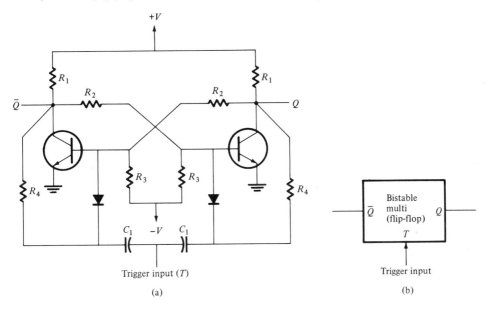

Figure 12.10 Bistable multivibrator.

As the name states, the flip-flop can *stay* with the Q output at $+5$ V (and $\bar{Q}$ at 0 V) *or* Q output at 0 V (and $\bar{Q}$ at $+5$ V) indefinitely, or as long as power is kept on. It may be triggered into the other states, opposite voltage levels, by an input trigger signal. Once triggered the circuit stays in the new state until triggered again. Either state can be maintained, as compared to the monostable circuit, which could stay in only one state indefinitely, or the astable circuit, which could not stay in any state for long. The uses of the flip-flop are many. It is basically a working storage device for temporary storage of a binary digit (called *bit*) and may also be used as a memory device. Within the main working of the computer it is used extensively. It can be used to form a counter where each flip-flop stage provides a power of two count. With two stages there can be four counts; with three stages, eight counts; and with n stages, 2^n counts. A counter can be used to accept input pulses and sum these by counting and can be used to count the number of steps in an operation. The bistable circuit may also be connected as a shift register, which allows moving a group of binary digits, forming what is called a word, from one computer location to the next, as from the arithmetic unit, etc., or as a parallel transfer register to move all bits at once.

A counter circuit for a count of eight is shown in Fig. 12.11 and a shift register is shown in Fig. 12.12.

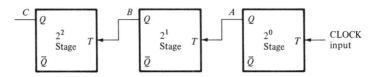

Figure 12.11 Three-state binary counter.

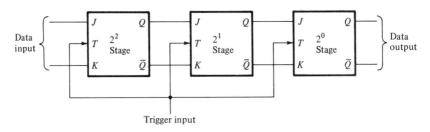

Figure 12.12 Three-stage shift register.

12.6 INTEGRATED CIRCUIT (IC) COMPUTER UNITS

Computer circuits are almost exclusively manufactured as integrated circuits or IC logic units (refer to Chapter 7). These ICs may be manufactured to contain as little as a couple of NAND or NOR gates, a half-dozen inverters, and one or two flip-flop circuits—these examples comprising small-scale integration (SSI). The IC unit may also be manufactured to contain a complete binary counter and a complete decoder of various types—these examples comprising medium-scale integration (MSI). A growing number of IC circuits are being manufactured as complete electronics, for example, for a calculator, electronic clock, computer memory, or computer CPU—this quite complicated electronic unit being a large-scale integrated (LSI) circuit.

Logic circuits built as IC units may be made using transistor-transistor logic (TTL) or field-effect transistors circuits (FET), as examples of two very popular techniques at present. Figure 12.13 shows a few types of IC logic gates—TTL, DTL, RTL, and FET—as examples of a few of the common techniques used presently. Various factors of size, switching speed, cost, and voltage compatibility are used in selecting which type of circuit to use.

IC flip-flop circuits made of these various types of construction are most often built as *JK* flip-flops (see Fig. 12.14). The *JK* designation refers to the input terminals provided to control the flip-flop operation when triggered. Complete specifications include the synchronous operation using the trigger (T) and J, K inputs, and the asynchronous operation using the direct set (preset), PS, and direct clear, CLR, inputs. In synchronous operation the inputs at the J and K terminals [there are four

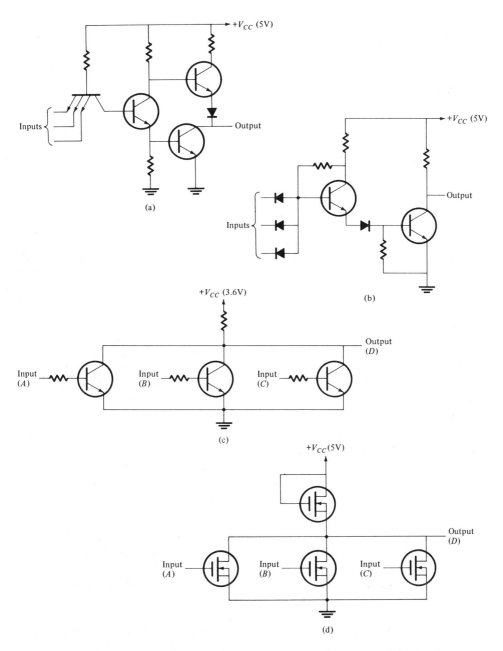

Figure 12.13 Digital IC logic gates: (a) TTL; (b) DTL; (c) RTL; (d) FET.

combinations of 1 and 0, as described in Fig. 12.14(b)] determine the resulting state of the flip-flop *after* a trigger pulse occurs so that the operation is synchronized by the trigger input pulse. For example, with J input low (0 V) and K input high $(+V)$ the circuit will be in the reset or clear state *after* the trigger pulse regardless of the

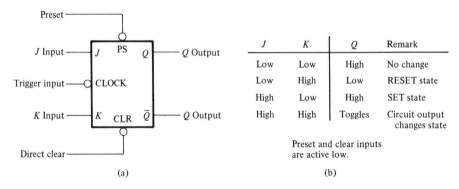

J	K	Q	Remark
Low	Low	High	No change
Low	High	Low	RESET state
High	Low	High	SET state
High	High	Toggles	Circuit output changes state

Preset and clear inputs are active low.

(a)   (b)

Figure 12.14 JK flip-flop. (a) logic diagram. (b) synchronous operation truth table.

state it was in before the trigger pulse. The PS input going low, on the other hand, would immediately set the flip-flop (and CLR going low would immediately reset the flip-flop)—no trigger pulse being necessary. This asynchronous operation takes place whenever an input (PS or CLR) goes low and needs no synchronizing trigger pulse to occur. Some typical operations using these *JK* flip-flops are shown in Fig. 12.15. Figure 12.15(a) shows a three-stage binary counter, Fig. 12.15(b) shows a decade or count-of-ten counter, and Fig. 12.15(c) shows a four-stage parallel transfer register.

While many SSI and MSI units are built using TTL circuits, LSI circuits are built using FET logic since the size and manufacturing yield are far better. A single LSI chip may contain a number of serial or parallel transfer registers, counters, logic arithmetic units, control units, and timing circuits. While we need some circuit operating description to use SSI and MSI units, we need far less specification for LSI units, which are so complete and specialized in application. A growing application of LSI technology is the microcomputer, with complete computer units being built on a single chip or on a couple of basic building block chips.

12.7 ARITHMETIC UNIT

The computer arithmetic unit (AU) carries out the operations of addition, subtraction, multiplication, and division as called for by the program. Other operations desired, such as taking the square root or cube root, cubing a number, doing differentiation or integration, etc., are all obtained from programs which use the four basic arithmetic operations to achieve the more complex function.

The data to be operated on are first read from an input unit into the memory unit. In using the arithmetic unit the numbers to be added, for example, are each brought into the arithmetic unit and then added together, the sum being left in the arithmetic unit. Actually, each computer may do this somewhat differently, but the basic operation of addition is carried out in the arithmetic unit. In computers designed primarily for scientific use the practice is to require an instruction to bring the first number into the AU and a second instruction to bring the second number into the AU and add it to the first number, leaving the answer in a location called an accu-

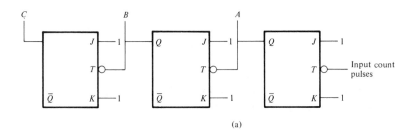

(a)

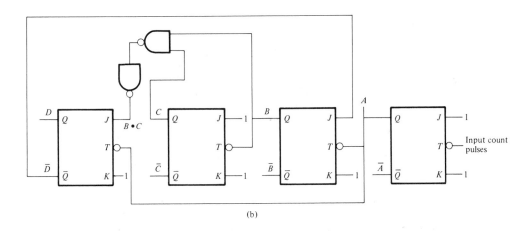

(b)

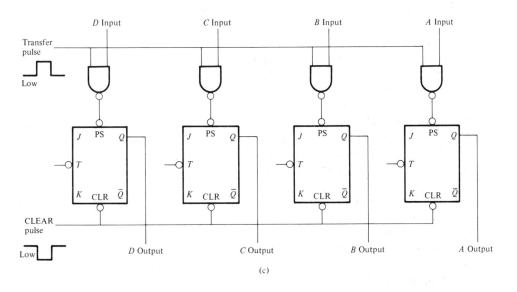

(c)

Figure 12.15 (a) 3-stage binary counter. (b) decade counter. (c) 4-stage parallel transfer register.

mulator. As the name accumulator implies, it is now possible to just give a command to add a third number, the resulting operation providing the sum of the first two and the third in the accumulator location. When the result is to be typed out or stored temporarily the accumulator value is stored in the memory unit. The basic operation, then, is to take the numbers from the memory unit and store the result back in the memory unit. A simple logic diagram of an arithmetic unit is shown in Fig. 12.16.

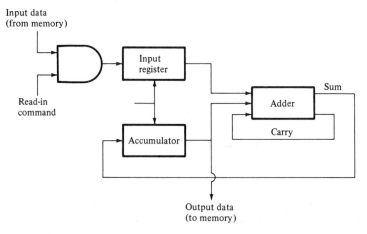

Figure 12.16 Arithmetic Unit, logic block diagram.

Two registers are shown, these being made of flip-flop circuits connected to form a shift register. The number of shift register stages is the same as the number of bits in a word for that particular computer. Word lengths of 12, 16, or 24 bits are common. The trigger pulses used to move the word into and out of the register are called shift pulses (SP), 16 shift pulses occurring to read a 16-bit word into the input register. As shown, the number in the input register and that in the accumulator are added together by the adder unit (similar to the full-adder circuit developed in Section 12.4), and the sum is shifted back into the accumulator. Thus, the input number is added to the number already in the accumulator, and the accumulated sum is placed back into the accumulator. When the sum obtained is to be read out the contents of the accumulator are shifted out and fed into the memory unit for storage. In some problems it would be helpful if when reading from the accumulator the accumulator contents are preserved. The shift-around accomplishes this by shifting the word out to the memory and also back into the accumulator register. Thus, as the accumulator word is being shifted out it is being shifted back into the left end of the register. At the end of the readout operation the accumulator word is stored in memory and is still present in the accumulator.

The adder unit has three inputs: the two numbers to be added and the carry. The two numbers are added one bit position at a time. As the number is shifted to the adder unit, bit by bit, the adder provides the sum of the bits for that position and the carry. The carry is stored by the adder (in a flip-flop) and used as input when the next-higher-position bits are added. The number must be shifted into the adder, lowest position first, and the carry is initially set to 0, since no previous carry has occurred. At the end of the add operation, the carry is also important. If the most significant position bits added (with the previous position carry) resulted in a carry, the answer is then of one position larger in size than the basic word length. Since the machine is designed to operate, say, with 16-bit words, the registers designed to handle 16-bit words, the memory to store 16-bit words, etc., the occurrence of the seventeenth bit is a special condition. Since it cannot be stored properly, the carry bit being 1 *after* all 16 positions of the number are added is an indication that there was an *overflow*. In some computers this will cause a stop in program operation; in most others it will cause a special error routine to take over.

It should be pointed out that many other arithmetic unit designs are implemented in various computers; in most the operation is carried out in parallel (all 16 bits added using 16 adders at one time) rather than *serially* (1 bit at a time) as illustrated here. The parallel operation has the disadvantage of greater cost but the important advantage of greater speed. With a 1-μsec add time the serial addition of a 16-bit word would take 16 μsec when done serially and only the 1 μsec done in parallel.

The multiplication operation carried out by the computer is actually implemented using the adder unit. Repeated addition and shift steps are performed to achieve the multiplication of two numbers. When multiplication is done the multiplicand is added to the product if the multiplier is 1 and zero added if the multiplier bit is 0. Then the multiplier word is shifted one position, as is the partial product obtained. This same procedure is again carried out as many times as the word size, until the complete product is obtained. Thus for a 16-bit word the addition and shift steps are carried out 16 times to obtain the multiplication of the two numbers. Obviously, multiplication operations take longer in a computer than additions or subtractions. Division is carried out using repeated subtraction and shift steps, again taking much longer than the addition or subtraction instruction to complete. With the speed of the computer this is still quite fast, multiplication times taking as little as 20 μsec, allowing 50,000 multiplications/sec.

Computer arithmetic units are becoming more complex in newer computers, some even having the operation of obtaining the square root as a basic instruction and others allowing operation on complex numbers. This allows faster operation, since the internal steps to carry out the operation are prewired (or stored in read-only memories), and the programmer has greater ease in using the computer since he can write instructions with these operations, rather than having to write programs for them. Even in these machines the basic use of the adder, the parallel or serial shifting of data, etc., are used.

12.8 MEMORY UNIT

The memory serves to store both the program and the data which the program acts on or generates. The most important aspect of the memory type selected is speed of operation. Other important factors are size, cost, and organization of the memory. The most popular memory has been the magnetic core, which uses little donut-shaped cores which can be magnetized in either a clockwise or counterclockwise flux direction to represent the 0 and 1 bit conditions. Core memory speeds are typically less than 1 μsec down to a few hundred nanoseconds. A core plane (arrangement of cores) is shown in Fig. 12.17. Another memory type is integrated circuit (IC) storage, which uses a large number of flip-flops organized into an array for storage and readout purposes. A number of newer memory types are being used in special areas because they have greater operating speed and smaller size than even the core memory, but none are widely used.

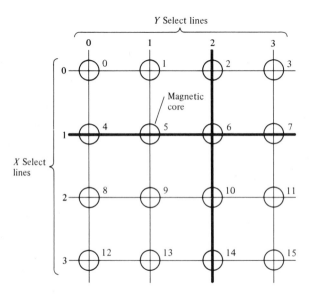

Figure 12.17 Core array showing selection scheme.

If each bit of a core plane had to be selected for read-in or readout using a separate wire, the total number of wires to handle and the number of selection gates would be too much to allow the size of core memories to reach into the hundreds of thousands to millions of cores that are now used. The scheme used allows selecting a particular core by the coincidence of what is called two half-select signals. In operation, the core memory is magnetized into either state by passing a current through a wire threading the core, opposite current directions causing the opposite flux states to occur. The core material used has a magnetization characteristic in which the amount of drive current must be above a certain value in order to "switch" the core. By using half this current value and two wires, so that both contribute a flux drive, the sum of which is sufficient to switch the core, a coindicent-current selection method may be employed. The arrangement of select wires and cores is shown in Fig. 12.17. If a current is placed into wires X_1 and Y_2, the core numbered 6 is the *only* one selected of the 16 shown. Cores 4, 5, 7 and 2, 6, 14 all receive half-select signals; the rest receive no signal. As stated, only core 6 receives two select signals and is the only core selected.

To read into or out of the core a certain order has to be followed. The basic operation of reading out of a core requires causing the core to switch and "sensing" whether the core switched from the 1 state or 0 state. The read operation requires writing a 0 into the selected core. If the core had previously stored a 1, the large flux change when switching from the 1 to 0 state is sensed on a pickup wire called a *sense* line, which has a voltage pulse induced by the changing flux. If, however, the core had previously stored a 0, the core doesn't switch state, and no voltage pulse is induced in the sense line. In either case, though, the core is at 0 *after* the read operation. Because the information must be destroyed in order to be read and since we would want the memory unit to allow a read *without* destroying the stored information, the basic memory cycle used is a read/write cycle, in which the bit just read out is then read back into the core again so that the net operation is to allow reading without loss of information.

The basic read/write cycle is used to advantage on writing into the core. Since a read operation always precedes a write, the core selected will have 0 stored at the time of the write operation, because the preceding read will have forced it to 0. The date input line is also called the *inhibit* line. If a 1 is to be written, the two half-select signals at write time cause the selected core to switch to the 1 state. If a 0 is to be written, the inhibit line has a pulse which opposes the same write pulses as for the 1, canceling out one half-select signal, so that the core cannot be switched. Since it was already in the 0 state due to the read, the result is a write of 0. In other words, the two half-select pulses to switch the core to 1 appear in either case, and an inhibit pulse appears for a 0 input and no inhibit pulse for a 1 input.

The last point to consider about the core memory is organization. One plane as considered so far represents one bit position of a word. If the memory stores 16-bit words, then there will be 16 core planes. If there are 1000 words in the memory, then each core plane has 1000 cores. Actually, the number of cores in a plane is usually some factor of 2 because of the use of binary address locations. A typical 4096-word memory with a 24-bit word length would have the following form. There would be 4096 cores on each plane and 24 planes. In total there would be 4096×24 cores, or 98,304 cores. This is a small-sized core memory, larger memories having hundreds of thousands to millions of words.

To specify a word in memory an *address* is used. The address for the 4096 core memory would require a 12-bit address word since 2^{12} is 4096. Of these 12 bits, 6 would be used for X select and 6 for Y select. For each selection part there would be 6 register stages to hold the address of the word called for, and along each side of the core plane there would be 2^6 selection lines or 64 half-select lines. For the complete plane there would be $64 + 64$ or 128 half-select lines. Only one sense wire is needed for each core plane since only the selected core will switch. Figure 12.18 shows the organization of a core memory plane, while Fig. 12.19 shows the organization of a full (3D) core memory system.

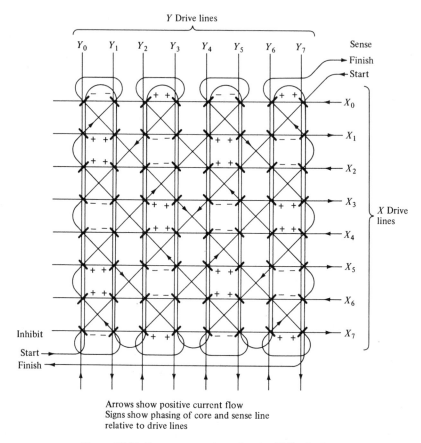

Figure 12.18 Core memory plane showing X-Y selection.

12.9 CONTROL UNIT

The heart of the computer is the control unit, which interprets each instruction, sends appropriate signals to the different units of the computer to carry out the operation, and then goes on to the next instruction. The program instruction contains two parts, an operation and an address. The control unit decodes the instruction part to determine which of the many possible instructions is presently being called for and then uses the address part to carry out the instruction on. For example, an instruction ADD address location 1075, meaning add the contents of memory word 1075 to whatever is presently in the arithmetic unit accumulator, is carried out by the memory unit and the arithmetic unit, the control unit providing all the necessary internal timing and control signals.

A number of types of control units are possible, the simplest having an instruction format of a single instruction and a single address. Some computers have two-address instructions, which provide information on the instruction desired, say, add; the address of the first number; and the address of the second number to add. Thus,

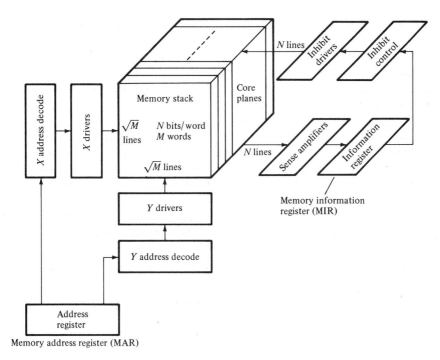

Figure 12.19 Core memory unit.

a single instruction will result in two numbers being added and the result left in the accumulator. A further complexity of control unit, but a reduction in the number of steps, is obtained with a three-address instruction, which would allow specifying the instruction, add, for example; the two address locations of the numbers to add; and the address location to place the resulting sum into. A four-address instruction is also found in use, allowing the additional specification of the address to obtain the next instruction from.

Parts of a control unit contain counters, logic gating, decoding gates, and flip-flop storage registers. The counters provide the timing steps which are used to carry out the different steps of each of the instructions with all the internal detail needed by the different computer units. The decoding gates take the many digits of the instruction and *decode* or convert them into signals on individual lines, for example, decoding a 3-bit instruction into one of eight lines of signal. The selected line provides the signal to the different parts of the computer to indicate which instruction is selected. During this particular instruction time, the control unit also provides all the smaller steps which are needed. The storage unit holds the instruction and address parts of the instruction and the next address to obtain the instruction from. A basic operating cycle in a computer is the FETCH/EXECUTE. During the FETCH part the present NEXT-INSTRUCTION ADDRESS is used, and the next instruction is picked up from the memory unit and brought into the control unit. Then the EXECUTE phase is performed, and the instruction is decoded and carried out. At the end of the FETCH operation the NEXT-INSTRUCTION ADDRESS, stored in a counter-register, is advanced by 1. At the end of an EXECUTE operation the program goes into a FETCH, and the address in the NEXT-INSTRUCTION ADDRESS register is used by the control unit to know which instruction to perform next. Thus, the program is advanced by the control unit, through instruction pickup and instruction operation, each instruction taken from the next memory instruction. This sequence may be altered by a branch instruction, in which the program instruction has the address of the next instruction to perform. During the EXECUTE phase, the address part of the instruction is used to replace the address of the NEXT-INSTRUCTION ADDRESS, and the program is advanced to the FETCH phase. The computer just carries out each step without regard to what just happened; the address in the NEXT-INSTRUCTION ADDRESS register is used to tell which address to read the instruction from. The address count is then increased by 1, and the program goes to the EXECUTE phase. The effect of a branch is to cause the program to continue reading instructions from a different location in memory so that, for example, the program can be directed to repeat certain instruction over again. A block diagram of a simple control unit showing the parts of the circuitry used for the FETCH/EXECUTE is shown in Fig. 12.20.

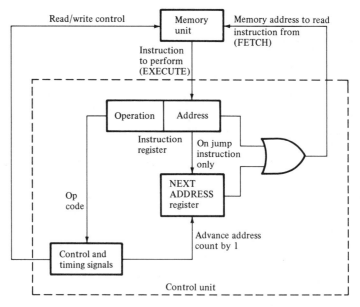

Figure 12.20 Control unit block diagram showing FETCH/EXECUTIVE.

12.10 INPUT-OUTPUT UNITS

The parts of a computer seen most by the user are those for putting information into and taking information from the computer. The punched card, for example, is the most common input medium, and punched-card readers are part of almost every commercial computer.

Punched Card

Since punched cards are so common, let's consider some details of the punched card itself and the punched-card reader, or punch. The basic arrangement on any 80-column card is that each column is a character position, such as a number, a letter, or any special character, and there are 12 rows which have the different possible binary positions used to code the character. To allow easy reading of the card, the 12-bit code selected is quite simple. Table 12.3 shows the code used, and Fig. 12.21

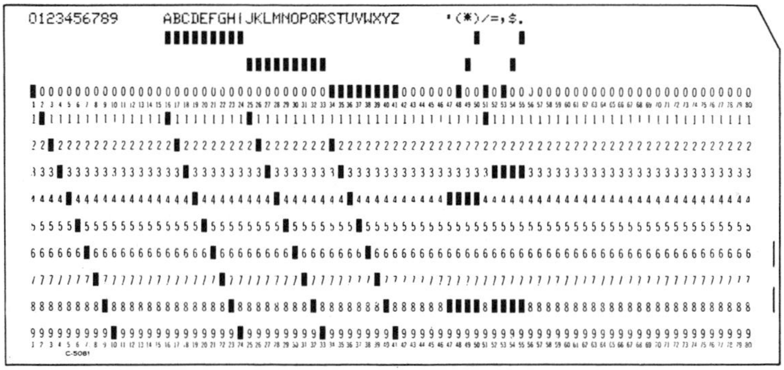

Figure 12.21 Punched card showing alphameric code.

shows a card punched with most of the possible characters. The rows from 0 down to 9 are numbered on the card, and a hole punched in row 6, for example, indicates the number 6. Only one character may be punched in any column. However, more than one hole may be punched, the combination of a hole in row 12 and row 1 representing the character A and holes in row 11 and row 7 the character P. The three rows 12, 11, and 0 are also called zones since they are used to modify the numbered rows 1–9 to

Table 12.3

Punched-Card Code Table

NUMERICAL ZONE (ROW) ONLY	ZONE 12 PLUS NUMERICAL ROW BELOW	ZONE 11 PLUS NUMERICAL ROW BELOW	ZONE 0 PLUS NUMERICAL ROW BELOW
0 = 0			
1 = 1	1 = A	1 = J	
2 = 2	2 = B	2 = K	2 = S
3 = 3	3 = C	3 = L	3 = T
4 = 4	4 = D	4 = M	4 = U
5 = 5	5 = E	5 = N	5 = V
6 = 6	6 = F	6 = O	6 = W
7 = 7	7 = G	7 = P	7 = X
8 = 8	8 = H	8 = Q	8 = Y
9 = 9	9 = I	9 = R	9 = Z

form the alphabetic characters. The combination of numeric and alphabetic characters is often referred to as alphanumeric, or more simply as alphameric characters.

Punched-card reader speeds range from about 100 to 1000 cards/min. The faster readers read the card along an entire row so that the data are not fully in until the 12 rows are read. Other readers read a column at a time but may achieve fast speeds when only a few columns are punched at the beginning of each card. Mechanical readers, such as pin or brush readers, are relatively slow as compared to optical readers, which have photocells sensing light sources on the opposite card side, the light passing through a punched hole causing the photocell to be excited. Card punch units are considerably slower, punching speeds of 200 cards/sec being typical.

Punched Paper Tape

Punched-paper-tape units are similar to card units in that either mechanical or photoelectric readers are used to sense holes punched. The codes used, however, are different since readability is not so important as the efficient use of the positions of the coding. Code levels (number of code bits per character) vary from 5 to 8 levels, the 8 level being more common because of teletype transmission, which uses the 8-level code, and also because of the adoption of either the EBCDIC or ASCII codes as standards for transmission and therefore input-output from computer units. Figure 12.22 shows a paper-tape reader and punch and Fig. 12.23 a punched section of tape showing the code used.

Magnetic Tape and Disk

Magnetic tape is used to store large amounts of information (as are magnetic disks); a typical tape reader unit is shown in Fig. 12.24. If eight-level codes are used, there are eight reader rows or positions across a cross section of tape. The tape is arranged somewhat differently, for efficient reading, than for the punched-card or tape readers. The much greater speed and capacity of the tape dictates that each larger group of data, arranged in records, is spaced with an interrecord gap separating them. Interrecord gaps are typically $\frac{3}{4}$ in., and records can be as long as desired. More efficient reading is obtained using larger records, since there are, then, less record gaps. When a record is found the entire record is read into core memory. Whereas punched cards are read as a single record of 80 characters and punched tape as separate characters, the magnetic tape (mag tape) is recorded as variable-length records. A typical operation with computers is reading data from punched cards onto magnetic tape and thereafter reading from the magnetic tape into the computer at more reasonable speeds for the computer units.

Figure 12.25 shows some details of a tape drive mechanism. The magnetic tape is threaded past drive/stop capstans and vacuum columns so that an amount of tape is hanging loose (free) from each reel. When tape is to be read (moved past the read/write heads) a drive pulley forces the tape against a rotating drive capstan, thereby pulling the loose tape past the read heads. The drive action of the two reels is separate

Figure 12.22 Paper tape units. (a) paper tape reader. (b)
paper tape punch.

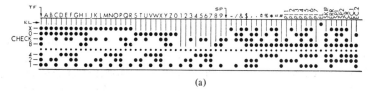

(a)

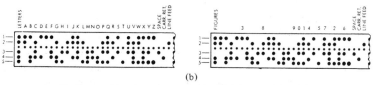

(b)

Figure 12.23 Paper tape showing (a) light channel code
(b) methods of reading punched tape.

Figure 12.24 Magnetic tape unit.

from the drive tape action. The two reels are driven independently to provide a preset amount of free or loose tape in the separate columns. In this way the tape drive "sees" only a loose piece of tape and can quickly start the tape in motion without drag, break, or stretch problems. This action can proceed in forward reading or reverse winding, or, in some computers, for forward and reverse reading and writing.

Figure 12.26 shows part of a magnetic disk unit. These disks are magnetically recorded on both sides in circumferential tracks with information obtained much faster than with magnetic tape. The read/write heads are usually on moving arms which can be stepped radially from track to track, a typical track arrangement being shown in Fig. 12.26(c).

High-Speed Printer

The job to output the large amount of data developed or stored in the computer is usually carried out on high-speed printer units which generate a full line of type at a time. Printing mechanisms include print wheels (one for each character position), print belts with a few sets of complete characters on a continuously rotating belt, and electrostatic character printers, as shown in Fig. 12.27.

The electrostatic printer forms a character as a series of dots by electrostatic discharge onto special printing paper. This method is presently the fastest, with printing units having 5000-line/min speeds. Print wheels and belt printers operate more conventionally with electrically actuated hammers striking to press the formed characters against paper and ink ribbon to generate the printed character. Separate hammers for each print position strike as the required character appears, and a line

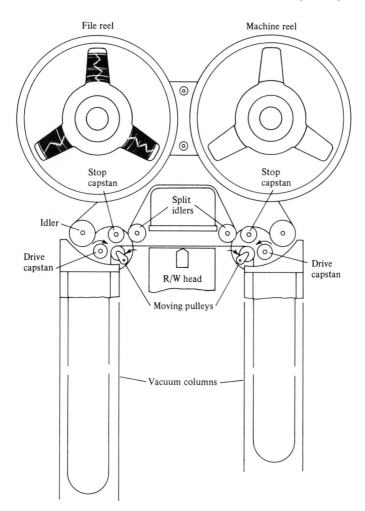

Figure 12.25 Diagram of tape transport unit.

feed occurs after all line characters have been printed. Rates of 1200–2000 lines/min can be obtained with these types of printers.

Terminals

A growing area of computer input-output devices is concerned with the small, inexpensive, individual user devices operating with a time-shared computer. Tens to hundreds of separate users can have "immediate" access to computer files and computer programs at these separate terminals. Actually, the computer operating at very fast speeds compared with the operating speeds of each terminal can sequentially, or on demand, sample the input data from each terminal and store the input infor-

Figure 12.26 Magnetic disk unit.

mation fast enough to handle these large number of users "at the same time"—called *time sharing*. Popular terminals for this time-shared operation include teletypes and CRT/keyboard terminals. Figure 12.28 shows both types of devices.

In typical operation a character of data is entered by pressing the appropriate character key. Electrical signals generated by the keyboard are transmitted directly to the computer or are modulated and sent over telephone lines by special telephone modems or couplers. Data are collected by the computer and stored, typically on magnetic disk, until a request for processing is received. Output data are then generated and sent directly to the terminal or storded temporarily on magnetic disk so that they can be sent to the terminal at a later time. Transmission rates over telephone lines are typically 10 or 30 characters/sec (line rates are usually given in bits per second or *baud*—110 or 300 baud being typical for teletype units)—or over specially conditioned telephone lines rates of 1200, 2400, or 4800 baud for CRT terminals up to 9600 baud for particular high-speed input-output devices.

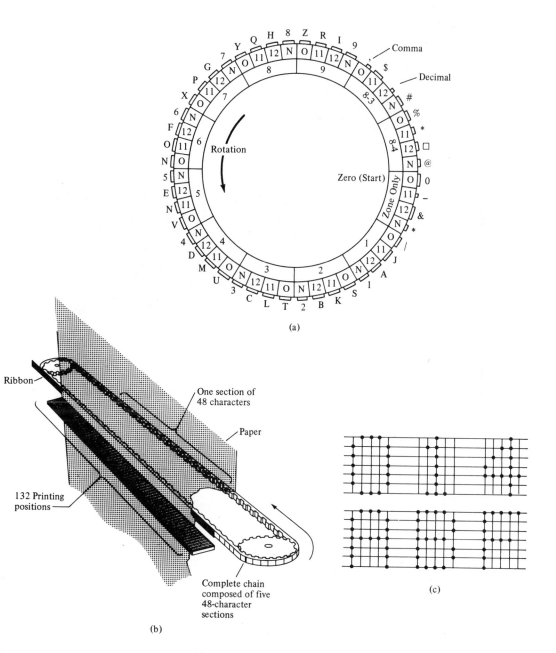

Figure 12.27 Printing mechanisms. (a) print wheel (b) print chain
(c) print matrix.

(a)

(b)

Figure 12.28 Time-share terminals. (a) teletype. (b) CRT. (Courtesy Hazeltine Corporation.)

PROBLEMS

§ 12.2

1. Convert the following binary numbers into decimal:
   (a) 110101   (b) 11100   (c) 101011

2. Convert the following decimal numbers into binary:
   (a) 63   (b) 76   (c) 120

3. Convert the following octal numbers into decimal:
   (a) 43   (b) 66   (c) 106

4. Convert the following binary numbers into octal:
   (a) 111001   (b) 101110111   (c) 100011010

5. Convert the following hexadecimal numbers into decimal:
   (a) 49   (b) A2   (c) 5AD

6. Convert the following binary numbers into hexadecimal:
   (a) 11010011   (b) 01011100   (c) 011111010101

7. Perform the following additions in the indicated base number system:

(a) $\left(\begin{array}{r} 110010 \\ + \ 01101 \end{array}\right)_2$ (b) $\left(\begin{array}{r} 236 \\ + \ 22 \end{array}\right)_8$ (c) $\left(\begin{array}{r} 2A \\ +372 \end{array}\right)_{16}$

8. Perform the following subtractions in the indicated base number system:

(a) $\left(\begin{array}{r} 110010 \\ - \ 10111 \end{array}\right)_2$ (b) $\left(\begin{array}{r} 463 \\ - \ 72 \end{array}\right)_8$ (c) $\left(\begin{array}{r} 1A9 \\ - \ 6B \end{array}\right)_{16}$

9. Express decimal 2984 in BCD form.

10. Express octal 273 in binary-coded octal form.

11. Express hexadecimal A29 in binary-coded hexadecimal form.

12. Express decimal 596 in BCD with an odd parity bit.

§ 12.4

13. Develop the truth table and obtain the logical expression of a circuit which provides an output when any one or two of three inputs are present.

14. Write the truth tables for the AND and OR functions.

15. Write the truth table for a NAND (AND followed by inverter) circuit.

16. Simplify the following logical expressions:

(a) $A\bar{B} + A\bar{B}C$ (b) $ABC + A\bar{B}C$ (c) $XY + XZ + XYZ$

17. Draw a logic diagram for $W = XYZ + \overline{XY}Z + Y\bar{Z}$ (don't simplify).

18. Draw a logic diagram for $(A\bar{B}C + AC) \cdot AB$ (don't simplify).

19. Write the logical expression for the logic circuits of Fig. 12.29.

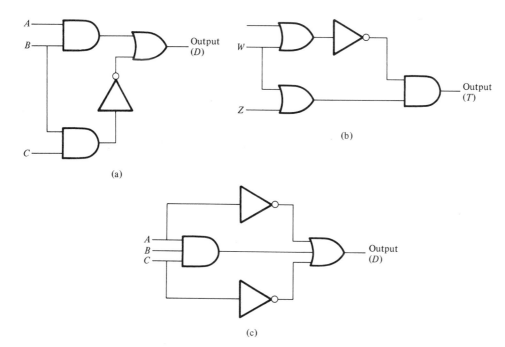

Figure 12.29 Logic circuits for problem 19.

20. Show how a full adder (sum of three inputs) can be made using two half-adders.

§ **12.5**

21. Draw the circuit of a bistable multivibrator circuit using BJTs.

22. Draw the circuit diagrams of astable and monostable multivibrator circuits using BJTs.

23. List the truth table for synchronous operation of a *JK* flip-flop.

24. How do preset and clear inputs to a TTL flip-flop operate?

13

Analog and Hybrid Computers

13.1 BASICS

Analog computers are made of continuous operating circuits, as compared to the discrete acting circuitry of the digital computer. Typically, a continuous voltage is developed by the analog circuitry which is analogous to or representative of the signal variation of a real device or system. The analog computer's voltage variation with time could, for example, be representative of the time variation of a control valve, or the movement of a spring in a particular system, or even the flow of a chemical in a special apparatus. In all these cases the system equations must be known, or at least proposed, and a model of the expected operation available. An analog connection is then put together to represent this real system, and the resulting signal variations within the analog computer circuitry should be, to some degree, representative of the variations in the real system.

Just as a program is necessary for the operation of a digital computer, a set of equations is necessary for the interconnection and operation of an analog computer. When a model or block diagram of the system is available the equations of the system can be obtained and the analog computer connected to represent this model. The circuitry of the analog computer contains the typical operations which can be specified by either integro-differential equations or a system model. There are integrating or differentiating circuits, adder or summer circuits, subtraction or inverting circuits, multiplier circuits, and function generator circuits, to name the most common. When using the digital computer many different programs can be written depending on the particular problem. With the analog computer many different circuit connections can be made using these basic circuits or building blocks to solve a given analog model or set of equations. Because of the basic nature of the analog computer—solving equations—it is mainly a tool of the mathematician or engineer and is used primarily to solve scientific problems. The digital computer, it will be remembered, can be programmed for any types of problems, business or scientific, with equal ease.

The accuracy of the solution obtained on an analog computer depends on the accuracy of the circuits used, typically 0.1 %, whereas the accuracy of a digital computer depends primarily on the number of bits used to handle the numbers, so that theoretically any desired accuracy is obtainable. In the type of problems solved by the analog computer the accuracy is generally quite suitable using circuits currently available. The speed with which a solution can be developed is also of concern. With the digital computer it depends on the basic memory speed and the speed of arithmetic operations, typically being on the order of fractions of a microsecond. However, the solution of a set of equations may require many steps so that the full solution may take milliseconds. The speed of the analog computer depends on the response time of its circuits, typically milliseconds to seconds of time. Where the digital computer requires a considerable number of numerical steps to solve the problem, the analog computer may be equally as fast, providing the direct solution for the wired circuit connection.

The analog computer (see Fig. 13.1) has one great advantage which often results in its being selected over the digital computer for solving a particular problem. This is the economy using a relatively inexpensive analog computer instead of a large general-purpose digital computer. If continuous solution of a set of equations is required, the use of a small, inexpensive analog computer might be more economical

Figure 13.1 Analog computer.

than tying up a large digital computer, capable of much more than is being used in this particular setup. On the other hand, it may also be more economical to solve a particular set of equations once, or relatively infrequently, using a large digital computer, which then goes on to do many other programs, rather than to require the use of an analog computer. Many factors go into the decision to use one or the other of these devices, but both have their value.

In some cases a combination of these two devices offers the best solution to a problem. This combined analog-digital computer, called a hybrid computer (see Fig. 13.2), could be basically an analog computer with a small amount of digital counters and logic controlling elements or a full digital and full analog computer with the required analog-to-digital and digital-to-analog converter circuitry needed to interconnect these two units. This hybrid configuration allows solving a problem economically on the analog circuitry, with the power of the digital machine to vary parameters, store results, and provide comprehensive presentation of the desired data.

Figure 13.2 Hybrid computer.

13.2 OPAMP CIRCUIT

An operational amplifier has, ideally, an infinite gain, infinite input impedance (open circuit), and zero output impedance. The OPAMP unit is basic to such analog circuits as the summer, integrator, and inverter. Figure 13.3 shows an OPAMP unit having two inputs and a single output. The two inputs are marked with *plus* ($+$) and *minus* ($-$) signs to indicate the relation between the input at that terminal and the resulting output voltage. A signal applied to the plus input appears in phase and amplified at the output, while that applied to the minus input is amplified but inverted at the output. While the basic OPAMP circuit has very high gain, the useful connection of an OPAMP circuit, as shown in Fig. 13.4(a), has a gain set by the external resistor components. To see how this is so, consider the ideal equivalent circuit of the OPAMP with the external resistor components as shown in Fig. 13.4(b).

We can determine the voltage V_o using superposition as follows. From the redrawn equivalent circuit of Fig. 13.4(c) the voltage V_i can be calculated as the result-

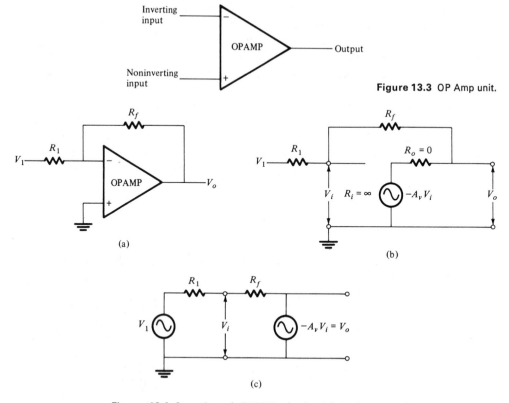

Figure 13.3 OP Amp unit.

(a)

(b)

(c)

Figure 13.4 Operation of OPAMP circuit: (a) basic connection (constant-gain multiplier); (b) ideal equivalent circuit; (c) redrawn equivalent circuit.

ing voltage due to source voltages V_1 and V_o. For source V_1 only (with V_o set to zero)

$$V_{i_1} = \frac{R_f}{R_1 + R_f} V_1 \tag{13.1}$$

For source V_o only (with V_1 set to zero)

$$V_{i_2} = \frac{R_1}{R_1 + R_f} V_o = \frac{R_1}{R_1 + R_f}(-A_v V_i) \tag{13.2}$$

where $V_o = -A_v V_i$, with A_v being the voltage gain of the OPAMP. Solving for V_i as the sum of the components due to those of Eqs. (13.1) and (13.2) results in

$$V_i = \frac{R_f}{R_f + (1 + A_v)R_1} V_i \tag{13.3}$$

For operation as an OPAMP, $A_v \gg 1$ is always true, and the choice of R_1 and R_f usually results in the condition $A_v R_1 \gg R_f$. With these approximations the voltage V_i in Eq. (13.3) can be simplified to

$$V_i = \frac{R_f}{A_v R_1} V_1 \tag{13.4}$$

Although the OPAMP gain A_v is important, especially that it be large enough for the approximations above, the useful circuit gain, from input V_1 to output V_o, calculated as V_o/V_1, is of greatest interest and use. Solving for V_o/V_1 we get

$$\frac{V_0}{V_1} = \frac{-A_v V_i}{V_1} = \frac{-A_v R_f V_1}{V_1} \frac{1}{A_v R_1} = -\frac{R_f}{R_1}$$

$$\boxed{\frac{V_0}{V_1} = -\frac{R_f}{R_1}} \qquad (13.5)$$

The result shows that the ratio of output to circuit input voltage is dependent *only* on the value of resistors R_1 and R_f, provided that A_v is very large. Using the high-gain OPAMP allows building a circuit whose circuit gain can be set exactly by passive components into, as will be shown, a large variety of useful configurations.

Virtual Ground

The output voltage is limited by the supply voltage, typically a few volts. Voltage gain for the amplifier, on the other hand, is typically very high. For example, an output voltage of 10 V with an amplifier gain of 10,000 would result from an input voltage of

$$V_i = \frac{V_0}{A_v} = \frac{10 \text{ V}}{10,000} = 1 \text{ mV}$$

The input voltage to the amplifier will typically be millivolts or even less for typical output voltages in the volt range with the high amplifier gain expected from OPAMPs used. If the circuit had an overall gain (V_o/V_1) of, say, 1, then the value of V_1 would be 10 for the above sample values. The value of the amplifier input voltage, V_i, which is not the same as the circuit input voltage, V_1, is quite small, and may be considered 0 V. (Note that although $V_i \simeq 0$ V, it is not *exactly* 0 V, since the output is the value V_i times the amplifier gain, A_v.)

The fact that $V_i \simeq 0$ V leads to the concept that at the input to the amplifier there exists a virtual short circuit or *virtual ground*. Since the input impedance of the amplifier is quite high, ideally infinite, there is also no current flow into the amplifier so that the virtual ground condition can be depicted as shown in Fig. 13.5. The heavy

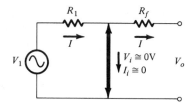

Figure 13.5 Virtual ground in an OPAMP circuit.

line is used to indicate that we may consider that a short circuit exists with $V_i \simeq 0$ V but that this is a virtual short in that no current flows through the short to ground. Current flows through resistors R_1 and R_f, as shown.

If we use the virtual ground concept, we can write equations for the current I as follows:

$$I = \frac{V_1}{R_1} = -\frac{V_o}{R_f}$$

which can be solved for V_o/V_1:

$$\frac{V_o}{V_1} = -\frac{R_f}{R_1}$$

The virtual ground concept, which depends on A_v being very large, allows simple solution for the overall voltage gain. It should be understood that although the circuit of Fig. 13.5 is not a physically real circuit, it does allow an easy means for solving the overall voltage gain.

Constant-gain multiplier. A basic form of the OPAMP circuit used in analog computers is an inverting constant-gain multiplier, as shown in Fig. 13.6. The operation of this circuit is summarized by the gain equation previously calculated in Eq. (13.5).

EXAMPLE 13.1

The circuit of Fig. 13.6 has $R_1 = 100$ K and $R_f = 0.5$ M. What is the output voltage for an input of $V_1 = -2$ V?

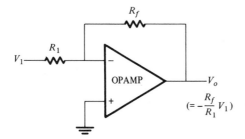

Figure 13.6 Inverting constant-gain multiplier.

Solution:

Using Eq. (13.5),

$$V_o = -\frac{R_f}{R_1}V_1 = -\frac{500\text{K}}{100\text{K}}(-2 \text{ V}) = +10 \text{ V}$$

Noninverting constant-gain multiplier. The connection of Fig. 13.7 shows an OPAMP circuit that operates as a noninverting constant-gain multiplier. To determine the voltage gain of the circuit we can apply the equivalent virtual ground representation in Fig. 13.7(b). Note that the voltage across R_1 is V_1 since $V_i \simeq 0$ V. This must be equal to the voltage due to the output, V_o, through a voltage divider of R_1 and R_f so that

$$V_1 = \frac{R_1}{R_1 + R_f}V_o$$

and

$$\boxed{\frac{V_o}{V_1} = \frac{R_1 + R_f}{R_1} = 1 + \frac{R_f}{R_1}} \qquad (13.6)$$

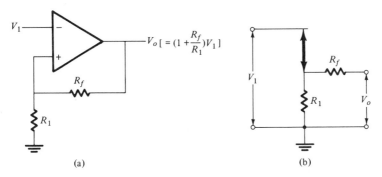

Figure 13.7 Noninverting constant-gain multiplier.

EXAMPLE 13.2

Calculate the output voltage of a noninverting constant-gain multiplier (as in Fig. 13.7) for the values $V_1 = 2$ V, $R_f = 0.5$ M, and $R_1 = 100$ K.

Solution:

Using Eq. (13.6),

$$V_o = 1 + \frac{R_f}{R_1} V_1 = 1 + \frac{500 \text{ K}}{100 \text{ K}} (2 \text{ V}) = +12 \text{ V}$$

Unity follower. A unity follower circuit, as in Fig. 13.8(a), provides a gain of 1 with no phase reversal. From the equivalent circuit with virtual ground [Fig. 13.8(b)] it is clear that

$$V_o = V_1 \tag{13.7}$$

and that the output is the same polarity and magnitude as the input. The circuit acts very much like an emitter-follower circuit except that the gain can be set much closer to being exactly unity.

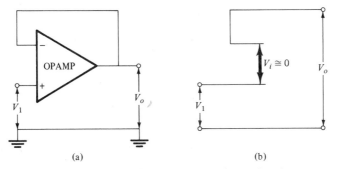

Figure 13.8 (a) Unity follower; (b) virtual ground equivalent circuit.

13.3 SUMMER AND INTEGRATOR CIRCUITS

By far the most useful of the OPAMP circuits used in analog computers are the summer and integrator circuits. The summer provides a means of doing addition or subtraction of signals, while the integrator circuit allows solving integro-differential equations such as usually occur in engineering and scientific studies.

Summer

Figure 13.9(a) shows a basic form of an OPAMP summer circuit, this one having three inputs. Each of these inputs is connected through a separate resistor component and can therefore be multiplied by a different constant-gain factor, as will be

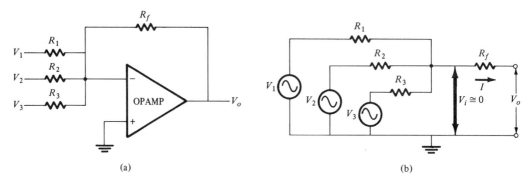

(a)   (b)

Figure 13.9 (a) Summer; (b) virtual ground equivalent circuit.

shown. Using the virtual ground equivalent circuit of Fig. 13.9(b), the current, I, can be expressed as

$$I = \frac{V_1}{R_1} + \frac{V_2}{R_2} + \frac{V_3}{R_3} = -\frac{V_o}{R_f}$$

which can be directly solved for V_o:

$$V_0 = -\left[\frac{R_f}{R_1}V_1 + \frac{R_f}{R_2}V_2 + \frac{R_f}{R_3}V_3\right] \tag{13.8}$$

In other words, each input adds a voltage term to the output as obtained for an inverting constant-gain circuit. If other than three inputs are used, each input adds a term to the output similar to the form in Eq. (13.8).

EXAMPLE 13.3

What is the resulting output voltage for an OPAMP summer for the following input voltages and resistor values: $V_1 = +2$ V, $V_2 = -4$ V, $V_3 = +5$ V, $R_1 = 500$ K, $R_2 = 250$ K, $R_3 = 1$ MΩ, and $R_f = 1$ M.

Solution:

Using Eq. (13.8),

$$V_o = -\left[\frac{1000 \text{ K}}{500 \text{ K}}(+2 \text{ V}) + \frac{1000 \text{ K}}{250 \text{ K}}(-4 \text{ V}) + \frac{1000 \text{ K}}{1000 \text{ K}}(+5 \text{ V})\right]$$

$$= -(+4 \text{ V} - 16 \text{ V} + 5 \text{ V}) = +7 \text{ V}$$

Figure 13.10(a) shows a practical summer circuit with gains of 1, 2, and 10 for inputs V_1, V_2, and V_3, respectively. Check, using Eq. (13.8). In many analog computer units the exact values of input of feedback resistors need not be specified if the input gains are indicated, as the simplified representation of Fig. 13.10(b) shows.

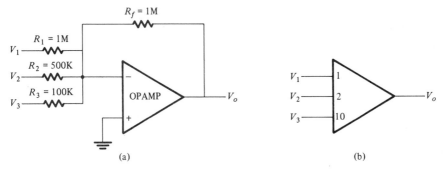

(a)   (b)

Figure 13.10 (a) Practical summer circuit; (b) simplified representation used in analog computer.

Integrator Circuit

The summer unit alone would allow addition or subtraction operation. An integrator circuit is also needed in solving differential equations. The constant-gain multiplier circuit of Fig. 13.4(a) can be converted to an integrator circuit by using a capacitor as feedback element rather than a resistor as shown in Fig. 13.11(a).

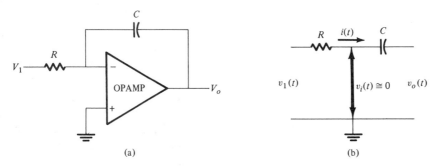

(a)   (b)

Figure 13.11 (a) Integrator circuit; (b) virtual ground equivalent circuit.

Using the virtual ground equivalent circuit of Fig. 13.11(b), the current can be expressed as

$$i(t) = \frac{v_1(t)}{R} = -C\frac{dv_o(t)}{dt}$$

From which

$$\frac{dv_o(t)}{dt} = -\frac{1}{RC}v_1(t)$$

Solving for $v_o(t)$ by integration,

$$v_o(t) = -\frac{1}{RC} \int v_1(t)\, dt \qquad (13.9)$$

If, for example, $R = 1\ \text{M}\Omega$ and $C = 1\ \mu\text{F}$, then

$$v_o(t) = -\frac{1}{10^6(10^{-6})} \int v_1(t)\, dt = -\int v_1(t)\, dt$$

showing that the output is the integral of the inverted input signal [see Fig. 13.12(a)]. If, for example, $R = 1\ \text{M}\Omega$ and $C = 0.1\ \mu\text{F}$, then Eq. (13.9) results in

$$v_o(t) = -10 \int v_1(t)\, dt$$

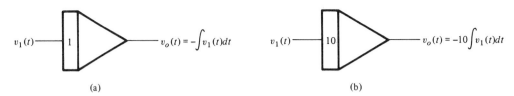

(a)   (b)

Figure 13.12 Integrator units (a) gain of 1; (b) gain of 10.

showing that the input can be both integrated *and* multiplied by a constant-gain factor and then inverted to produce the output signal [see Fig. 13.12(b)].

Referring to the practical integration circuit of Fig. 13.11(a), it can be seen that with the output left unconnected or connected to a very high impedance load a voltage across capacitor C would be maintained for a long time after the input signal is removed. In fact, even with no input signal a voltage might result across capacitor C as a result of some very small practical current after a long period of time. Usual practice is to connect a short across capacitor C when the circuit is not operating on the input signal. Figure 13.13(a) shows this initial condition of 0 V across capacitor C, which is practically the same as placing 0 V at the output. Figure 13.13(b) shows the 0-V initial condition in the usual representation in analog computers. Figure 13.13(c) shows how an initial condition (other than 0 V) can be placed in the output. As shown, a battery voltage of 5 V is placed across capacitor C so that the output voltage is initially 5 V when the switch controlling the initial condition across capacitor C is opened at the start of an operating cycle. The usual analog computer representation of initial conditions on an integrator is shown in Fig. 13.13(d).

13.4 SPECIAL DIODE CIRCUITS

Special circuit connections can be used for a great variety of purposes. Of particular significance are those using diodes for limiting the range of the output to achieve special actions. Consider the output-input relation of a basic constant-gain inverting amplifier unit as shown in Fig. 13.14(a). We know that the output is the input inverter and multiplied by a constant-gain factor. This can be represented graphically as shown

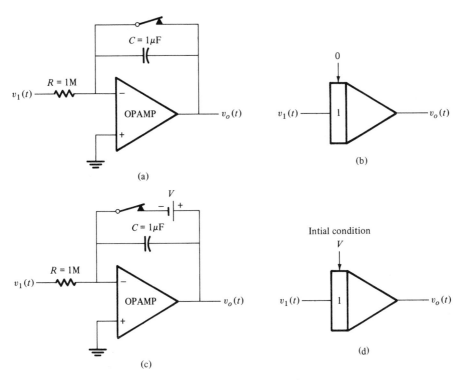

Figure 13.13 Integrating amplifier.

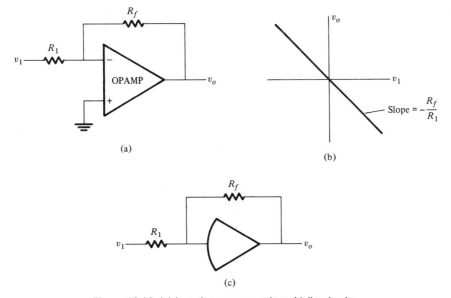

Figure 13.14 (a) Inverting constant-gain multiplier circuit; (b) output-input relationship; (c) simplified circuit representation of inverting constant gain multiplier.

in Fig. 13.14(b). A simplified circuit representation can also be used where the OPAMP inputs are not marked for polarity and the reference plus (+)—input connection to ground—is not shown. To indicate this simplified OPAMP connection the circuit of Fig. 13.14(c) can be used.

An example of how a diode can be used to provide special circuit action is shown by the diode limiter circuit in Fig. 13.15 along with the resulting output-input relation.

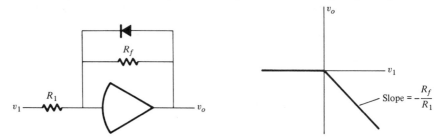

Figure 13.15 Diode limiter circuit.

As shown, the diode across feedback resistor R_f prevents the output from going positive (the diode would conduct and clamp the output to 0 V). Recall from the virtual ground concept that the junction of R_f and R_1 at the input to the OPAMP is considered to be at 0 V, so that the diode will conduct if the output goes above 0 V (ideally).

When the input goes positive the output goes negative, and with the diode now reverse-biased the output is the inverted input multiplied by the constant gain factor of R_f/R_1.

In the circuit of Fig. 13.16 a pair of Zener diodes are connected back to back, as shown. With a breakdown voltage of V_Z the output will be clamped so that neither positive nor negative voltage levels can exceed V_Z in magnitude. The input-output relationship in Fig. 13.16 shows that beyond the range of V_Z the output voltage remains at the clamped level, whereas within that range the circuit operates as an inverting constant-gain multiplier for both positive and negative voltages.

Another practical limiter circuit using diodes is shown in Fig. 13.17. In this example two amplifiers with unity gain are used along with two diodes to provide an output which is the absolute value of the input. Thus, whether the input is negative

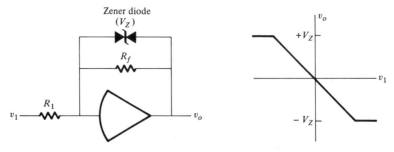

Figure 13.16 Limiter circuit using zener diode pair.

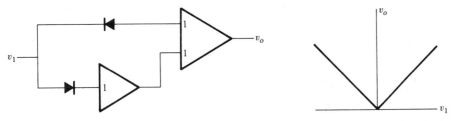

Figure 13.17 Absolute value circuit.

or positive the output is always positive. When the input is negative it back-biases the lower diode and results in an output for the upper part of the circuit. Conversely, a positive input voltage back-biases the upper diode and operates the lower part of the circuit to result in a positive output voltage.

Function Generator

Another use of diodes in special analog circuits is in the design of a function generator—a special circuit which can generate almost any type of input-output relationship on a piecewise basis. The circuit of Fig. 13.18 is an example of a function generator circuit. Both positive and negative slopes can be generated with breakpoints

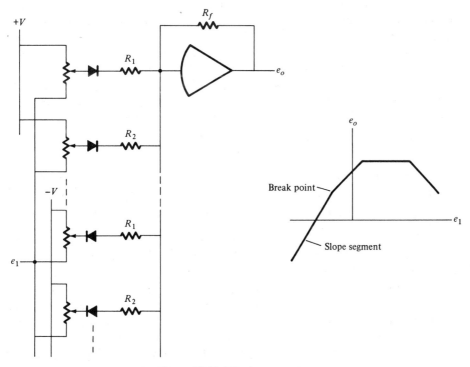

Figure 13.18 Function generator.

of slope change set by a potentiometer. The number of potentiometer diode-resistor inputs sets the number of line segments in the input-output relationships. The larger the number of such breakpoints, the smoother the resulting input-output curve. Generally, one such function generator may be used in an analog computer to provide a particular input signal for computer processing. More advanced problems may require a number of such function generators.

Multiplier

Another important part of analog computers is a circuit for multiplying two variable signals (as opposed to a constant-gain multiplier circuit). One means of producing the multiplication operation using only electronic circuitry involves the use of square-law circuits. A square-law circuit has essentially a parabolic input-output relation, which could be obtained from a function generator but is usually more economically obtained using special diode or thermistor circuitry. Figure 13.19 shows a multiplier unit which is capable of providing the product of independent signals x and y for any polarity of either signal. Such four-quadrant multiplier circuits are the most generally useful, while simpler multipliers are also available where signal polarity of one or both signals is limited.

In the circuit of Fig. 13.18(a) a biased diode network is used to do the operation of squaring a signal applied to it as a result of the diode network's parabolic output current to input voltage relationship. This action only squares a single signal. One way to obtain multiplication of two separate signals is to square the sum of these signals resulting in $(x + y)^2 = x^2 + 2xy + y^2$, then generate a second squared signal from $(x - y)^2 = x^2 - 2xy + y^2$, and finally subtract these resulting signals. Since addition (or subtraction) is easily possible with analog circuitry, the use of this technique allows obtaining the desired product of signals x and y. To combine x and y signals by adding them using equal input resistors, as in Fig. 13.19, the resulting sum obtained is $(x + y)/2$. The resulting product can then be obtained as

$$\left[\left(\frac{x + y}{2} \right)^2 - \left(\frac{x - y}{2} \right)^2 \right] = xy$$

If only positive x and y signals were allowed, then only a pair of resistor diode networks to sum x and y and a pair to sum x and $-y$ would be needed. For the full range of positive and negative voltages for both x and y input signals, the complete multiplier unit of Fig. 13.19 is required.

The resulting output of $v_o = -xy/100$ is inverted from the product $xy/100$, but a simple inverter can form the noninverted output. For four-quadrant multiplication the product can go both positive and negative, where the negative sign indicates a 180° phase reversal of the product signal. The magnitude of the resulting signal from the diode network (and then the output) is specifically reduced by a factor dependent on the maximum output voltage. In the circuit shown the resulting factor of 100 allows that with values of x and y less than the circuit reference voltage, the output is also less than the reference value. Figure 13.19(b) shows the multiplier symbol for use in an analog computer. Note that it is necessary to provide both x, y, and their

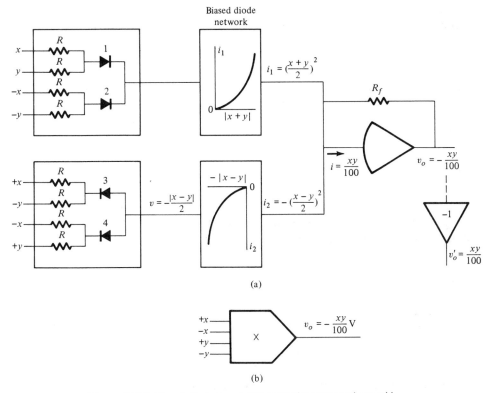

(b)

Figure 13.19 Biased-diode square law networks connected as multi-plier. (a) detailed circuit; (b) multiplier symbol.

inverted signals as input to the multiplier and that the resulting product is the inverted value of xy divided by the scale factor of 100.

Divider Circuit

The square-law multiplier circuit can also be used in such a way as to provide a division of one independent variable signal by another. One such form of divider circuit is shown in Fig. 13.20. The inputs to the square-law multiplier circuit are y

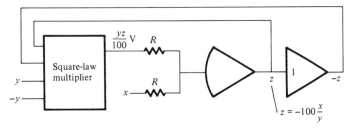

Figure 13.20 Divider circuit.

and z. The remaining analog circuitry sums $yz/100$ and x and then amplifies this sum by a very high value so that the value of z (used then as one input of the square-law multiplier circuit) is

$$z = A\left(\frac{yz}{100} + x\right)$$

which can be manipulated to the form

$$(100 - Ay)z = 100Ax$$

If A is very large we can rewrite the above expression as

$$-Ayz = 100Ax$$

so that

$$z = -\frac{100x}{y}$$

and we see the divider operation of the circuit. The only restriction on using this type of divider circuit is that y be larger in magnitude than x at all times.

13.5 SAMPLE PROBLEMS

As an example of how the various analog computer circuits can be used to solve specific problems, consider the differential equation of a mass-spring-damper system:

$$M\frac{d^2x}{dt^2} + B\frac{dx}{dt} + Kx = F$$

which can be written in a more usual dot notation:

$$M\ddot{x} + B\dot{x} + Kx = F$$

(where each dot over the variable term indicates a degree of differentiation). The equation describes the motion with respect to time of a system such as that of Fig. 13.21(a). The distance x is moved by mass M, from the left wall due to a driving force F with the mass acted on by a spring having a spring constant K, and the movement of the mass damped by the viscous friction B.

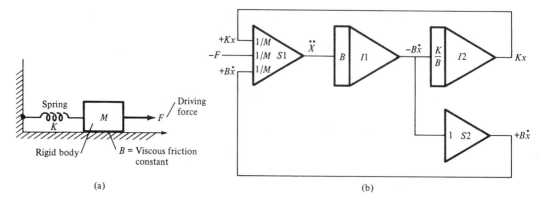

(a)   (b)

Figure 13.21   Mass-spring-damper system.

Rewriting the equation to solve for $\ddot{x}$ gives

$$\ddot{x} = -\frac{B}{M}\dot{x} - \frac{K}{M}x + \frac{F}{M} = -\frac{1}{M}(B\dot{x} + Kx - F)$$

Summing unit $S1$ sums the three signals expressed above, multiplies each by $1/M$, and provides an inverted output which is equal to $\ddot{x}$. Integrator $I1$ then operates on $\ddot{x}$ to provide $-B\dot{x}$ as output. Since one input to $S1$ is $+B\dot{x}$, the output of $I1$ is fed back to $S1$ through a gain of one inverting amplifier. Integrator $I2$ multiplies the output of $I1$ by a factor of B/K and provides an output of $+Kx$ as required for input to $S1$. The solution is obtained as the variation of the voltage from $I2$ (Kx can be scaled down to x by the recording or display instrument). With the integrator initial conditions set (in this case to 0 V each)—the forcing function F connected as input to $S1$—an operate cycle can occur in which the variation of the displayed voltage representing the signal x is the solution to the problem for the set of values of B, K, and M. Changing any one or more of these values would result in a new solution and can be observed by repeating an operating cycle.

The same analog computer setup can be used to solve for the series RLC circuit of Fig. 13.22(a) since the differential equation is the same form of second-order equation:

$$v(t) = L\frac{d^2q}{dt^2} + R\frac{dq}{dt} + \frac{1}{C}q$$

where the change, q, is the variable to be solved for. Figure 13.22(b) shows the same analog computer setup as in Fig. 13.21 with the appropriately defined input gain settings.

The analog circuit of Fig. 13.21 can thus be viewed as the general setup to solve a second-order differential equation with the only requirement that of identifying the coefficient parameters of the actual system.

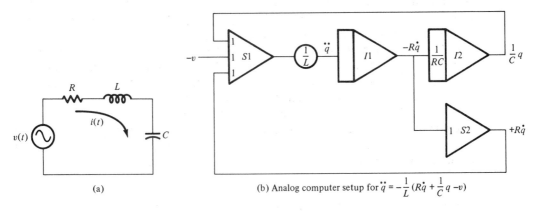

(a)

(b) Analog computer setup for $\ddot{q} = -\frac{1}{L}(R\dot{q} + \frac{1}{C}q - v)$

Figure 13.22 Series RLC circuit and analog representation.

Analogs

This idea of electrical circuit analogs for mechanical quantity can be used to investigate the action of various mechanical systems with the easier setup and parameter variations obtained using the electrically analogous equivalent circuit. Table 13.1 shows a number of common mechanical quantities and their electrically equivalent quantities using a mass-inductance, or a mass-capacitance, analogy.

Table 13.1

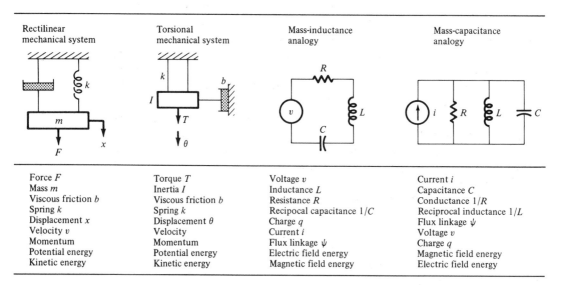

Rectilinear mechanical system	Torsional mechanical system	Mass-inductance analogy	Mass-capacitance analogy
Force F	Torque T	Voltage v	Current i
Mass m	Inertia I	Inductance L	Capacitance C
Viscous friction b	Viscous friction b	Resistance R	Conductance $1/R$
Spring k	Spring k	Recipocal capacitance $1/C$	Reciprocal inductance $1/L$
Displacement x	Displacement θ	Charge q	Flux linkage ψ
Velocity v	Velocity	Current i	Voltage v
Momentum	Momentum	Flux linkage ψ	Charge q
Potential energy	Potential energy	Electric field energy	Magnetic field energy
Kinetic energy	Kinetic energy	Magnetic field energy	Electric field energy

Basic Driving Functions

In many problems the input used to drive or operate the system is either a step, ramp, sinusoidal, or damped sinusoidal function. Some examples of circuits used to generate these functions are shown in Fig. 13.23 along with the developed time relationships. In Fig. 13.23(a) the amplitude of the signal is set by the initial condition of -5 V, while the delay time constant of -2 is set by the integrator gain factor of 2. In Fig. 13.23(b) the ramp slope is set by the value of the input step voltage -10 V (and the integrator gain). The circuit shown starts at 0 V and goes positive until it saturates (overloads) the amplifier output. The starting point could be shifted to, say, -10 V using an initial condition of 10 V. The sinusoidal outputs developed by the setup of Fig. 13.23(c) are sinudoidal with the x_1 and x_2 signals 90° out of phase. Potentiometers marked ω can be used to set the oscillator frequency, while the sinusoidal amplitude is set by the initial condition voltage, A (and gain values of amplifiers, if desired).

13.6 ANALOG/DIGITAL CONVERSION TECHNIQUES

While some problems are best solved only by a digital computer and others only be analog computation, a large number of practical problems are best handled by some combination of these two techniques. Accordingly we shall consider techniques for

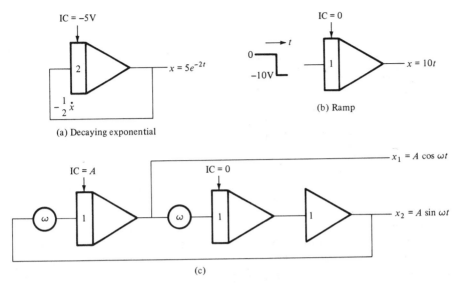

Figure 13.23 Generation of selected driving functions.

converting analog signals into digital form (analog-to-digital, or A/D, conversion) and for converting digital data back into analog signals (D/A conversion). In a typical application a number of different operations in a manufacturing plant or on a plane provide analog output signals to indicate their operation. Each of these analog form signals can then be converted by either separate A/D convertors or by one convertor using multiplex time switching. The resulting digital values can be handled by a single digital computer, these different data values being stored in the computer memory until used. After some programmed digital computation, digital values can be converted back into analog form for one or more output channels.

dc Voltage A/D Conversion

A popular analog signal is a dc voltage which can be converted into a digital count by a circuit as shown in Fig. 13.24. A special circuit is used to provide timing and control signals, these being a start count pulse to initialize the binary counter to zero (reset the counter), start a constant voltage ramp signal applied to a dc voltage comparator, and provide clock pulses which will advance the counter through its full range during one full sweep of the ramp voltage signal. The counter continues to advance by one count for each clock pulse until the dc voltage comparator provides a stop count pulse. The count is stopped when the ramp voltage rises up to the value of the unknown dc input voltage, the binary count at that time being proportional to the input dc voltage. If, for example, a 10-stage counter is driven at a clock rate of 1 MHz, it can go through its full 1024 steps in about 1 msec so that a conversion could be completed in, at most, about 1 msec, and at least 1000 such conversions could be carried out each second.

The resolution to which the dc voltage is converted depends on the number of

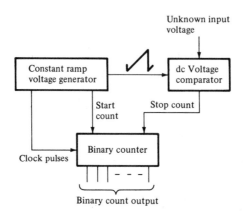

Figure 13.24 A/D conversion for dc voltage.

counter stages, about 1 part in 1000 resolution in the above example. The accuracy of the conversion, on the other hand, depends on the accuracy of the dc voltage comparator and on the linearity of the constant ramp voltage generator. One way to avoid the problem of ramp voltage linearity is to use a stepwise increasing voltage from a ladder network, as shown in Fig. 13.25(a).

After a reset pulse to clear the binary count to zero, clock pulses drive the binary counter at the rate set by the clock frequency. The parallel output bits of the binary counter then drive a special circuit called a ladder network, which provides a voltage output which increases or steps up for each binary count increase as shown in Fig. 13.25(b). When the stepped output voltage just equals or exceeds the input dc voltage a stop count pulse is provided by the dc voltage comparator, and the binary count equivalent of the input dc voltage is then available as a parallel binary value from the counter.

Ladder network. The ladder network, so-called because of its physical appearance as shown in Fig. 13.26, would have 10 inputs for a 10-stage counter, each input being logical-1 (*one*) or logical-0 (*zero*), depending on the binary count. For a count of all 0s the output is 0 V; for an input of all 1s the count is approximately V_{ref}, the precise reference voltage; and for any other count the output is a binary or proportional voltage equivalent. It is useful to consider the 2^9 bit as having a value of 0.5 ($\frac{1}{2}$) if that bit is 1 and similarly the 2^8 as having a value $\frac{1}{4}$, 2^7 a value $\frac{1}{8}$, etc. A more systematic way to specify the above is that the 2^9 bit has value $2^9/2^{10}$ or $\frac{1}{2}$, the 2^8 bit has value $2^8/2^{10}$ or $\frac{1}{4}$, the 2^7 bit has value $2^7/2^{10}$ or $\frac{1}{8}$, until the last bit having value $2^0/2^{10}$ or $\frac{1}{1024}$ in this example.

The ladder voltage in this case can be stepped by $V'_{ref}/1,024$ V for each binary count, until the dc voltage comparator shows the ladder voltage to be equal to or to just exceed the unknown input dc voltage. For any binary count each of the 10 inputs to the ladder network is either 0 V or V_{ref}. Using superposition (or other analysis techniques) allows calculating the resulting output voltage. Because of the particular arrangement of the ladder resistors and their proportional value (R and $2R$), each input of V_{ref} adds to its binary value over 2^{10}, times the input of V_{ref}, while each input of 0 V adds no voltage to the output summed voltage. Thus a binary count of

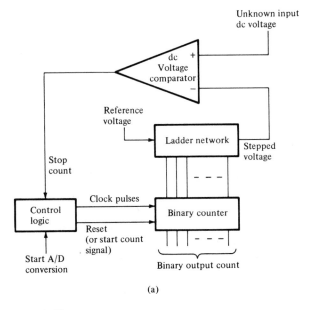

(a)

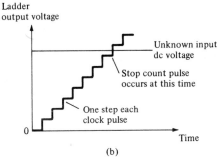

(b)

Figure 13.25 A/D dc voltage conversion using ladder network.

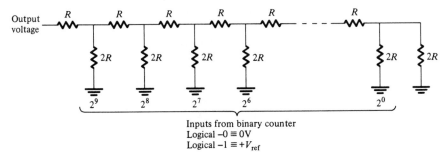

Figure 13.26 Ladder network.

0000010110 would result in an output voltage of

$$\frac{2^4}{2^{10}}V_{ref} + \frac{2^2}{2^{10}}V_{ref} + \frac{2^1}{2^{10}}V_{ref} = \frac{16+4+2}{1024}V_{ref} = \frac{22}{1024}V_{ref}$$

(or its binary equivalent value of 22 over the full count 1024 times the reference voltage).

Successive approximation converter. The previous method of conversion required the binary counter to go from a count of all 0s (reset) to all 1s for the longest conversion operation. On the average, the conversion time will be half the longest time required to get a full binary count. For the example of the 10-stage counter and the 1-MHz clock the average conversion time would be about 0.5 msec. If the same converter were to be multiplexed among a number of inputs and if 1000 conversions/sec were required for each input, the conversion time of the A/D unit would soon become too long. When greater conversion speed is required (less time per conversion) a technique of successive approximations may be used, among others. In this technique each bit of the 10-bit count value is handled on each clock pulse, so that a 10-stage counter would take only 10 clock steps for a full conversion time. With a 1-MHz clock the conversion time would then be the same 10 μsec each A/D conversion instead of the average 500 μsec previously discussed.

Figure 13.27 shows a block diagram of how a successive approximation converter can be built. The conversion method is carried out by comparing the unknown voltage against the generated ladder network voltage for each of the values $\frac{1}{2}V_{ref}$, $\frac{1}{4}V_{ref}$, $\frac{1}{8}V_{ref}$, etc., up to $\frac{1}{1024}V_{ref}$ for the example of a 10-bit conversion. The conversion circuitry consists of a shift register through which the single binary 1 is shifted, signifying which bit the conversion is being carried out on. An output register drives the ladder network.

Starting with the most significant bit (MSB), a 1 is set into the MSB position of

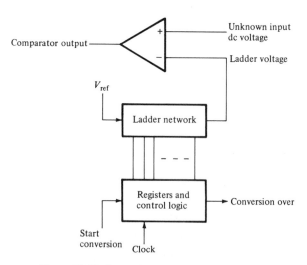

Figure 13.27 Successive approximation converter.

the output register and the output of the ladder network is $\frac{1}{2}V_{ref}$. If the input voltage is greater than the generated ladder output voltage (as indicated by the comparator output voltage), the MSB of the output register is left ON (otherwise it is turned OFF) by the control logic circuitry. On the next clock pulse, a binary 1 is set into the next lower output register stage and $\frac{1}{4}V_{ref}$ is added onto the output voltage. If the unknown dc input voltage is greater than this generated voltage, then the corresponding output register bit is left ON; otherwise it is turned OFF. This process continues for descending order bits until all bit positions have been examined (10 steps for a 10-bit conversion). The output register contains the binary equivalent value at the end of this 10-step conversion process. As a binary 1 moves down the shift register stages, it indicates which bit is presently being operated on. When the binary 1 is shifted out of the least significant bit position the conversion is finished.

Shaft position A/D conversion. In addition to converting a dc voltage into an equivalent binary value it is often useful to monitor some positional rotation of a shaft and convert this shaft rotational position into an equivalent binary value. The amount of positional movements of an airplane wing flap or rudder, of a valve opening, or of a navigational device rotation are examples where information occurs by nature of a shaft rotating some number of degrees. One popular technique incorporates either a brush or optical encoding disk attached to the shaft to be monitored. The encoding disk provides a direct binary output giving the digital output value of the shaft position. However, the output code from such a disk is most often in Gray code rather than the more common straight binary code. This is due to the desirable property of the Gray code (also a property of many other codes) that only 1 bit changes from one position to the next. This restricted code feature is important for reliable code reading, especially for any reasonable accuracy—say 10 bits or more. Mechanical limitations of reading the code value off the disk, especially for reasonably small disk sizes, necessitate the use of a code such as the Gray code being set onto the disk attached to the shaft. Thus, an instantaneous conversion is possible by directly reading the binary value of shaft position—this binary value in Gray code form.[*]

When, as usual, it is desired to have positional value converted into straight binary code, a relatively simple scheme, such as that shown in Fig. 13.28, can be used. The Gray code data are taken in serial form and used to toggle a flip-flop, which acts

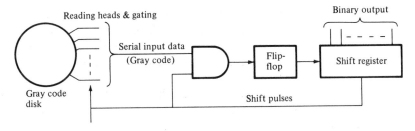

Figure 13.28 Gray-to-binary code converter.

[*]Refer to *Introduction to Digital Computer Technology*, by L. Nashelsky, Wiley, New York, 1972, pp. 86–90, for a description of the Gray code and its conversion from and into straight binary code.

as the converter. The resulting binary code value is then shifted into a storage register which provides the full binary value as parallel outputs after the conversion is completed. Conversion time is short—being only the time, for example, for shifting 10 bits at a 1-MHz rate, or 10 μsec. With such short conversion time it would be no difficulty to multiplex the input from many different Gray code disks using only the single code converter.

Digital-to-Analog (D/A) Conversion

Having converted analog data into digital form for use by the digital computer or special digital circuitry it is often desirable to convert some resultant digital values back into analog form for control or display purposes. The means for converting digital-to-analog dc voltage has already been presented. Using a ladder network, as shown in Fig. 13.29, it is easy to change a binary value in the data buffer register into

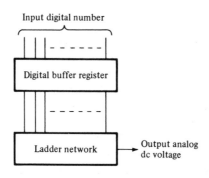

Figure 13.29 Dc voltage digital-to-analog converter (DAC).

an analog dc voltage. This dc digital-to-analog converter, or DAC, provides an output voltage almost simultaneously when the input value is shifted into the buffer register so that conversion is almost instantaneous.

A single DAC unit could be multiplexed to provide output to a number of output channels using special *hold* circuits as shown in Fig. 13.30. A digital value is transferred into a DAC buffer register whose dc voltage output is then connected to a hold circuit. As the name implies a dc voltage applied is then held after the multiplex operation moves the DAC output to the following output channels. This hold circuit may be as simple as a large capacitor which maintains or holds the dc voltage applied across it or some more involved electronic circuit which not only holds the dc voltage applied but allows connection to some external device with negligible loading. As the multiplex operation continues, different digital values are transferred into the DAC buffer register with each respective dc voltage held by that channel's hold circuit. After one complete cycle through all appropriate channels the DAC returns to operate the first channel again, updating the dc voltage applied to the hold circuit.

Shaft position DAC. A digital value can be converted into analog form as the position of a shaft. A digital value can then determine the rotational position of a shaft attached to some output device. Mechanical positioning of the shaft can be carried

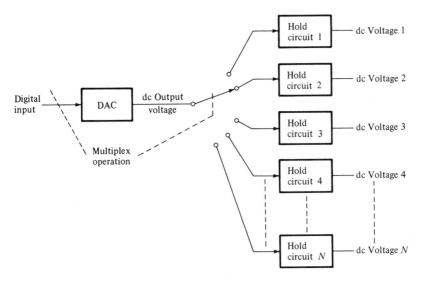

Figure 13.30 Multiplexed DAC output dc voltage.

out by a control motor, by a servomechanism motor (servomotor), or often by a stepping motor. The last is more directly operated by digital circuitry since it can be moved one step (each step being a fixed angular rotation of some fraction of a degree) for each stepping pulse applied to its electronic circuitry. Thus, the conversion from a digital value into shaft position requires generating the number of pulses for the number of steps the shaft must rotate.

In a full operating environment the present position of the shaft is obtained by an A/D converter; the computer determines the desired new position of the shaft and generates a number of pulses to move it to the desired position (see Fig. 13.31). This movement can be either clockwise or counterclockwise, as needed. In the sample system of Fig. 13.31 the shaft position is obtained by a shaft encoder A/D converter, and a binary positional value is obtained. This may be one value among a number

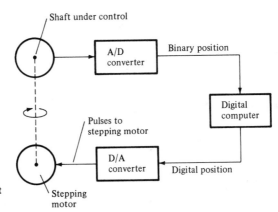

Figure 13.31 Computer control of shaft position.

that the computer obtains. Then, after some prescribed calculations, a digital position command is generated, and the D/A converter unit generates the number of
pulses required to step the motor and thereby the shaft attached to it, to the new
required position.

PROBLEMS

§ 13.1

1. List some of the basic parts of an analog computer.

2. What are some of the differences between analog and digital computers?

3. Can an analog computer be used in place of a digital computer?

§ 13.2

4. What are the ideal characteristics of an OPAMP?

5. What is meant by virtual ground?

6. Calculate the output voltage of the constant-gain multiplier circuit in Fig. 13.32.

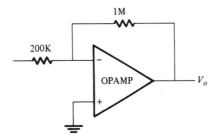

Figure 13.32 OPAMP circuit for problem 6.

7. Calculate the output voltage of the summer circuit shown in Fig. 13.33.

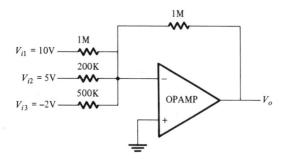

Figure 13.33 OPAMP summer for problem 7.

§ 13.3

8. What is the output voltage expression for the circuit of Fig. 13.34?

9. Describe (and sketch) the output voltage for the circuit in Fig. 13.35.

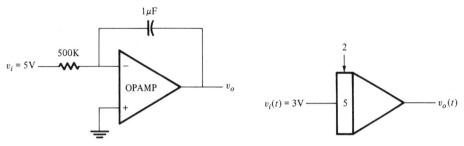

Figure 13.34 Circuit for problem 8.

Figure 13.35 Circuit for problem 9.

§ 13.4

10. Sketch the output waveform for the circuit of Fig. 13.36.

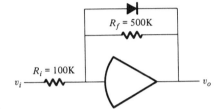

Figure 13.36 Circuit for problem 10.

11. Describe the basic operation of a function generator.

12. Describe the basic operation of a square-law multiplier (such as that of Fig. 13.19).

13. How does the divider circuit of Fig. 13.20 operate?

§ 13.5

14. Write the expression of the analog computer setup shown in Fig. 13.37.

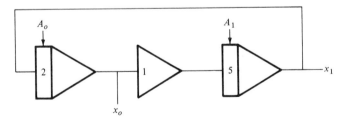

Figure 13.37 Circuit for problem 14.

15. Repeat Problem 14 for Fig. 13.38.

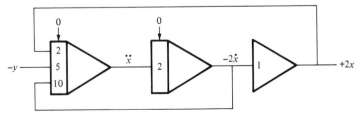

Figure 13.38 Circuit for problem 15.

§ 13.6

16. Describe the basic difference between a staircase-type A/D converter as shown in Fig. 13.25 and the successive approximation type shown in Fig. 13.27.

17. What output voltage results for an eight-stage A/D converter using a ladder network with $V_{ref} = 12$ V and a binary count of 10110011.

18. How long would the maximum conversion time be for an 8-bit A/D converter using a 10-μsec clock in a staircase-type converter.

19. Why is Gray code used on many optical shaft-encoded disks?

20. What is the purpose of the hold circuit used with D/A converters?

14

Power Supplies

14.1 INTRODUCTION

There is a great deal more to the basic laboratory supply than the face of some recently designed supplies might suggest. The manufacturers are reaching a level of sophistication that permits a set of controls on the face of the instrument that are clear in purpose and function and an output significantly less affected by the load applied to the supply. However, be aware that the old adage "you get what you pay for" applies in this area also. Even though two supplies may be rated $0 \rightarrow 40$ V dc with a current range $0 \rightarrow 500$ mA, there are other factors to consider such as voltage regulation, internal impedance, and, probably most important, reliability in the design. The last of the three is not a quantity normally included in the advertising literature but is one that can be of significant importance for a number of obvious reasons. The full use of ICs and digital readout has been the direction of design in recent years, resulting in a number of beautifully packaged supplies for all applications.

14.2 dc SUPPLIES

The basic components of a dc power supply appear in Fig. 14.1.

The input signal is first transformed to the desired voltage level and then passed through a full-wave rectifier, resulting in the indicated waveform described in Chapter 6. The 1000-μF capacitor is a filtering element to be considered in the next section that will result in an increased dc component as shown in the figure.

The preregulator ensures a fixed current I_{C_3} for the transistor Q_2 (like a current source). For proper regulation the current I_{B_1} should be sensitive only to the variations of Q_2 carried by a feedback path from the difference amplifier which is sensing the output voltage V_L. By designing the preregulator so that the current I_{C_3} is basically

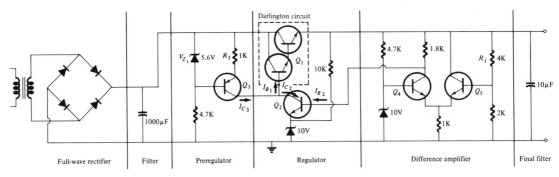

Figure 14.1 Complete voltage-regulated power supply.

independent of the variations in V_L the base current I_{B_1} will be sensitive only to changes in the behavior of Q_2. That is, an increase in I_{B_2} due to a change in V_L will cause a change in I_{C_2} which will be noted directly by I_{B_1} if I_{C_3} is held fixed (Kirchhoff's current law). If a simple resistor were to replace Q_3, the base of the Darlington amplifier would lose a full measure of its sensitivity to variations in the behavior of Q_3 since the collector current of Q_3 (I_{C_3}) would also vary with changes in V_L. Numerically, if we assume $V_{BE} = 0.6$ V, then

$$V_{R_3} = V_Z - 0.6 = 5.6 - 0.6 = 5 \text{ V}$$

and

$$I_{R_3} = I_{E_3} \simeq I_{C_3} = \frac{5}{1 \text{ K}} = \textbf{5 mA}$$

The output V_{C_4} of the difference amplifier is sensitive to the difference in the signals applied to the base terminals of the back-to-back transistors. Since V_{B_4} is fixed by the Zener diode, V_{C_4} is directly sensitive to variations in V_{B_5} controlled by the level of the output voltage. The collector voltage V_{C_4} appears at the base of Q_2 and therefore will obviously affect the ON condition of the transistor. Let us now see if we can follow the effect of an undesirable change in V_L (due to a change in load) on the behavior of the network. An increase in V_L (undesired) will cause the base voltage V_{B_5} to rise and cause an increase in the collector potential V_{C_4}. This will in turn cause V_{B_2} to rise and result in an increase in the collector current I_{C_2}. Through Kirchhoff's current law, if I_{C_3} is fixed, $I_{B_1} = I_{C_3} - I_{C_2}$ must drop in magnitude for an increase in I_{C_2}. The result is a drop in I_{C_1} and correspondingly in V_L to offset the undesired increase. The reverse would be true for decreasing levels of V_L.

The Darlington configuration was employed to increase the sensitivity of I_C to variations in I_B. Recall that for this configuration $I_C = \beta^2 I_B$ and not simply $I_C = \beta I_B$, as encountered for single-transistor configurations.

The ability of the network to maintain V_L at a fixed level is a measure of its regulation characteristics. Naturally, a supply that maintains a set voltage is desirable over one that changes its terminal voltage with each slight change in load. Voltage regulation will be examined in detail in a section to follow.

The 10-μF capacitor is a final filter on the supply voltage. The output voltage V_L can be varied (with maintained regulation) by changing the resistance R_1, which draws current from the load and therefore determines the load voltage through Ohm's law.

The above design is unfortunately more sophisticated than found in some commercially available supplies. However, we must also keep in mind that the high regulation characteristics are not required for all applications. In fact, some will only incorporate the rectifier and filtering section to obtain a dc level. Filters and other types of regulators will be examined in detail in the sections to follow.

14.3 RECTIFICATION AND FILTERS

The process of rectification was introduced in Chapter 6. In Fig. 14.2 the resulting waveforms for a half- and full-wave rectifier with a sinusoidal input are provided. The dc (or average value) of each waveform is given by

$$\boxed{\begin{aligned} V_{dc}(\text{half-wave}) &= 0.318 V_{peak} \\ V_{dc}(\text{full-wave}) &= 0.636 V_{peak} \end{aligned}} \tag{14.1}$$

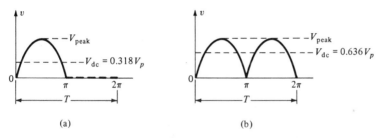

(a)   (b)

Figure 14.2 Rectified waveforms (a) half-wave; (b) full-wave.

The sole function of the rectifying circuit has therefore been to convert the sinusoidal input with zero average value to one with an average value of sufficient magnitude. The purpose of the filter in Fig. 14.1 was to increase the dc level above that established solely by the half-wave or full-wave rectifier. The term filter comes from the fact that the additional components will filter out some of the ac signal (variations in level) appearing in the rectified signals. In some applications, such as the charging of a battery, the output of a half- or full-wave rectifier may be a satisfactory level, but for an electronic amplifier the variations in (or ripple in) level will be carried through and distort the signal to be amplified. This is especially true for any audio or video equipment such as a stereo or TV receiver.

The simplest filter is the capacitive filter appearing in Fig. 14.3 with a load R_L and a full-wave rectifier. The output waveform as shown in the figure still has some "ripple," but it is certainly significantly less than that of the pure rectified output. During the interval of time denoted T_1 the capacitor C_1 is charging to the peak value of the full-wave rectified signal. During the interval denoted T_2 the voltage across the capacitor is greater than the input pulse signal, and the diode is reverse-biased (open circuit). During this same interval the capacitor will find itself discharging

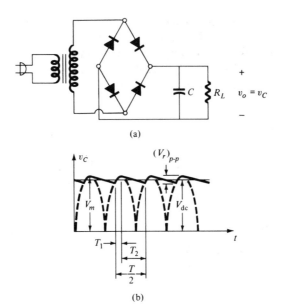

(a)

(b)

Figure 14.3 Capacitor filter: (a) capacitor filter circuit; (b) output voltage waveform.

through the load R_L at a time constant determined by

$$\tau = R_L C \tag{14.2}$$

You will recall from Chapter 2 that the charging and discharging period can be approximated by five time constants (5τ). Note that the discharge does not follow the input but discharges at a much slower rate. In fact an increase in C will cause an increase in the time constant and an improvement in the dc level. The value of C is limited by its bulkiness and the high resulting diode currents that must flow during the charging periods since the period of charging will likewise be decreasing.

The dc level of the signal is determined by

$$V_{dc} = V_m - \frac{V_{r(p-p)}}{2} \tag{14.3}$$

It can be shown using integral calculus that the rms of the ripple is given by

$$V_r(\text{rms}) = \frac{V_{r(p-p)}}{2\sqrt{3}} \tag{14.4}$$

A quantity called the *ripple factor* gives an immediate indication of the ripple content in a waveform whether it be filtered or not.

It is determined by

$$r\% = \frac{V_r(\text{rms})}{V_{dc}} \times 100\% \tag{14.5}$$

Obviously, since V_{dc} should be a maximum and V_r (rms) a minimum, the smaller the ripple factor, the better the output. In terms of the parameters of the network the ripple is given by

$$r = \frac{2.4}{R_L C} \qquad \text{(14.6)}$$

EXAMPLE 14.1

A capacitive filter provides a 5-K load with a 20-mA dc load current. If the peak-to-peak value of the ripple voltage is 4 V, determine

1. V_r (rms).
2. V_{dc}.
3. Percentage ripple.
4. C.

Solution:

1.
$$V_r(\text{rms}) = \frac{V_{r(p-p)}}{2\sqrt{3}} = \frac{4}{2\sqrt{3}} = \frac{2}{\sqrt{3}} = \textbf{1.156 V}$$

2.
$$V_{dc} = I_{dc}R_L = (20 \times 10^{-3})(5 \times 10^3) = \textbf{100 V}$$

3.
$$\% \text{ ripple} = \frac{V_r\,(\text{rms})}{V_{dc}} \times 100\% = \frac{1.156}{100} \times 100\% = \textbf{1.156}\%$$

4. From Eq. (14.6),

$$C = \frac{2.4}{rR_L} = \frac{2.4}{(0.01156)(5 \times 10^3)} = \frac{2.4}{57.8} = \textbf{0.0415 F}$$

A measurable improvement in the simple capacitive filter can be obtained by adding an additional RC filter as shown in Fig. 14.4.

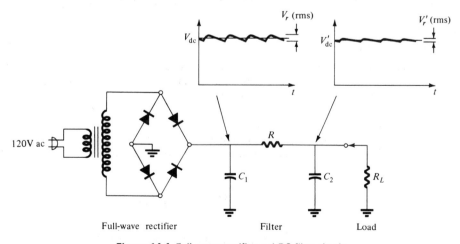

Figure 14.4 Full-wave rectifier and RC filter circuit.

The resistor R is chosen sufficiently small compared to R_L to ensure that the major portion of the input dc level appears across the load. Through an application of the voltage-divider rule (capacitors open circuit for dc),

$$V_L(\text{dc}) = \frac{R_L(V_{\text{dc}})}{R_L + R} \qquad (14.7)$$

where V_{dc} is the dc level provided by the simple capacitive filter. The capacitor C_2 is chosen so that its reactance X_{C_2} is low compared to the resistor R and R_L. This will permit the following approximations:

$$X_{C_2} \| R_L \simeq X_{C_2} \qquad (14.8)$$

and

$$V_L(\text{rms}) \simeq \frac{X_{C_2}[V_r(\text{rms})]}{(X_{C_2} + R) \simeq R}$$

$$V_L(\text{rms}) \simeq \frac{X_{C_2}}{R}[V_r(\text{rms})] \qquad (14.9)$$

The reactance X_{C_2} is determined by $X_{C_2} = \dfrac{1}{2\pi f C}$. For a half-wave-rectified signal the frequency is that of the applied sinusoidal signal. For a full-wave rectifier since two pulses appear for every cycle of the input, the frequency must be doubled. That is,

$$\begin{array}{ll} f = f_{\text{input}} & \text{(half-wave)} \\ f = 2f_{\text{input}} & \text{(full-wave)} \end{array} \qquad (14.10)$$

Equation (14.9) indicates that for $X_{C_2} \ll R$ the output ripple will be significantly reduced. The loss in dc level is considerably less, and an improved ripple factor is obtained as shown in Fig. 14.6.

EXAMPLE 14.2

An RC filter with $R = 0.5$ K and $C_2 = 26.5$ μF is added to a simple capacitive filter with a 5-K load. The measured values obtained are $V_{\text{dc}} = 120$ V and $V_r(\text{rms}) = 2$ V from a full-wave rectifier with a 60-Hz input.

1. Determine the new dc load voltage.
2. Calculate the rms value of the ripple voltage.
3. Find the new ripple factor and compare to one having the input levels.

Solution:

1.
$$V_L(\text{dc}) = \frac{5\text{ K}(120)}{5\text{ K} + 0.5\text{ K}} = 109.09\text{ V}$$

2.
$$X_{C_2} = \frac{1}{2\pi f C} = \frac{1}{(6.28)(120)(26.5 \times 10^{-6})} = 50\ \Omega$$

$$V_L(\text{rms}) \simeq \frac{X_{C_2}}{R}[V_r(\text{rms})] = \frac{50}{5000}(2) = 2 \times 10^{-2} = \textbf{20 mV}$$

3.
$$r\% = \frac{V_L(\text{rms})}{V_L(\text{dc})} \times 100\% = \frac{0.02}{109.09} \times 100\% = \textbf{0.0183}\%$$

$$r\% = \frac{V_r(\text{rms})}{V_{\text{dc}}} \times 100\% = \frac{2}{120} \times 100\% = \textbf{1.67}\% \gg \textbf{0.0183}\%$$

The dc loss across the resistor R in the previous filter can be reduced somewhat by replacing the resistor R by a coil as shown in Fig. 14.5. This particular filter con-

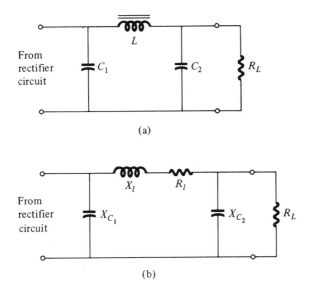

Figure 14.5 π-type filter circuit: (a) components; (b) impedances.

figuration is referred to as a π-type filter for reasons obvious from its appearance. For dc conditions, the coil is ideally a short, and all the dc voltage appearing across the first capacitor will find its way directly to the load. Of course, there will be some dc resistance associated with the coil, but then Eq. (14.7) can simply be applied. For ac conditions, X_{C_2} is still chosen sufficiently small at the applied frequency to permit the approximation $X_{C_2} \ll R_L$ and $X_{C_2} \parallel R_L \simeq X_{C_2}$. To ensure that the major portion of the ripple appears across X_L and not the load the following condition must be satisfied: $X_L \gg X_{C_2}$. Then

$$V_L(\text{rms}) \simeq \frac{X_{C_2}[V_r(\text{rms})]}{|X_L - X_{C_2}| \cong X_L}$$

$$\boxed{V_L(\text{rms}) \simeq \frac{X_{C_2}}{X_L}[V_r(\text{rms})]} \qquad (14.11)$$

The last type of filter to be introduced is called the *L-type* or *Choke input* filter, as evidenced by its appearance in Fig. 14.6.

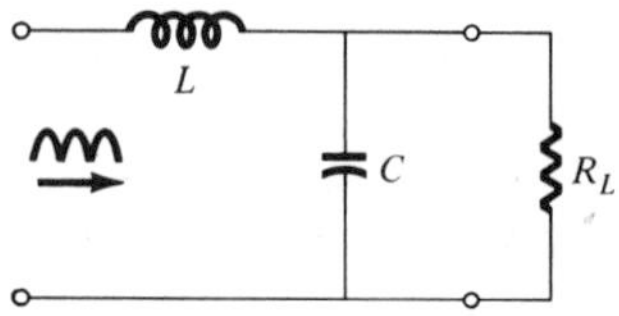

Figure 14.6 L-type filter.

The input capacitor C_1 was removed to improve the current limitations associated with the diodes of the rectifier network. Diodes of lower current ratings can be used with this filter for the same input-rectified signal. Of course, there is a reduction in the output dc level since the first filtering stage has been eliminated. The analysis of this filter is very similar to those described previously.

14.4 MULTIPLIER NETWORKS

There are diode-capacitive techniques for increasing the peak rectified voltage without increasing the peak transformer voltage. One such network, called a voltage doubler, appears in Fig. 14.7.

During the positive portion of the input signal, D_1 will be conducting and D_2 will be reverse-biased, causing capacitor C_1 to charge to V_m as indicated in Fig. 14.8. During the negative portion of the input D_2 will be conducting and D_1 nonconducting, as shown in Fig. 14.9.

But now the output peak voltage is the sum of that applied and the voltage held by the capacitor C_1, resulting in $2V_m$, as indicated in Fig. 14.9.

An extension of the doubler will result in the tripler and quadrupler of Fig. 14.10. The description of its behavior is simply a continuation of that described above for the doubler.

In each of the above networks, the PIV rating of each diode must be at least $2V_m$.

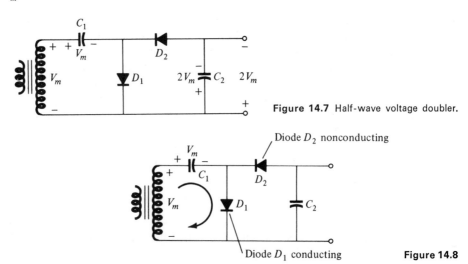

Figure 14.7 Half-wave voltage doubler.

Diode D_2 nonconducting

Diode D_1 conducting

Figure 14.8

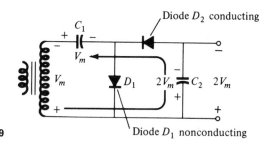

Figure 14.9

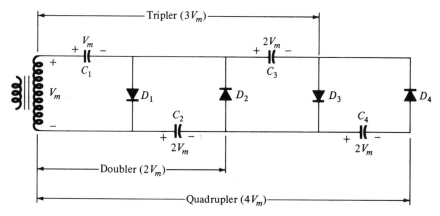

Figure 14.10 Voltage tripler and quadrupler.

14.5 REGULATION

The regulation of a supply, whether it be current or voltage, is a measure of how much the terminal rating will vary with load. Ideally, the dc voltage supply should supply a fixed voltage to a circuit independent of the variations in the circuit. In other words, when we set a dc supply to 20 V we want that voltage to remain fixed no matter how we vary the load connected to the supply. Unfortunately, however, the ideal is unattainable, but we can come very close to it if we pick the right supply for the application in mind. The terminal voltage of most supplies will decrease with increase in load current as shown in Fig. 14.11 with the no-load level (open-circuit terminal conditions) greater than the full-load (rated current) condition. In percent, the voltage regulation is determined by

$$\text{VR}\,\% = \frac{V_{\text{NL}} - V_{\text{FL}}}{V_{\text{FL}}} \times 100\,\% \qquad (14.12)$$

The voltage regulation is one of those quantities normally found on the specification sheet of the supply. Naturally the smaller its value (indicating the least slope in the characteristics of Fig. 14.11), the better the supply for the same voltage and current range.

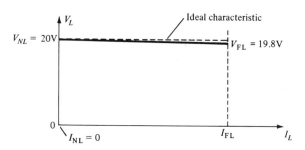

Figure 14.11 Terminal characteristics of a dc supply.

EXAMPLE 14.3

Determine the voltage regulation for the supply having the characteristics of Fig. 14.11.

Solution:

$$VR\% = \frac{V_{NL} - V_{FL}}{V_{FL}} \times 100\% = \frac{20 - 19.8}{19.8} \times 100\% = \frac{0.2}{19.8} \times 100\% = \mathbf{1.01\%}$$

For the supply of Fig. 14.12, the load voltage V_L can never equal the open-circuit voltage of 20 V due to the drop across the internal resistance of the supply. However, so long as the load applied is significantly larger than the internal resistance the terminal voltage will vary only slightly with change in load. For the case of Fig. 14.12 with $R_L = 5$ K,

$$V_L = \frac{5 \text{ K } (20)}{5 \text{ K} + 50} = 19.8 \text{ V} \simeq 20 \text{ V}$$

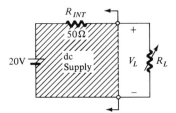

Figure 14.12 Circuit demonstrating the need for voltage and current regulators.

However, for $R_L = 100\ \Omega$,

$$V_L = \frac{100(20)}{100 + 50} = 13.33 \text{ V} \neq 20 \text{ V}$$

The internal impedance is a typically available piece of data on the specification sheet of the supply. It should obviously be a consideration for the types of loads to be encountered for a particular application.

There are, fundamentally, two techniques for ensuring that the terminal voltage is not sensitive to loads as the example of Fig. 14.12. One employs a series regulator such as shown in Fig. 14.13(a), while the other employs a shunt regulator such as appears in Fig. 14.13(b).

In series regulation the series element is a variable resistor whose resistance level is sensitive to variations in the load voltage. For increasing values of V_L the

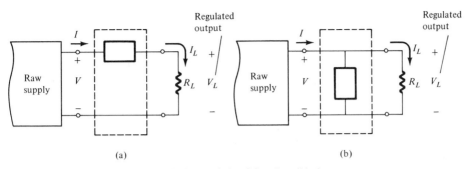

Figure 14.13 Regulation (a) series; (b) shunt.

resistance will increase, with the reverse occurring for decreasing values of V_L. For example, if V_L should start to drop due to a change in load, the resistance of the regulator will drop. The voltage drop across it will be less, and V_L will be stabilized at its original higher level.

Shunt regulators operate in the reverse manner in that an increase in V_L will cause a drop in the shunt regulator resistance and vice versa. For example, if the voltage should start to rise, the shunt resistance will drop, drawing current away from the parallel load, and cause V_L to stabilize at its original lower value.

There is a wide variety of techniques for effecting the series and shunt regulation defined above. One of the simplest is the *Zener shunt regulator* appearing in Fig. 14.14. As long as the input voltage is sufficiently high, the Zener diode will be in

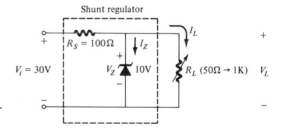

Figure 14.14 Zener diode shunt regulator.

the ON state and have a voltage drop of 10 V across it. However, for a set input voltage, if R_L is too small, 10 V will not appear across the Zener diode, and it will not be in the ON state. Before conduction the Zener diode is for all practical purposes an open circuit. Using this equivalent for the diode we find that the condition for firing is determined by

$$V_Z = 10 = \frac{R_L V_i}{R_L + R_S}$$

and

$$R_{L_{min}} = \frac{V_Z R_s}{V_i - V_z} \qquad\qquad (14.13)$$

For this example,

$$R_{L_{min}} = \frac{10(100)}{(30 - 10)} = 50 \ \Omega$$

and

$$I_{max} = \frac{V_Z}{R_{min}} = \frac{10}{50} = 200 \ \text{mA} = I_{L_{max}}$$

Therefore, for R_L (50 $\Omega \rightarrow$ 1 K) the diode is in the ON state and $V_L = 10$ V. For the ideal characteristics above where the internal resistance of the Zener diode is ignored the regulation is obviously 0%. However, in the practical world there is some resistance associated with the Zener diode, and even though it may only be 1 or 2 Ω, a regulation of 2 or 3% may result. Incidentally, the difference in current levels between that supplied and that which reaches the load is absorbed by the Zener diode. Recall the almost vertical characteristics of the device in Chapter 6.

Three reference potentials can be made available using three Zener diodes in the series manner indicated in Fig. 14.15. The reference level is determined by the series addition of Zener voltage levels. That is,

$$V_{R_1} = V_{Z_1} + V_{Z_2} + V_{Z_3} = 25 \ \text{V}$$
$$V_{R_2} = V_{Z_2} + V_{Z_3} = 18.2 \ \text{V}$$
$$V_{R_3} = V_{Z_3} = 10 \ \text{V}$$

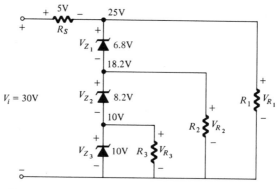

Figure 14.15 Zener regulator having three reference potentials.

A shunt regulator employing a thermistor appears in Fig. 14.16. A decrease in V_L will cause a corresponding decrease in the current through the thermistor. This decrease in thermistor current will reduce the heating level in the device and since it has a negative temperature coefficient will cause an increase in its terminal resistance. The resulting resistance of the parallel branches will therefore increase, causing an increase in V_L through simple voltage-divider action to offset the decreasing attitude.

Transistors are also used in series regulators as shown in Fig. 14.17 and previously in Fig. 14.1. The base to emitter voltage in Fig. 14.17 is determined by $V_{BE} = V_Z - V_L$. For increasing values of V_L, V_{BE} will decrease, and the base and collector currents

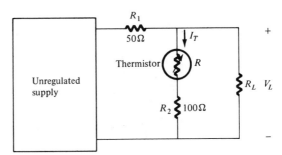

Figure 14.16 Thermistor shunt regulator.

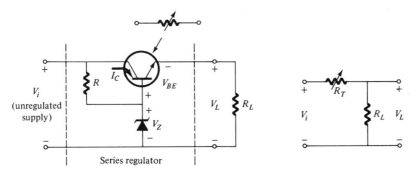

Figure 14.17 Transistor series voltage regulator.

will follow suit. Decreasing values of I_C are associated with increasing collector to emitter resistance (series variable resistor), causing a greater series drop across the transistor and a maintaining of V_L at its initial lower value.

The shunt transistor regulator appears in Fig. 14.18. In this case, V_L is always greater than V_Z and $V_{BE} = V_L - V_Z$. For increasing values of V_L, V_{BE} will increase, causing an increase in I_B and I_C. Increasing values of I_C will draw current from the load, and its terminal voltage will drop to its initial value. Or in terms of resistance the terminal (C-E) resistance of the transistor will drop, causing a drop in the total resistance of the parallel branches and a drop in V_L.

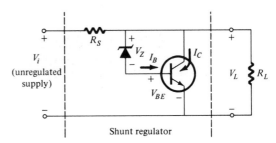

Figure 14.18 Transistor shunt voltage regulator.

Shunt regulator

14.6 LABORATORY SUPPLIES

An IC (dual-in-line package) regulated dc power supply appears in Fig. 14.19. The unit shown is a $0 \rightarrow 20$-V dc, $0 \rightarrow 0.65$-A meter. Its ripple content is 1 mV peak to

Figure 14.19 I-C regulated dc supply.

peak or 250 μV rms. For a 20-V output this would result in a percentage ripple of

$$\% r = \frac{V_r(\text{rms})}{V_{\text{dc}}} \times 100\% = \frac{250 \times 10^{-6}}{20} \times 100\% = 12.5 \times 10^{-4}\% = \mathbf{0.00125\%}$$

The load regulation is given as 4 mV, indicating that from no-load to full-load conditions the terminal voltage will vary only 4 mV. For the 20-V level this would provide a voltage regulation of

$$\text{VR}\% = \frac{V_{\text{NL}} - V_{\text{FL}}}{V_{\text{FL}}} \times 100\% = \frac{20 - (20 - 4\text{ mV})}{(20 - 4\text{ mV})} \times 100\%$$

$$= \frac{20 - 19.996}{19.996} \times 100\% = \mathbf{0.02\%}$$

A dc supply with a range of $0 \longrightarrow 250$ V dc appears in Fig. 14.20. The load regulation for this supply is given as 0.0005% + 100 μV for load variations from 0 to full load. For an output of 200 V this would represent a variation in terminal voltage of (0.0005/100)200 + 100 μV = 1000 μV + 100 μV = 1100 μV (0.0011 V) from no-load to full-load conditions. Its output level is set by the five-digit system as shown in the figure. The maximum current is determined by temperature levels. At 30°C it is 0.1 A, while at 60°C it is 0.07 A. It can be set for constant current and

Figure 14.20 High voltage dc laboratory supply.

result in less than 2-mA regulation for input variations of 105 to 132 V_{ac} and from 0 to rated V_{dc} load voltage change.

The internal construction of a dc supply appears in Fig. 14.21. Note the use of a printed circuit board for packaging and increased availability of components. This particular unit is designed specifically for rack mounting.

Power kits are available for constructing your own power supply such as shown in Fig. 14.22. After choosing the desired output dc voltage and current levels the components are provided for the construction of the supply. The series regulator units of Fig. 14.23, which are also available in design packages such as the above, employ a dual-in-line package IC for the regulation control on the output (also provided).

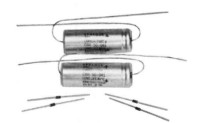

Figure 14.21 Internal construction of a dc supply.

Figure 14.22 Power kit.

14.7 dc/ac INVERTORS AND dc/dc CONVERTORS

Our interest thus far has been to convert an ac signal into one with an average or dc level. The block diagram of Fig. 14.24 indicates the sequence of operations required to convert a dc input to a higher dc level or an ac sinusoidal voltage. The latter process is called an *inversion* operation, while the former is a *conversion* process. Note that for either process the input dc must first be "chopped up" as shown in the figure using any one of the techniques listed. It can then be transformed to a higher level through transformer action, and it can then follow one of the two paths indicated

Figure 14.23 Series regulated unit.

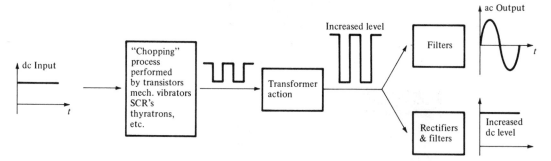

Figure 14.24 Inversion and conversion process.

for the type of output desired. Recall that transformers can operate only on a changing signal, thereby requiring the chopping action indicated. In the inversion process, the filter (LC) will provide the smoothing action to convert the input waveform to one more resembling the sinusoidal pattern.

One network configuration capable of performing the chopping function on the dc input appears in Fig. 14.25. Although the two transistors chosen have the same stock number and manufacturer, they can never be *exactly* alike in characteristics. When the dc input is applied the basic biasing for each transistor by the resistors R_1 and R_2 and dc voltage V_{CC} will turn that transistor on with the least base to emitter resistance. The other, as we shall see, will enter the nonconduction state. Let us assume for discussion sake that Q_1 is turning on. The resulting collector current

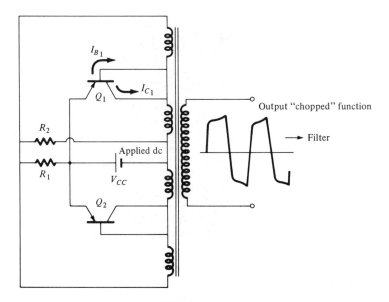

Figure 14.25 Inverter network.

for that transistor will pass through the primary winding of the transformer and since it is a changing level will induce a voltage across the other "feedback" primary windings through the flux induced in the core. For the ON transistor this induced voltage at the base of the device will turn the device heavily on, causing a sharp increase in I_{B_1} and I_{C_1}. The induced voltage at the base of the other transistor has a polarity to further turn that device off. For the ON transistor, the collector current will reach a sufficiently high level to saturate the core. That is, the flux in the core is an absolute maximum, and any further increase in primary current will have almost no effect on the flux in the core. When this occurs, the induced voltage in the primary windings is zero since Faraday's law $[e = N(d\phi/dt)]$ depends on a changing flux. The ON transistor will find itself starting to turn off, causing a decrease in collector current. A decreasing collector current in the primary will induce a voltage of reverse polarity in the base windings, further turning Q_1 off and starting to turn Q_2 on. Eventually, Q_2 will enter the full ON state with Q_1 off. However, when Q_2 saturates the primary as Q_1 did, the process will reverse itself. The reversing collector currents will result in reversing induced voltages in the secondary as shown in Fig. 14.15. The nonlinearity of the magnetic core of the transformer will result in the major portion of the distortion appearing in the almost square-wave signal. For an ac output a choke input filter will smooth the edges for an acceptable output sinusoidal signal. For the dc output the signal is rectified and filtered as described in this chapter. The turns ratio of the transformer will determine the output swing of the "square" wave as compared to the primary voltage.

A commercially available dc/dc convertor appears in Fig. 14.26 with its terminal data.

Figure 14.26 dc-dc converter.

PROBLEMS

§ 14.3

1. (a) A half-wave rectified signal has a peak value of 170 V. Determine the average (dc) value.

(b) The average level of a full-wave signal is 60 V. Determine the peak value of the signal.

2. Sketch (from memory) a half- and full-wave rectifier.

3. (a) Sketch the output (V_o) of the network of Fig. 14.27.

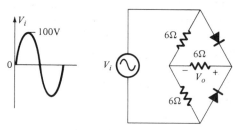

Figure 14.27

(b) Find the average value of the output waveform of part (a).

4. (a) The output waveform from a simple capacitive filter has a peak value (V_m) of 160 V and a peak-to-peak ripple voltage of 6 V. Determine the dc level.

(b) Determine the rms value of the ripple voltage of part (a).

(c) Calculate the ripple in percent.

(d) For $R_L = 1$ K, determine the capacitance C using the result of part (c).

5. A capacitive filter delivers 5 mA (dc) to a 2-K load. If the ripple is 4%, determine

(a) V_{dc}.

(b) The rms value of the ripple voltage.

(c) The peak-to-peak value of the ripple.

(d) The capacitance C.

6. A simple capacitive filter with a 2-K load is augmented with an RC filter with $R = 0.1$ K and $C_2 = 40$ μF. Measurements indicate that $V_{dc} = 60$ V and that $V_r(\text{rms}) = 1$ V from a half-wave rectifier with a 60-Hz input. Determine
   (a) The new dc load voltage.
   (b) The rms value of the output ripple.
   (c) The ripple factor.

7. Design an RC filter (find R and C) to have a ripple of 1% at $V_{L(dc)} = 100$ V for a 4-K load. A full-wave rectifier provides a dc level of 120 V and $V_r(\text{rms}) = 4$ V. The applied frequency is 60 Hz.

8. A π-type filter has an input voltage of 150 V dc across the input capacitor. If the filter inductance is 8 H (internal dc resistance of 400 Ω), calculate the output dc voltage at a load current of 50 mA.

9. Determine the dc output voltage of a π filter having 120 V (dc) across the input filter capacitor for a 4-K load. The inductor has a dc resistance of 500 Ω.

10. If the input rms ripple voltage to a π filter is 2 V, determine the load rms ripple voltage if $L = 5$ H (dc resistance of the coil $= 200$ Ω) and $C_2 = 5$ μF. A full-wave rectified input was developed from a 60 Hz input.

11. (a) Calculate the output dc voltage across a 2-K load preceded by an L-type filter ($L = 5$ H, $R_l = 200$ Ω, $C = 10$ μF) if the full-wave rectified input has a 200-V average dc level.
   (b) Determine the dc load current for part (a).
   (c) If $f = 60$ Hz (input to full-wave rectifier), determine the percent ripple at the load if the rms ripple voltage at the input to the filter is 4 V.

§ 14.4

12. Sketch (from memory) a voltage doubler.

13. Determine the voltage regulation of a supply that has a full-load output voltage of 120 V with a change in output level of 2 V from no-load to full-load conditions.

14. (a) The internal resistance of a dc laboratory supply is 50 Ω. If the no-load voltage is 40 V and a 4-K load is connected to the output terminals, determine the voltage across the load.
   (b) Calculate the voltage regulation of the supply if the load represents full-load conditions.

15. Repeat Problem 14, part (a), for a 500-Ω load.

16. Determine the voltages V_R and V_L in the network of Fig. 14.28.

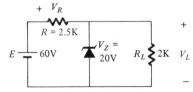

Figure 14.28

17. Repeat Problem 16 if $R_L = 1$ K.

15

Electronic Instrumentation

15.1 GENERAL INSTRUMENTATION

The measurement of various physical occurrences is one of the most important aspects of instrumentation devices. These measurements must be made accurately, without appreciably affecting the system under measurement, and provide display in readable or useful form. Measurement of a physical quantity requires some type of *transducer* which either provides direct indication of the measurement quantity or provides some other physical quantity, usually electrical, for display purposes.

A measuring device or instrument may be described by the general arrangement of Fig. 15.1. The measured quantity or *parameter* (e.g., temperature, pressure, fluid

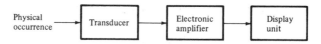

Figure 15.1 General instrument block diagram.

flow) causes the transducer to provide a voltage which can then be amplified, if necessary, by an electronic circuit to bring it to a level to drive a display device such as a meter deflection. An alternative display which is growing in popularity is the direct digital readout, requiring circuitry to convert the amplified signal from analog form into digital form to operate a digital display unit.

There are many types of transducers available for converting various parameters into an electrical quantity which can then be used to operate a display unit. A microphone, for example, converts audio or acoustic energy into electrical form. A strain-gage transducer converts an applied force into electrical form, while a thermocouple can be used to provide an electrical quantity proportional to temperature. In each case such quantities as transducer linearity, range of operation, amplitude, output electrical operating environment, and physical size must all be considered in selecting a suitable device.

Basic dc, ac, and magnetic measurements have already been covered in Chapters 2–4. In the present chapter we shall concentrate more on electronic devices or measurements. Electronic voltmeters are covered, followed by the very important cathode-ray oscilloscope electronic instrument. A few electronic signal-generating circuits are then discussed and special waveform measuring instruments are described.

15.2 ELECTRONIC VOLTMETERS

Purely passive measuring meters are small, economical, and usually quite portable as compared to a vacuum-tube voltmeter (VTVM). The advent of solid-state devices, especially field-effect transistors (FETs) and integrated circuit (IC) amplifier units, has made possible electronic voltmeters which compare favorably in terms of small size, portability, and relatively low cost. An electronic voltmeter can have a number of advantages over a purely passive meter; some of these advantages include relative isolation from the circuit under measurement, high sensitivity, and a great variety of display formats including digital readout and automatic scale ranging. Figure 15.2 shows how electronic voltmeters can be generally organized. In Fig. 15.2(a) the input voltage to be measured is applied to range attenuators, allowing

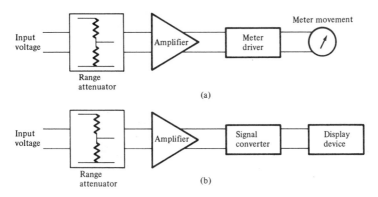

Figure 15.2 Electronic voltmeter, block representations.

measurement of a wide range of voltage. The rescaled voltage is then applied to an amplifier to handle low-level signals or to provide isolation from the circuit under measurement. The amplified signal can then be applied to the display meter through a meter-driver circuit, if necessary. For other than meter movement displays, the arrangement of Fig. 15.2(b) shows the inclusion of some signal conversion circuitry to alter the voltage reading into a form suitable for the type of display used. If the display is digital, then the signal conversion circuitry would include a voltage to digital converter and digital display driver circuitry. In more elaborate units solid-state switching of the range attenuators would allow automatic scale ranging so that voltages over a very wide range can be measured without unnecessary scale setting.

Electronic voltmeters can be broadly classified as either ac or dc. The amplifier circuits are built using either direct-coupled dc amplifiers for economical dc voltmeters operating down into the millivolt range or chopper-type dc amplifiers for more stable operation down into microvolt range.

Range Setting

Precision resistors in a voltage-divider chain are commonly used to set the scale or range of the meter. In the sample circuit shown in Fig. 15.3 the input impedance of the meter is essentially the fixed value of probe and divider chain resistances—about 12 MΩ in this case. The voltage to be measured is applied to the probe. The input voltage connected to the FET amplifier is taken through the range switch and depends on the divider chain setting. At the 3-V setting the divider action of 8 and 12 MΩ total resistance results in a calibrated full-scale 3 V. With the range switch moved to the 30-V position, for example, the divider resistance to ground, 800 K, over the divider chain resistance of 8 MΩ requires an input voltage 10 times as large as in the 3-V switch position to produce the same full-scale deflection.

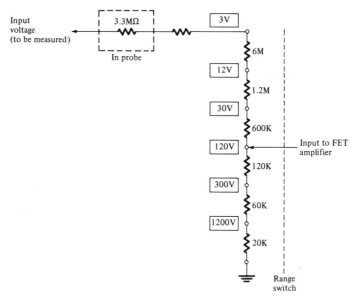

Figure 15.3 Range setting of voltmeter using voltage divider resistor chain.

ac Voltage Measurement

While the divider chain shown above can be directly used in dc voltage measurement, additional wave shaping must be done in measuring ac voltage. Figure 15.4 shows how wave shaping a sinusoidal signal can be accomplished to provide a dc voltage for meter deflection. In Fig. 15.4(a) the rectifier-filter circuit provides a dc

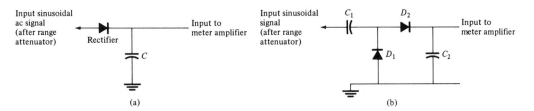

Figure 15.4 Ac voltage measurement. (a) rms; (b) peak-to-peak.

voltage which is the peak sinusoidal voltage. Proper calibration can then be made so that either peak or rms ($0.707V_{peak}$) is read on the meter scale. In Fig. 15.4(b) a voltage doubler circuit is used so that the dc voltage is $2V_{peak}$ or V_{p-p}, which can then be read directly on the meter. Proper range attenuation of the input sinusoidal signal before wave shaping keeps the voltage applied to the wave-shaping circuit at about the same level regardless of the input signal magnitude. It should be clear from the circuits shown in Fig. 15.4 that the readings obtained apply to purely sinusoidal signals only. Other types of signals are best handled by meter types, discussed later, or by CROs.

Resistance Measurement

The measurement of resistance is obtained using an external (or internal) battery in series with a resistor divider chain as shown in Fig. 15.5. The unknown resistance to be measured is applied at a point in the attenuator chain set by the range switch, usually in scale steps which are multiples of 10. With no resistance connected to the ohms input the internal battery results in full-scale deflection (usually with some external calibration available.) When the ohms input is connected to ground the 0-V input results in no deflection of the meter. Then a resistance connected across the

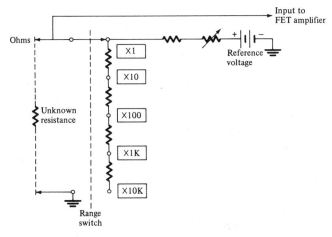

Figure 15.5 Resistance measurement circuit connection showing divider chain and reference voltage.

input leads results in a voltage division of the reference battery voltage, and less than full-scale deflection is obtained. The nature of this divider action is to produce a non-linear scale from 0- to ∞-Ω reading. When the reading is too far off the center region the use of the range switch to change scales is appropriate. As shown in Fig. 15.5 a number of divider resistors are used in series to change the scale of full-scale deflection by factors of 10 from $\times 1$ (times one) to $\times 10$ K (times 10 kilohms) in this example.

Electronic Voltmeter Using dc Amplifier

Electronic voltmeters built using FET and bipolar transistors allow for battery-operated portable or small-sized meters having good sensitivity, high input impedance, and high reliability. Solid-state voltmeters are thus replacing VTVMs as basic measuring instruments. As a simple example of a transistor voltmeter, consider the balanced bridge circuit shown in Fig. 15.6. The FET provides a very high input

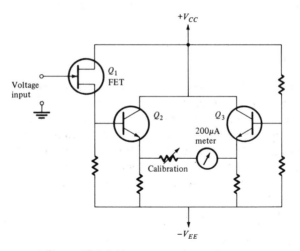

Figure 15.6 Bridge-type transistor voltmeter.

impedance, typically greater than 10 MΩ, with the input dc voltage coupled directly to the base of bipolar transistor Q_2. The FET source-follower stage allows driving the balanced bridge circuit of transistors Q_2 and Q_3 while providing the desired high impedance to the input. The current through a 200-μA meter movement in this example can be calibrated to provide a desired full-scale voltage reading. With 0-V input the bridge should be balanced with no net current through the meter movement.

Another type of comparison meter is the differential voltmeter. One very accurate means of measuring a voltage involves comparing the unknown voltage to a well-defined reference voltage. A precision potentiometer (see Fig. 15.7) allows adjusting the amount of reference voltage to balance against the unknown input voltage so as to obtain a *null* or 0-V condition of the meter movement. When the reference voltage obtained from the divider results in a null condition the setting of the precision divider

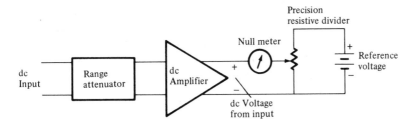

Figure 15.7 Differential voltmeter.

can be read directly as the value of the unknown dc voltage measured. The meter movement sensitivity, resistor divider precision, and reference voltage accuracy determine the overall accuracy of the differential voltmeter.

Measurement of dc voltages in the microvolt range would require very elaborate low-drift dc amplifiers to increase the signal voltage to a sufficient level to operate a meter movement. To avoid this drift problem a chopper amplifier, containing a chopper or modulator circuit to convert the low-level dc into an ac pulse waveform (Fig. 15.8), a conventional ac amplifier to increase the pulse waveform level, and a dc modulator-filter network to convert the signal back to a proportional dc voltage are used. The entire modulator-ac amplifier-demodulator-filter network is then referred to as a single chopper dc amplifier stage. The chopper modulator-demodulator circuit can be made using FET switches which connect either the input dc voltage or ground, at a rate set by an oscillator circuit, to achieve a chopped signal. The demodulator is also an FET switching circuit operated synchronously with the input chopper to provide a nonzero average signal to be filtered.

The full dc voltmeter includes ranging attenuators and often a null comparator, as covered above, to allow for good measurement accuracy. In addition to the FET-type chopper circuits, mechanical vibrators and photoelectric choppers are also commonly used.

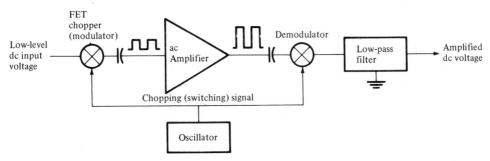

Figure 15.8 Chopper-stabilized dc amplifier.

Digital Voltmeters

With the increased availability of medium- and large-scale integrated circuits and the low cost of these units the digital voltmeter (DVM) has grown rapidly in use. Even with the previous large size and cost the DVM was popular because of the ease of reading measured values. Digital display not only reduces reading error and increases reading speed; it can also provide the data in suitable digital form to feed other digital data collection circuitry for additional processing and recording. DVM units can be built to include many special features such as automatic floating decimal point, polarity sign, scale factor, or range display and range switching, for example.

Digital voltmeters can be generally classified by method of conversion, which include

1. Linear ramp.
2. Staircase ramp.
3. Integrating.
4. Servo-balancing potentiometer.

The *linear ramp* DVM uses the time it takes a linear ramp voltage to just equal or exceed the unknown input voltage to determine the input voltage magnitude. A counter, operated at a fixed clock frequency, is scaled so that the resulting count value indicates the voltage value. The counter binary value then operates some form of decimal-indicating display so that direct visual readout is provided. Figure 15.9(a) shows a linear ramp signal and gating interval control signal both started at the same time and continuing until the ramp voltage reaches the value of the input dc voltage. The circuitry to achieve this operation, shown in Fig. 15.9(b), provides a digital clock to operate the unit, some logic control gating for starting the conversion operating, a comparator to determine when the voltages are equal (or ramp voltage just exceeds the input), and a digital counter and display to provide visual output of the converted voltage value. The conversion accuracy depends on the linearity of the ramp signal generated and on the accuracy of the comparator circuit. The number of counter stages determines the precision to which the readout value can be set. Some instruments allow front panel adjustment of the sampling rate, while others used a fixed rate, typically 5 Hz.

A *staircase ramp* signal is used in place of the linear ramp to bypass the difficulty in generating an accurate linear ramp signal. The waveforms in Fig. 15.10(a) show a ramp-like voltage which increases by a set amount at a preset stepping rate. When this staircase-like voltage just equals or exceeds the unknown input dc voltage a comparator circuit will indicate this condition so that the conversion operation is stopped. A block diagram of a possible digital voltmeter using this staircase-ramp comparator voltage is shown in Fig. 15.10(b). The input dc voltage is connected to one input of a comparator circuit after range attenuation and amplification to handle a wide range of dc voltage level. A sample rate oscillator controls how often a conversion is carried out and initiates the conversion action. When conversion starts the digital counter is reset, and then clock pulses are gated from an oscillator to operate the digital counter. As the count increases the output voltage of the special ladder network is the desired staircase waveform, which is compared to the input dc voltage. The ladder network is an arrangement of precision resistors (see Chapter 13) which act as a digital-to-analog converter, changing the digital count

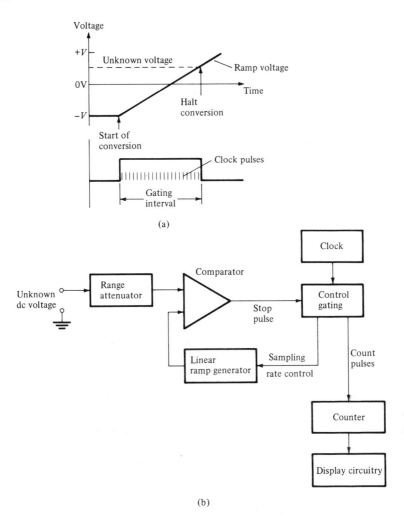

Figure 15.9 Linear ramp type DVM.

into a proportional dc output voltage. The comparator halts the count action when the staircase voltage just equals or exceeds the input. At that time the counter value is proportional to the input dc voltage and is used to operate display circuitry so that a direct decimal readout is provided.

An *integrating-type* DVM measures the average of the input signal over a fixed measuring period rather than sampling the voltage at the end of the conversion operation. This kind of conversion is capable of accurate measurement in the presence of noise because of the integrating action. A sample DVM is shown in Fig. 15.11. The A/D circuit is essentially a voltage to frequency converter providing a higher rate of pulses for larger dc voltages. Thus, as the dc voltage increases or decreases the integrator pulse generator circuitry provides a varied pulse rate. Over a fixed measuring period (manually adjustable with the sample rate control circuitry)

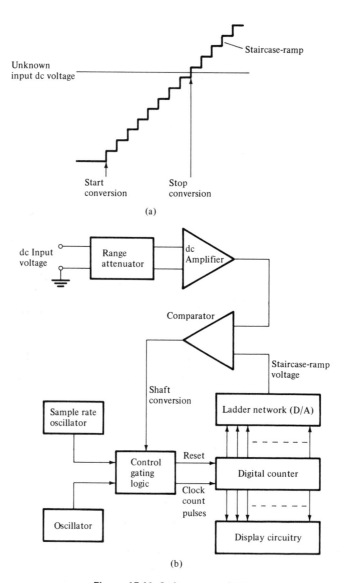

Figure 15.10 Staircase ramp DVM.

the pulses from the integrator-pulse generator is proportional to the average voltage during the period. Control gating logic provides counter reset, gated clock pulses, and readout display timing.

A *servo-balancing potentiometer*-type DVM, as shown in Fig. 15.12, drives a mechanical servomotor and digital readout dials. The circuit is continuous acting, with the servomotor being driven to maintain a null condition with the input dc voltage. The nulling voltage is obtained from a fixed precision reference supply and precision potentiometer controlled by the servomotor. The slow speed of the servo

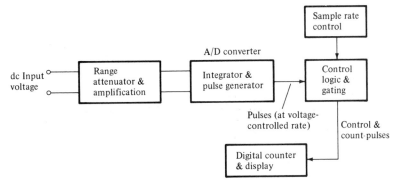

Figure 15.11 Integrator type DVM.

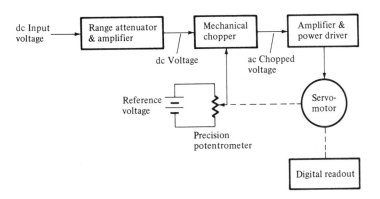

Figure 15.12 Servo-balancing potentiometer type DVM.

system requires measuring times of a few seconds rather than the fractions of a second with the fully electronic circuitry covered previously.

15.3 CATHODE-RAY OSCILLOSCOPE

The cathode-ray oscilloscope (CRO) is by far the most useful electronic measurement instrument. Without the viewing ability made possible by the CRO it would be virtually impossible to measure various waveform characteristics—amplitudes, duration times, and phase relations. CROs are made in a variety of presentations, although the basic operation of all is the same. All CROs use a cathode-ray tube (CRT) as the viewing element. The CRT construction, shown in Fig. 15.13, contains an electron gun to generate a stream of electrons traveling at high velocity, pairs of horizontal and vertical deflection plates to move the beam of electrons in each direction, and a phosphor screen at the tube's end to provide a visible image. The deflection plates in an oscilloscope use electrostatic deflection to control the electron beam. The phosphor coating used depends on the intended CRO area of application, with some phosphors having long persistence (continues to glow for a few milliseconds after

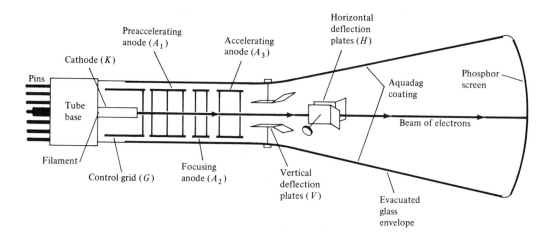

Figure 15.13 Cathode ray tube: basic construction.

beam is removed) for slow or nonrepetitive signals and others having short persistence for high-frequency signal viewing. Special CRTs are used in memory-type scopes having effective persistence in the range of hours.

Basic CRO Operation

The CRT is the output element of the oscilloscope instrument having basic parts as shown in Fig. 15.14. Voltages of hundreds of volts are usually necessary at the deflection plates to move the viewed beam across the visible screen area. The vertical and horizontal attenuators and amplifiers are therefore needed to scale the input signal up or down for a desired viewing range. For repetitive viewing of the

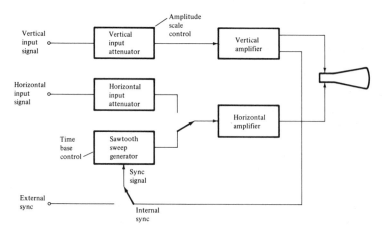

Figure 15.14 Cathode ray oscilloscope, general block diagram.

input signal a synchronizing signal (discussed later) is obtained either internally from the vertical signal itself or from some externally connected signal. The horizontal deflection signal can be either an input signal for Lissajous patterns or, more often, an internal sweep signal to control the amount of signal viewed, as discussed next.

CRO Deflection and Sweep Operation

To obtain a visual pattern of the input signal an internally generated sawtooth sweep voltage is applied to the horizontal deflection plates. The sawtooth voltage rises at an adjustable rate from a negative voltage sufficient to deflect the electron beam horizontally to the extreme left of the tube up to a positive voltage deflecting the beam across the tube face to the extreme right. The resulting display is a straight line, as shown in Fig. 15.15, if the sweep rate is fast enough so that the tube persistence

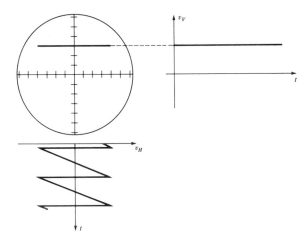

Figure 15.15 Scope display for dc vertical signal and linear horizontal sweep signal.

maintains the image. The sawtooth, as shown, is a repetitive signal so that the beam is repeatedly swept across the tube, producing a steady, strong image.

An input signal applied to the vertical deflection plates would produce a vertical straight line by causing the beam to deflect up and down (see Fig. 15.16). Any signal would give this same display when no horizontal sweep signal is present. To obtain a visual display showing the form of the vertical input signal requires both horizontal sweep signal *and* vertical input signal, as shown in Fig. 15.17. Figure 15.17(a) shows the result of a sawtooth sweep signal set to the same frequency rate as the input signal, while Fig. 15.17(b) shows the resulting display for a sweep rate one-half as fast as the input signal. Notice that adjusting the internal sawtooth sweep rate varies the resulting display, this display being a fraction of the input signal across the full display area up to many cycles. To ensure a resulting steady picture each display sweep should start at the same point in the input signal cycle, a process of *synchronization* being necessary.

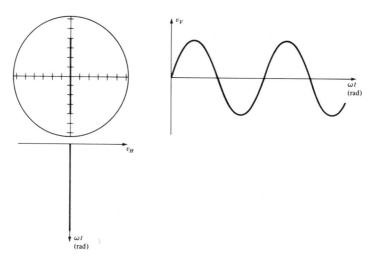

Figure 15.16 Resulting scope display for sinusoidal vertical input and no horizontal input.

Synchronization

The sweep process described above results in repeated displays of a portion of the vertical input signal dependent on the sweep rate of the horizontal sawtooth voltage. In practice it would be practically impossible to adjust the sawtooth rate to remain a fixed multiple of the vertical signal frequency. When such synchronization is attained the steady display due to a repeated identical trace results as shown in Fig. 15.18. Since the signals come from two independent sources which will individually vary somewhat in frequency and more so in relation to one another, the retrace will start at slightly different times, as shown in Fig. 15.19, resulting in an apparent drift of the displayed signal waveform.

A means of stabilizing this display is obtained using a triggering circuit to initiate the start of a sweep rather than a fixed sawtooth sweep rate. In the standard oscilloscope a trigger signal is derived from the input signal at a selected point in the signal cycle and used to initiate each sweep as shown in Fig. 15.20. The horizontal signal goes through a single sweep, retraces to the left of the screen, and then remains there until the trigger signal initiates another sweep. In this way the display remains a steady picture even if the signal frequency varies slightly since each new retrace occurs at the same place in the input signal cycle.

Display Modes

It is often desirable to be able to view more than one signal waveform at the same time. A number of modes of presentation are possible using a single electron gun CRT by means of electronic switching techniques. Figure 15.21 shows two common modes of presentation. In Fig. 15.21(a) a full sweep of the signal applied

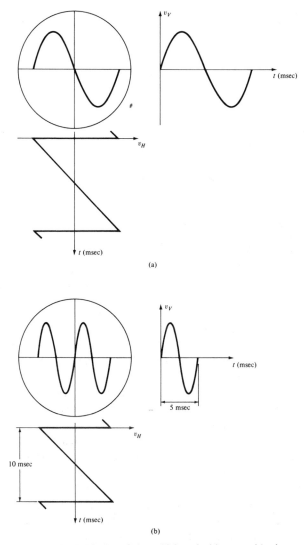

Figure 15.17 Display of sinusoidal vertical input and horizontal sweep input (a) display of vertical input signal using linear sweep signal for horizontal deflection; (b) scope display for a sinusoidal vertical input and a horizontal sweep speed equal to one-half that of the vertical signal.

to vertical input *A* is shown after which electronic switching connects an input *B* to the vertical channel and a second, or *alternate*, sweep takes place. As long as the signals being viewed have a frequency greater than a few thousand hertz the top sweep will still persist (on usual medium-persistence CRTs) while the bottom signal is being traced out, with the net result of two waveforms appearing "at the same time" insofar as the viewer can see. When two signals of frequency too low for good viewing

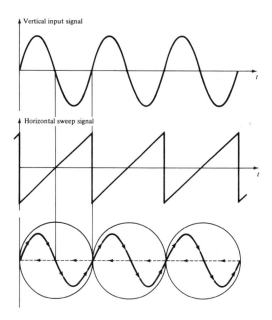

Figure 15.18 Steady scope display—input and sweep signals synchronized.

using alternate mode are to be displayed a *chopped mode* of presentation, shown in Fig. 15.21(b), can be used. In the chopped mode the electronic switch moves very rapidly back and forth between input *A* and input *B* during a single sweep so that both appear "at the same time." The vertical chopping lines are normally quite faint for low input signal frequencies.

A sophisticated use of triggered sweeps is the delayed sweep mode of showing part of a basic sweep in expanded view. Figure 15.22(a) shows a typical pulse waveform. Using the delayed sweep action a portion of this main sweep waveform, say the third pulse, can be selected to be viewed in expanded form, as in Fig. 15.22(b). The electronic circuitry to achieve this delayed sweep mode of display is shown in block diagram form in Fig. 15.22(c). A main trigger signal initiates the main sweep to give the overall main view as in Fig. 15.22(a). A separate delayed trigger, which can be externally adjusted, is used to initiate the delayed sweep at some time after the start of the main sweep. In the present example the delay could be set so that the delayed sweep starts when the third pulse occurs. If the main sweep is then selected, the waveform of Fig. 15.22(a) is displayed. If the delayed sweep is selected, however, then the view of Fig. 15.22(b) is shown.

Measurements Using Calibrated CRO Scales

One important use of CROs is to view various signal waveforms to obtain some picture of what is happening in the electronic circuit under investigation. A second very important use is for measurement of various amplitude, frequency, phase displacements or time intervals within a signal or between signals. Calibrated vertical

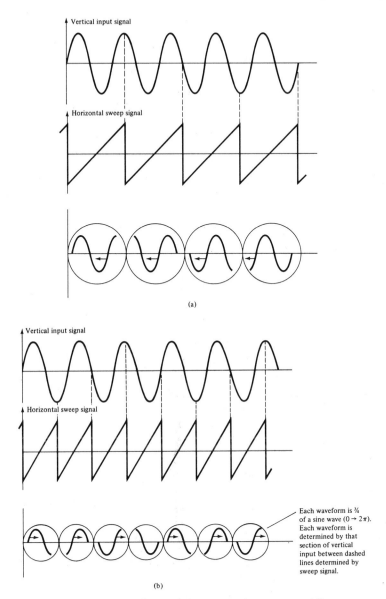

Figure 15.19 (a) Sweep frequency too low—apparent drift
to left; (b) sweep frequency too high—apparent drift to right.

gain and horizontal time base, use of various fixed-scale ranges, and scaled axes on
the scope face all allow for these various measurements.

Amplitude measurement. The vertical scale of a scope is usually calibrated in units
of volts per centimeter (V/cm) or volts per box as shown in the sample amplitude
measurement of Fig. 15.23. Having centered the display, the peak of the signal is

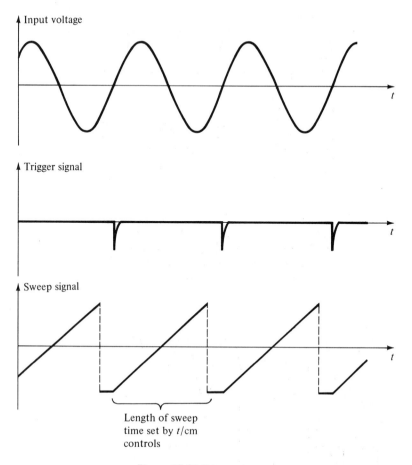

Figure 15.20 Triggered sweep.

seen to be 3.6 boxes or 3.6 cm. If the scope were then set at 1 V/cm, the voltage measured would be

$$1 \text{ V/cm} \times 3.6 \text{ cm} = 3.6 \text{ V}$$

If, however, the scope scale setting were 0.05 V/cm, then the voltage measured would
be

$$0.05 \text{ V/cm} \times 3.6 \text{ cm} = 0.18 \text{ V}$$

The range used is usually selected to obtain a reasonably large display (within the
screen limitations, typically 8–10 cm).

A sinusoidal signal can be measured as in Fig. 15.23 using instruments other
than a CRO. Voltage measurements of pulse waveforms, shown, for example, in
Fig. 15.24, depend on CRO display and calibrated scales for proper measurement.
The various amplitude values are read off the scope face in centimeter units, with
the scope scale setting then determining the exact voltage values. Some more sophisticated scopes provide display of the selected scale setting in volts per centimeter

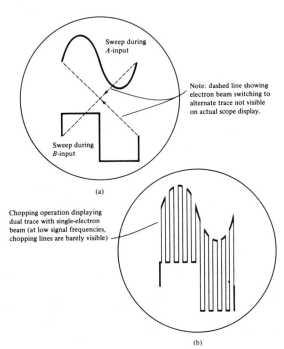

Note: dashed line showing electron beam switching to alternate trace not visible on actual scope display.

Sweep during *A*-input

Sweep during *B*-input

(a)

Chopping operation displaying dual trace with single-electron beam (at low signal frequencies, chopping lines are barely visible)

(b)

Figure 15.21 ALTERNATE and CHOP-PED mode displays for dual-trace operation: (a) ALTERNATE mode for dual-trace using single electron beam; (b) CHOPPED mode for dual-trace using single electron beam.

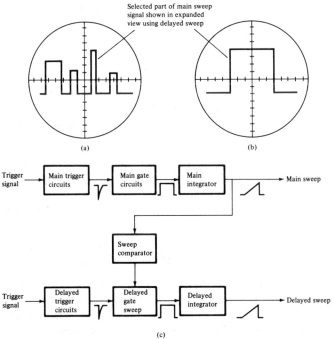

Selected part of main sweep signal shown in expanded view using delayed sweep

(a)

(b)

Trigger signal → Main trigger circuits → Main gate circuits → Main integrator → Main sweep

Sweep comparator

Trigger signal → Delayed trigger circuits → Delayed gate sweep → Delayed integrator → Delayed sweep

(c)

Figure 15.22 Operation of delayed sweep: (a) main sweep presentation; (b) delayed sweep presentation; (c) delayed sweep operation-block diagram.

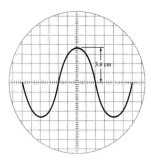

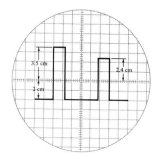

Figure 15.23 Scope scale showing amplitude measurement.

Figure 15.24 Measurement of pulse-type
waveform amplitudes.

so that the display screen contains all the necessary information to obtain various
voltage values.

Time measurement. Time intervals can be measured using the calibrated horizontal
scale shown in Fig. 15.25. In this example the dual trace display allows measurement
not only of the pulse width of two separate signals but also the time between the start
of each pulse as shown. If the scale setting were 20 μsec/cm in the present example,
the time intervals of each pulse would be

$$1.2 \text{ cm} \times 20 \ \mu\text{sec/cm} = 24 \ \mu\text{sec}$$

$$2.3 \text{ cm} \times 20 \ \mu\text{sec/cm} = 46 \ \mu\text{sec}$$

and the time between start of each pulse would be

$$2.1 \text{ cm} \times 20 \ \mu\text{sec/cm} = 42 \ \mu\text{sec}$$

Frequency measurement. It is also possible to use the calibrated time scale (horizontal
sweep scale) of the scope to calculate the frequency of an observed signal. After
measuring the time interval of a full cycle of the signal the signal frequency is calculated using $f = 1/T$.

Phase shift measurement. The scope's calibrated time scale can also be used to
calculate phase shift between two signals of the same frequency. If the two signals
are sinusoidal as shown in the dual trace presentation of Fig. 15.26, the phase shift
between the signals can be measured since a cycle of signal is 360° and any portion
of that cycle is then

$$\theta = \text{Phase shift} = \frac{\text{No. of cm of phase shift}}{\text{No. of cm of full cycle}} \times 360°$$

The above technique would apply to any type of repetitive signal. The technique is
easily applied on a dual-trace scope, but care must be taken that each signal sweep
is triggered by the *same* reference signal. On some scopes either the *A* or *B* input is
the only trigger input; otherwise use of an external trigger input is necessary to ensure
that a true phase shift relation is observed.

On single trace scopes it is possible to view each signal at a time by changing
input connections while maintaining a reference external trigger. For sinusoidal

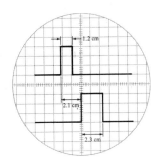

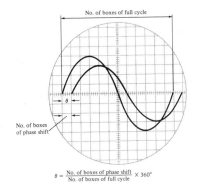

$$\theta = \frac{\text{No. of boxes of phase shift}}{\text{No. of boxes of full cycle}} \times 360°$$

Figure 15.25 Pulse waveforms showing time measurements using calibrated sweep scales.

Figure 15.26 Phase shift measurement using horizontal scope scale.

signals a special technique known as Lissajous figures can also be applied. This technique* applies one signal to the vertical input and the other to the horizontal input (in place of the sweep) to develop a Lissajous pattern from which the amount of phase shift can be calculated.

15.4 SIGNAL GENERATION

Various types of signal generators are necessary for the design, testing, and operation of electronic circuits. These generators include sinusoidal waveform generators over a wide range of frequency, pulse-type generators of variable amplitude, pulse width, and repetition rate, and special function generators for ramp, sawtooth, and other similar signals. For example, Table 15.1 provides a listing of the more common

Table 15.1

Signal Generator Frequency Bands

BAND	APPROXIMATE FREQUENCY RANGE
Low frequency	0.05–100 Hz
Audio frequency (AF)	20 Hz–20 kHz
AM radio frequency	0.5–1.6 MHz
High frequency (HF)	1.5–30 MHz
FM radio frequency	87–108 MHz
Very high frequency (VHF)	30–300 MHz
Ultra high frequency (UHF)	300–3000 MHz
Microwave	Above 3000 MHz

*For details on the use of Lissajous figures to measure phase shift or frequency, refer to *Electronic Devices and Circuit Theory*, by R. Boylestad and L. Nashelsky, Prentice-Hall, Inc., Englewood Cliffs, N.J., 1972, pp. 752–759.

frequency bands for which commercial signal generators are readily available. Signal generators may be specified within any of the above bands, may fall into a number of bands, may be more selective than the ranges listed in Table 15.1, or may be of broader or wider range. Sinusoidal oscillator circuits used depend very much on the frequency range intended, on the factor of narrow or broad frequency band, and on the required frequency stability of the signal. A variety of oscillator circuits, covered in Chapter 10, could be used in a sinusoidal signal generator. A standard signal generator would then include a number of basic parts as shown in Fig. 15.27.

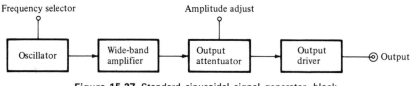

Figure 15.27 Standard sinusoidal signal generator, block diagram.

An oscillator circuit with provision for adjusting the frequency provides the desired signal. The signal directly obtained from the oscillator can then be amplified to some maximum output level and then passed through output attenuators which allow adjusting or selecting the output signal level desired. An output driver amplifier, such as an emitter follower, can be provided to enable the output signal to operate a wide range of output devices.

Pulse generators are quite popular, especially with digital-type circuitry. A basic generator, such as that shown in Fig. 15.28, can provide pulses at any desired frequency within the range of the particular unit, with additional adjustments of duty cycle (pulse width) and pulse amplitude. A pair of opposite polarity current sources is switched into a capacitor to provide opposite polarity ramp signals to a Schmitt trigger circuit. The output of the Schmitt trigger circuit is the desired pulse waveform which can be provided externally through a pulse driver and amplitude attenuator. The output pulse repetition rate and duty cycle are set by separately adjusting the two current sources and by the value of capacitor used. This type of circuit allows a greater adjustment range than a simpler nonstable multivibrator which could provide the same output pulse waveform.

Function Generators

A more versatile instrument than either the sinusoidal generator or pulse generator is one providing a variety of output wave shapes including sinusoidal, ramp, pulse, and sawtooth. Figure 15.29 shows the block design of a basic function generator providing sinusoidal, sawtooth, and pulse signals. A generator providing these various signal types at frequencies from a fraction of a hertz to hundreds of kilohertz with adjustable output amplitudes is a handy signal source.

A selector switch can be used to provide as output either triangular (sawtooth), pulse, or sinusoidal signal. This versatility of signal type makes the function gen-

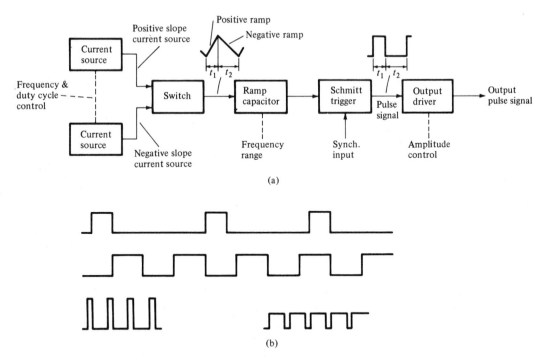

(a)

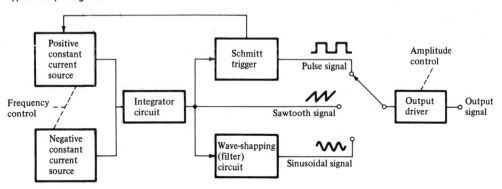

(b)

Figure 15.28 Basic pulse signal generator (a) block diagram; (b) typical output signals.

Figure 15.29 Function generator, typical block diagram.

erator, as described here, a generally useful signal source for a wide variety of uses at a relatively inexpensive price.

Wave Analyzer

A wave analyzer is a special instrument to measure the amplitude component of a single frequency of some complex or distorted waveform. In effect, the analyzer instrument filters out all but the desired frequency component, acting as a voltmeter for that signal component. The popular audio-range frequency analyzer, operating in the range of 20 Hz to 20 kHz, can be implemented using high-Q active filtering as shown in the block diagram of Fig. 15.30. After initial range selection using an input attenuator and driver amplifier, the signal, made up of many frequency components, is applied to an adjustable frequency high-Q active filter circuit which is highly tuned

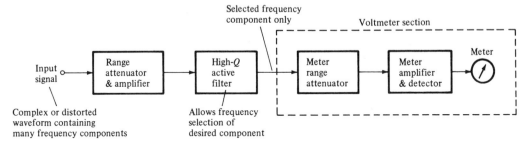

Figure 15.30 Audio-range wave analyzer, functional block diagram.

to attenuate all signal frequencies off the frequency selected. The output of this filter circuit is then a voltage whose amplitude is the relative component at the selected frequency. The signal then is applied to a voltmeter circuit containing range selection, rectification, and meter driver circuitry to provide either a voltage or dB reading. Manual selection of a number of frequency components (typically, harmonics of some fundamental frequency) allows readings from which analysis of the complex waveform can be made. Distorted audio signals can likewise be tested and the amount of signal distortion measured using the wave analyzer.

While the analyzer of Fig. 15.30 is applicable to the audio range, signals in the megahertz frequency range usually require a heterodyning-type wave analyzer as shown in the block diagram of Fig. 15.31.

Tuning a local oscillator whose signal mixes or beats with the input signal (after range attenuation and amplification) results in an intermediate frequency (i.f.) signal for a selected frequency component. The setting of the local oscillator results in heterodyning a desired frequency component into the 30-MHz i.f. frequency. This i.f. signal is then amplified and again mixed with a precise 30-MHz crystal oscillator signal to extract the desired signal component centered on zero frequency. The filtered component of the input signal then drives a voltmeter section and output meter to display the voltage component in either volts or dB units.

Signal distortion can be measured for each harmonic component of the distorted signal by separately measuring the second, third, fourth, etc., harmonic components. Each harmonic component relative to the fundamental signal level can be used to calculate the total distortion using

$$D = \sqrt{D_2^2 + D_3^2 + D_4^2 + \cdots}$$

where $D =$ the total distortion, $D_2 =$ second harmonic distortion, $D_3 =$ third harmonic distortion, $D_4 =$ fourth harmonic distortion, etc.

Harmonic distortion analyzers can be used to separately obtain the amount of

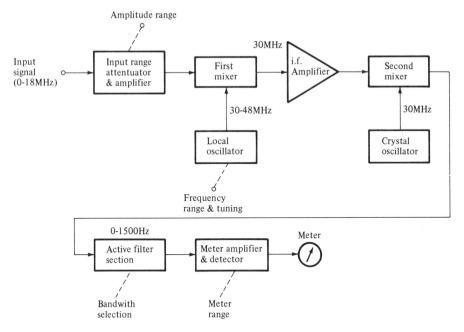

Figure 15.31 Heterodyning wave analyzer, functional block diagram.

the second, third, fourth, etc., harmonic components present in any distorted signal and then the total harmonic distortion can be calculated, if desired. As with wave analyzers, harmonic distortion analyzers can be built to directly filter the selected harmonic component or to use heterodyning to separate out the harmonic components for high-frequency signals.

Spectrum Analyzer

A complex or distorted signal can be mathematically described by its fundamental and harmonic components. The spectrum analyzer instrument provides a display of this total frequency spectrum showing the relative harmonic components. A total display is obtained from an automatic *sweeping* of the frequency range of interest rather than separate manual adjustment of each component as previously described. The resulting spectral information is quite useful in determining information about the input signal. Typically, instruments are built to operate in either the audio range or on rf signals, the latter being the more important since rf signals include the radar and navigation areas. Rf spectrum analyzers can be set to display a signal's frequency makeup for a portion of the i.f. spectrum.

A basic spectrum analyzer block diagram, as shown in Fig. 15.32, shows how the resulting spectrum of an rf signal can be obtained and displayed on a CRT. The rf input signal is heterodyned or mixed with a local oscillator signal, resulting in an i.f. signal, which is then amplified and applied to a detector and video amplifier

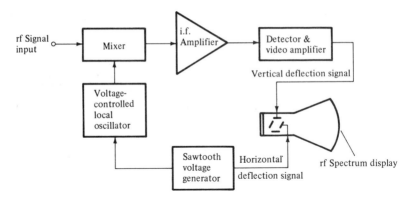

Figure 15.32 Basic spectrum analyzer, block diagram.

to provide a vertical deflection signal to a CRT. A sawtooth signal provides not only the necessary horizontal sweep signal but also sweeps the local oscillator frequency which acts to sweep the input signal.

PROBLEMS

§ 15.3

1. A sinusoidal signal displayed on a CRO has a peak amplitude of 5.5 cm when the scope vertical gain setting is 2 V/cm. Calculate the peak and rms values of the signal.

2. Draw the CRO display for a 12-V, rms sinusoidal waveform displayed at a vertical scale setting of 2 V/cm.

3. What are the peak-to-peak and rms voltage values for the sinusoidal waveform of Fig. 15.33?

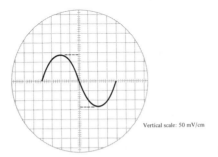

Vertical scale: 50 mV/cm

Figure 15.33 Waveform for problem 3.

4. A square-wave signal displayed on a CRO has a peak amplitude of 2.5 V. How many centimeters of peak signal amplitude are observed at a scale setting of 0.5 V/cm?

5. A pulse-type signal is observed on a CRO to have a width of 5.8 cm. Calculate the pulse width time for the following sweep scale settings: (a) 10 msec/cm, (b) 500 μsec/cm, and (c) 5 μsec/cm.

6. Two pulse-type signals are observed on a CRO at a scale setting of 50 μsec/cm. If the pulse widths are measured as 1.5 and 4.2 cm, respectively, and both start at the same time, calculate the time width of each pulse and the time delay between the end of the pulses.

7. Calculate the signal frequency of a square-wave signal having a half-cycle width of 4.5 cm at a scale setting of 20 μsec/cm.

8. A 2000-Hz signal is observed on a CRO. How many centimeters should be observed for four cycles of display at a scale setting of 10 msec/cm?

9. For the sinusoidal CRO display of Fig. 15.34, calculate (a) V_{p-p} and V_{rms}; (b) the time for one complete cycle, T; and (c) the frequency of waveform signal, f.

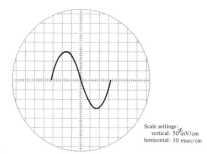

Scale settings:
vertical: 50 mV/cm
horizontal: 10 msec/cm

Figure 15.34 Waveform for problem 9.

10. For the CRO display of Fig. 15.35, calculate (a) V_{p-p}, (b) the time for three cycles, and (c) the pulse repetition rate, f.

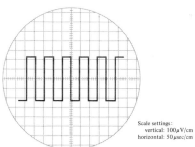

Scale settings:
vertical: 100μV/cm
horizontal: 50μsec/cm

Figure 15.35 Waveform for problem 10.

11. A full cycle of a sinusoidal signal is set to 6 cm. A second sinusoidal signal (same frequency) is observed to be phase-shifted by 2.5 cm. Calculate the phase shift in degrees.

12. How many centimeters displacement will be observed for a phase shift of 45° between two sinusoidal signals (same frequency) if one cycle of signal is observed to be 6 cm?

Appendices

Magnetic Parameter Conversions

	ENGLISH	CGS	MKS
Φ	10^8 lines	= 10^8 Maxwells	= 1 Weber
B	6.452×10^4 lines/in.^{2}	= 10^4 Maxwells/cm^2	= 1 Weber/m^2
		(10^4 Gauss)	
μ_0	3.20 lines/A·in.	= 1 Gauss/Oersted	= $4\pi \times 10^{-7}$ Weber/A·m
F	NI	0.4πNI	NI
	(At)	(Gilberts)	(At)
	(1 At = 1.256 Gilberts)		
H	NI/l	0.4πNI/l	NI/l
	(At/in.)	Gilbert/cm	At/m
		(Oersted)	

Greek Alphabet and Common Designations

Name	Capital	Lower case	Used to designate
alpha	A	α	Angles, area, coefficients
beta	B	β	Angles, flux density, coefficients
gamma	Γ	γ	Conductivity, specific gravity
delta	Δ	δ	Variation, density
epsilon	E	ϵ	Base of natural logarithms
zeta	Z	ζ	Impedance, coefficients, coordinates
eta	H	η	Hysteresis coefficient, efficiency
theta	Θ	θ	Temperature, phase angle
iota	I	ι	
kappa	K	κ	Dielectric constant, susceptibility
lambda	Λ	λ	Wave length
mu	M	μ	Micro, amplification factor, permeability
nu	N	ν	Reluctivity
xi	Ξ	ξ	
omicron	O	o	
pi	Π	π	Ratio of circumference to diameter = 3.1416
rho	P	ρ	Resistivity
sigma	Σ	σ	Sign of summation
tau	T	τ	Time constant, time phase displacement
upsilon	Υ	υ	
phi	Φ	ϕ	Magnetic flux, angles
chi	X	χ	
psi	Ψ	ψ	Dielectric flux, phase difference
omega	Ω	ω	Capital: ohms; lower case: angular velocity

Common Logarithms

Formulae:
$$\log ab = \log a + \log b$$
$$\log \frac{a}{b} = \log a - \log b$$
$$\log a^n = n \log a$$

no.	0	1	2	3	4	5	6	7	8	9
0	....	0000	3010	4771	6021	6990	7782	8451	9031	9542
1	0000	0414	0792	1139	1461	1761	2041	2304	2553	2788
2	3010	3222	3424	3617	3802	3979	4150	4314	4472	4624
3	4771	4914	5051	5185	5315	5441	5563	5682	5798	5911
4	6021	6128	6232	6335	6435	6532	6628	6721	6812	6902
5	6990	7076	7160	7243	7324	7404	7482	7559	7634	7709
6	7782	7853	7924	7993	8062	8129	8195	8261	8325	8388
7	8451	8513	8573	8633	8692	8751	8808	8865	8921	8976
8	9031	9085	9138	9191	9243	9294	9345	9395	9445	9494
9	9542	9590	9638	9685	9731	9777	9823	9868	9912	9956
10	0000	0043	0086	0128	0170	0212	0253	0294	0334	0374
11	0414	0453	0492	0531	0569	0607	0645	0682	0719	0755
12	0792	0828	0864	0899	0934	0969	1004	1038	1072	1106
13	1139	1173	1206	1239	1271	1303	1335	1367	1399	1430
14	1461	1492	1523	1553	1584	1614	1644	1673	1703	1732
15	1761	1790	1818	1847	1875	1903	1931	1959	1987	2014
16	2041	2068	2095	2122	2148	2175	2201	2227	2253	2279
17	2304	2330	2355	2380	2405	2430	2455	2480	2504	2529
18	2553	2577	2601	2625	2648	2672	2695	2718	2742	2765
19	2788	2810	2833	2856	2878	2900	2923	2945	2967	2989
20	3010	3032	3054	3075	3096	3118	3139	3160	3181	3201
21	3222	3243	3263	3284	3304	3324	3345	3365	3385	3404
22	3424	3444	3464	3483	3502	3522	3541	3560	3579	3598
23	3617	3636	3655	3674	3692	3711	3729	3747	3766	3784
24	3802	3820	3838	3856	3874	3892	3909	3927	3945	3962
25	3979	3997	4014	4031	4048	4065	4082	4099	4116	4133
26	4150	4166	4183	4200	4216	4232	4249	4265	4281	4298
27	4314	4330	4346	4362	4378	4393	4409	4425	4440	4456
28	4472	4487	4502	4518	4533	4548	4564	4579	4594	4609
29	4624	4639	4654	4669	4683	4698	4713	4728	4742	4757
30	4771	4786	4800	4814	4829	4843	4857	4871	4886	4900
31	4914	4928	4942	4955	4969	4983	4997	5011	5024	5038
32	5051	5065	5079	5092	5105	5119	5132	5145	5159	5172
33	5185	5198	5211	5224	5237	5250	5263	5276	5289	5302
34	5315	5328	5340	5353	5366	5378	5391	5403	5416	5428
35	5441	5453	5465	5478	5490	5502	5514	5527	5539	5551
36	5563	5575	5587	5599	5611	5623	5635	5647	5658	5670
37	5682	5694	5705	5717	5729	5740	5752	5763	5775	5786
38	5798	5809	5821	5832	5843	5855	5866	5877	5888	5899
39	5911	5922	5933	5944	5955	5966	5977	5988	5999	6010
40	6021	6031	6042	6053	6064	6075	6085	6096	6107	6117
41	6128	6138	6149	6160	6170	6180	6191	6201	6212	6222
42	6232	6243	6253	6263	6274	6284	6294	6304	6314	6325
43	6335	6345	6355	6365	6375	6385	6395	6405	6415	6425
44	6435	6444	6454	6464	6474	6484	6493	6503	6513	6522
45	6532	6542	6551	6561	6571	6580	6590	6599	6609	6618
46	6628	6637	6646	6656	6665	6675	6684	6693	6702	6712
47	6721	6730	6739	6749	6758	6767	6776	6785	6794	6803
48	6812	6821	6830	6839	6848	6857	6866	6875	6884	6893
49	6902	6911	6920	6928	6937	6946	6955	6964	6972	6981
50	6990	6998	7007	7016	7024	7033	7042	7050	7059	7067
no.	0	1	2	3	4	5	6	7	8	9

Common Logarithms (continued)

no.	0	1	2	3	4	5	6	7	8	9
50	6990	6998	7007	7016	7024	7033	7042	7050	7059	7067
51	7076	7084	7093	7101	7110	7118	7126	7135	7143	7152
52	7160	7168	7177	7185	7193	7202	7210	7218	7226	7235
53	7243	7251	7259	7267	7275	7284	7292	7300	7308	7316
54	7324	7332	7340	7348	7356	7364	7372	7380	7388	7396
55	7404	7412	7419	7427	7435	7443	7451	7459	7466	7474
56	7482	7490	7497	7505	7513	7520	7528	7536	7543	7551
57	7559	7566	7574	7582	7589	7597	7604	7612	7619	7627
58	7634	7642	7649	7657	7664	7672	7679	7686	7694	7701
59	7709	7716	7723	7731	7738	7745	7752	7760	7767	7774
60	7782	7789	7796	7803	7810	7818	7825	7832	7839	7846
61	7853	7860	7868	7875	7882	7889	7895	7903	7910	7917
62	7924	7931	7938	7945	7952	7959	7966	7973	7980	7987
63	7993	8000	8007	8014	8021	8028	8035	8041	8048	8055
64	8062	8069	8075	8082	8089	8096	8102	8109	8116	8122
65	8129	8136	8142	8149	8156	8162	8169	8176	8182	8189
66	8195	8202	8209	8215	8222	8228	8235	8241	8248	8254
67	8261	8267	8274	8280	8287	8293	8299	8306	8312	8319
68	8325	8331	8338	8344	8351	8357	8363	8370	8376	8382
69	8388	8395	8401	8407	8414	8420	8426	8432	8439	8445
70	8451	8457	8463	8470	8476	8482	8488	8494	8500	8506
71	8513	8519	8525	8531	8537	8543	8549	8555	8561	8567
72	8573	8579	8585	8591	8597	8603	8609	8615	8621	8627
73	8633	8639	8645	8651	8657	8663	8669	8675	8681	8686
74	8692	8698	8704	8710	8716	8722	8727	8733	8739	8745
75	8751	8756	8762	8768	8774	8779	8785	8791	8797	8802
76	8808	8814	8820	8825	8831	8837	8842	8848	8854	8859
77	8865	8871	8876	8882	8887	8893	8899	8904	8910	8915
78	8921	8927	8932	8938	8943	8949	8954	8960	8965	8971
79	8976	8982	8987	8993	8998	9004	9009	9015	9020	9025
80	9031	9036	9042	9047	9053	9058	9063	9069	9074	9079
81	9085	9090	9096	9101	9106	9112	9117	9122	9128	9133
82	9138	9143	9149	9154	9159	9165	9170	9175	9180	9186
83	9191	9196	9201	9206	9212	9217	9222	9227	9232	9238
84	9243	9248	9253	9258	9263	9269	9274	9279	9284	9289
85	9294	9299	9304	9309	9315	9320	9235	9330	9335	9340
86	9345	9350	9355	9360	9365	9370	9375	9380	9385	9390
87	9395	9400	9405	9410	9415	9420	9425	9430	9435	9440
88	9445	9450	9455	9460	9465	9469	9474	9479	9484	9489
89	9494	9499	9504	9509	9513	9518	9523	9528	9533	9538
90	9542	9547	9552	9557	9562	9566	9571	9576	9581	9586
91	9590	9595	9600	9605	9609	9614	9619	9624	9628	9633
92	9638	9643	9647	9652	9657	9661	9666	9671	9675	9680
93	9685	9689	9694	9699	9703	9708	9713	9717	9722	9727
94	9731	9736	9741	9745	9750	9754	9759	9763	9768	9773
95	9777	9782	9786	9791	9795	9800	9805	9809	9814	9818
96	9823	9827	9832	9836	9841	9845	9850	9854	9859	9863
97	9868	9872	9877	9881	9886	9890	9894	9899	9903	9908
98	9912	9917	9921	9926	9930	9934	9939	9943	9948	9952
99	9956	9961	9965	9969	9974	9978	9983	9987	9991	9996
100	0000	0004	0009	0013	0017	0022	0026	0030	0035	0039
no.	0	1	2	3	4	5	6	7	8	9

D

Natural Trigonometric Functions for Angles in Degrees and Decimals

Deg.	Sin	Tan	Cot	Cos	Deg.
0.0	.00000	.00000	∞	1.0000	90.0
.1	.00175	.00175	573.0	1.0000	89.9
.2	.00349	.00349	286.5	1.0000	.8
.3	.00524	.00524	191.0	1.0000	.7
.4	.00698	.00698	143.24	1.0000	.6
.5	.00873	.00873	114.59	1.0000	.5
.6	.01047	.01047	95.49	0.9999	.4
.7	.01222	.01222	81.85	.9999	.3
.8	.01396	.01396	71.62	.9999	.2
.9	.01571	.01571	63.66	.9999	89.1
1.0	.01745	.01746	57.29	0.9998	89.0
.1	.01920	.01920	52.08	.9998	88.9
.2	.02094	.02095	47.74	.9998	.8
.3	.02269	.02269	44.07	.9997	.7
.4	.02443	.02444	40.92	.9997	.6
.5	.02618	.02619	38.19	.9997	.5
.6	.02792	.02793	35.80	.9996	.4
.7	.02967	.02968	33.69	.9996	.3
.8	.03141	.03143	31.82	.9995	.2
.9	.03316	.03317	30.14	.9995	88.1
2.0	.03490	.03492	28.64	0.9994	88.0
.1	.03664	.03667	27.27	.9993	87.9
.2	.03839	.03842	26.03	.9993	.8
.3	.04013	.04016	24.90	.9992	.7
.4	.04188	.04191	23.86	.9991	.6
.5	.04362	.04366	22.90	.9990	.5
.6	.04536	.04541	22.02	.9990	.4
.7	.04711	.04716	21.20	.9989	.3
.8	.04885	.04891	20 45	.9988	.2
.9	.05059	.05066	19.74	.9987	87.1
3.0	.05234	.05241	19.081	0.9986	87.0
.1	.05408	.05416	18.464	.9985	86.9
.2	.05582	.05591	17.886	.9984	.8
.3	.05756	.05766	17.343	.9983	.7
.4	.05931	.05941	16.832	.9982	.6
.5	.06105	.06116	16 350	.9981	.5
.6	.06279	.06291	15.895	.9980	.4
.7	.06453	.06467	15.464	.9979	.3
.8	.06627	.06642	15.056	.9978	.2
.9	.06802	.06817	14.669	.9977	86.1
4.0	.06976	.06993	14.301	0.9976	86.0
.1	.07150	.07163	13.951	.9974	85.9
.2	.07324	.07344	13.617	.9973	.8
.3	.07498	.07519	13.300	.9972	.7
.4	.07672	.07695	12.996	.9971	.6
.5	.07846	.07870	12.706	.9969	.5
.6	.08020	.08046	12.429	.9968	.4
.7	.08194	.08221	12.163	.9966	.3
.8	.08368	.08397	11.909	.9965	.2
.9	.08542	.08573	11.664	.9963	85.1
5.0	.08716	.08749	11.430	0.9962	85.0
.1	.08889	.08925	11.205	.9960	84.9
.2	.09063	.09101	10.988	.9959	.8
.3	.09237	.09277	10.780	.9957	.7
.4	.09411	.09453	10.579	.9956	.6
.5	.09585	.09629	10.385	.9954	.5
.6	.09758	.09805	10.199	.9952	.4
.7	.09932	.09981	10.019	.9951	.3
.8	.10106	.10158	9.845	.9949	.2
.9	.10279	.10334	9.677	.9947	84.1
6.0	.10453	.10510	9.514	0.9945	84.0
.1	.10626	.10687	9.357	.9943	83.9
.2	.10800	.10863	9.205	.9942	.8
.3	.10973	.11040	9.058	.9940	.7
.4	.11147	.11217	8.915	.9938	.6
.5	.11320	.11394	8.777	.9936	.5
.6	.11494	.11570	8.643	.9934	.4
.7	.11667	.11747	8.513	.9932	.3
.8	.11840	.11924	8.386	.9930	.2
.9	.12014	.12101	8 264	.9928	83.1
7.0	.12187	.12278	8.144	0.9925	83.0
Deg.	**Cos**	**Cot**	**Tan**	**Sin**	**Deg.**

Deg.	Sin	Tan	Cot	Cos	Deg.
7.0	.12187	.12278	8.144	0.9925	83.0
.1	.12360	.12456	8.028	.9923	82.9
.2	.12533	.12633	7.916	.9921	.8
.3	.12706	.12810	7.806	.9919	.7
.4	.12880	.12988	7.700	.9917	.6
.5	.13053	.13165	7.596	.9914	.5
.6	.13226	.13343	7.495	.9912	.4
.7	.13399	.13521	7.396	.9910	.3
.8	.13572	.13698	7.300	.9907	.2
.9	.13744	.13876	7.207	.9905	82.1
8.0	.13917	.14054	7.115	0.9903	82.0
.1	.14090	.14232	7.026	.9900	81.9
.2	.14263	.14410	6.940	.9898	.8
.3	.14436	.14588	6.855	.9895	.7
.4	.14608	.14767	6.772	.9893	.6
.5	.14781	.14945	6.691	.9890	.5
.6	.14954	.15124	6.612	.9888	.4
.7	.15126	.15302	6.535	.9885	.3
.8	.15299	.15481	6.460	.9882	.2
.9	.15471	.15660	6.386	.9880	81.1
9.0	.15643	.15838	6.314	0.9877	81.0
.1	.15816	.16017	6.243	.9874	80.9
.2	.15988	.16196	6.174	.9871	.8
.3	.16160	.16376	6.107	.9869	.7
.4	.16333	.16555	6.041	.9866	.6
.5	.16505	.16734	5.976	.9863	.5
.6	.16677	.16914	5.912	.9860	.4
.7	.16849	.17093	5.850	.9857	.3
.8	.17021	.17273	5.789	.9854	.2
.9	.17193	.17453	5.730	.9851	80.1
10.0	.1736	.1763	5.671	0.9848	80.0
.1	.1754	.1781	5.614	.9845	79.9
.2	.1771	.1799	5.558	.9842	.8
.3	.1788	.1817	5.503	.9839	.7
.4	.1805	.1835	5.449	.9836	.6
.5	.1822	.1853	5.396	.9833	.5
.6	.1840	.1871	5.343	.9829	.4
.7	.1857	.1890	5.292	.9826	.3
.8	.1874	.1908	5.242	.9823	.2
.9	.1891	.1926	5.193	.9820	79.1
11.0	.1908	.1944	5.145	0.9816	79.0
.1	.1925	.1962	5.097	.9813	78.9
.2	.1942	.1980	5.050	.9810	.8
.3	.1959	.1998	5.005	.9806	.7
.4	.1977	.2016	4.959	.9803	.6
.5	.1994	.2035	4.915	.9799	.5
.6	.2011	.2053	4.872	.9796	.4
.7	.2028	.2071	4.829	.9792	.3
.8	.2045	.2089	4.787	.9789	.2
.9	.2062	.2107	4.745	.9785	78.1
12.0	.2079	.2126	4.705	0.9781	78.0
.1	.2096	.2144	4.665	.9778	77.9
.2	.2113	.2162	4.625	.9774	.8
.3	.2130	.2180	4.586	.9770	.7
.4	.2147	.2199	4.548	.9767	.6
.5	.2164	.2217	4.511	.9763	.5
.6	.2181	.2235	4.474	.9759	.4
.7	.2198	.2254	4.437	.9755	.3
.8	.2215	.2272	4.402	.9751	.2
.9	.2233	.2290	4.366	.9748	77.1
13.0	.2250	.2309	4.331	0.9744	77.0
.1	.2267	.2327	4.297	.9740	76.9
.2	.2284	.2345	4.264	.9736	.8
.3	.2300	.2364	4.230	.9732	.7
.4	.2317	.2382	4.198	.9728	.6
.5	.2334	.2401	4.165	.9724	.5
.6	.2351	.2419	4.134	.9720	.4
.7	.2368	.2438	4.102	.9715	.3
.8	.2385	.2456	4.071	.9711	.2
.9	.2402	.2475	4.041	.9707	76.1
14.0	.2419	.2493	4.011	0.9703	76.0
Deg.	**Cos**	**Cot**	**Tan**	**Sin**	**Deg.**

Deg.	Sin	Tan	Cot	Cos	Deg.
14.0	.2419	.2493	4.011	0.9703	76.0
.1	.2436	.2512	3.981	.9699	75.9
.2	.2453	.2530	3.952	.9694	.8
.3	.2470	.2549	3.923	.9690	.7
.4	.2487	.2568	3.895	.9686	.6
.5	.2504	.2586	3.867	.9681	.5
.6	.2521	.2605	3.839	.9677	.4
.7	.2538	.2623	3.812	.9673	.3
.8	.2554	.2642	3.785	.9668	.2
.9	.2571	.2661	3.758	.9664	75.1
15.0	0.2588	0.2679	3.732	0.9659	75.0
.1	.2605	.2698	3.706	.9655	74.9
.2	.2622	.2717	3.681	.9650	.8
.3	.2639	.2736	3.655	.9646	.7
.4	.2656	.2754	3.630	.9641	.6
.5	.2672	.2773	3.606	.9636	.5
.6	.2689	.2792	3.582	.9632	.4
.7	.2706	.2811	3.558	.9627	.3
.8	.2723	.2830	3.534	.9622	.2
.9	.2740	.2849	3.511	.9617	74.1
16.0	0.2756	0.2867	3.487	0.9613	74.0
.1	.2773	.2886	3.465	.9608	73.9
.2	.2790	.2905	3.442	.9603	.8
.3	.2807	.2924	3.420	.9598	.7
.4	.2823	.2943	3.398	.9593	.6
.5	.2840	.2962	3.376	9588	.5
.6	.2857	.2981	3.354	.9583	.4
.7	.2874	.3000	3.333	.9578	.3
.8	.2890	.3019	3.312	.9573	.2
.9	.2907	.3038	3.291	.9568	73.1
17.0	.2924	.3057	3.271	0.9563	73.0
.1	.2940	.3076	3.251	.9558	72.9
.2	.2957	.3096	3.230	.9553	.8
.3	.2974	.3115	3.211	.9548	.7
.4	.2990	.3134	3.191	.9542	.6
.5	.3007	.3153	3.172	.9537	.5
.6	.3024	.3172	3.152	.9532	.4
.7	.3040	.3191	3.133	.9527	.3
.8	.3057	.3211	3.115	.9521	.2
.9	.3074	.3230	3.096	.9516	72.1
18.0	.3090	.3249	3.078	0.9511	72.0
.1	.3107	.3269	3.060	.9505	71.9
.2	.3123	.3288	3.042	.9500	.8
.3	.3140	.3307	3.024	.9494	.7
.4	.3156	.3327	3.006	.9489	.6
.5	.3173	.3346	2.989	.9483	.5
.6	.3190	.3365	2.971	.9478	.4
.7	.3206	.3385	2.954	.9472	.3
.8	.3223	.3404	2.937	.9466	.2
.9	.3239	.3424	2.921	.9461	71.1
19.0	.3256	.3443	2.904	0.9455	71.0
.1	.3272	.3463	2.888	.9449	70.9
.2	.3289	.3482	2.872	.9444	.8
.3	.3305	.3502	2.856	.9438	.7
.4	.3322	.3522	2.840	.9432	.6
.5	.3338	.3541	2.824	.9426	.5
.6	.3355	.3561	2.808	.9421	.4
.7	.3371	.3581	2.793	.9415	.3
.8	.3387	.3600	2.778	.9409	.2
.9	.3404	.3620	2.762	.9403	70.1
20.0	.3420	.3640	2.747	0.9397	70.0
.1	.3437	.3659	2.733	.9391	69.9
.2	.3453	.3679	2.718	.9385	.8
.3	.3469	.3699	2.703	.9379	.7
.4	.3486	.3719	2.689	.9373	.6
.5	.3502	.3739	2.675	.9367	.5
.6	.3518	.3759	2.660	.9361	.4
.7	.3535	.3779	2.646	.9354	.3
.8	.3551	.3799	2.633	.9348	.2
.9	.3567	.3819	2.619	.9342	69.1
21.0	0.3584	0.3839	2.605	0.9336	69.0
Deg.	**Cos**	**Cot**	**Tan**	**Sin**	**Deg.**

Deg.	Sin	Tan	Cot	Cos	Deg.
21.0	0.3584	0.3839	2.605	0.9336	**69.0**
.1	.3600	.3859	2.592	.9330	68.9
.2	.3616	.3879	2.578	.9323	.8
.3	.3633	.3899	2.565	.9317	.7
.4	.3649	.3919	2.552	.9311	.6
.5	.3665	.3939	2.539	.9304	.5
.6	.3681	.3959	2.526	.9298	.4
.7	.3697	.3979	2.513	.9291	.3
.8	.3714	.4000	2.500	.9285	.2
.9	.3730	.4020	2.488	.9278	68.1
22.0	0.3746	0.4040	2.475	0.9272	**68.0**
.1	.3762	.4061	2.463	.9265	67.9
.2	.3778	.4081	2.450	.9259	.8
.3	.3795	.4101	2.438	.9252	.7
.4	.3811	.4122	2.426	.9245	.6
.5	.3827	.4142	2.414	.9239	.5
.6	.3843	.4163	2.402	.9232	.4
.7	.3859	.4183	2.391	.9225	.3
.8	.3875	.4204	2.379	.9219	.2
.9	.3891	.4224	2.367	.9212	67.1
23.0	0.3907	0.4245	2.356	0.9205	**67.0**
.1	.3923	.4265	2.344	.9198	66.9
.2	.3939	.4286	2.333	.9191	.8
.3	.3955	.4307	2.322	.9184	.7
.4	.3971	.4327	2.311	.9178	.6
.5	.3987	.4348	2.300	.9171	.5
.6	.4003	.4369	2.289	.9164	.4
.7	.4019	.4390	2.278	.9157	.3
.8	.4035	.4411	2.267	.9150	.2
.9	.4051	.4431	2.257	.9143	66.1
24.0	0.4067	0.4452	2.246	0.9135	**66.0**
.1	.4083	.4473	2.236	.9128	65.9
.2	.4099	.4494	2.225	.9121	.8
.3	.4115	.4515	2.215	.9114	.7
.4	.4131	.4536	2.204	.9107	.6
.5	.4147	.4557	2.194	.9100	.5
.6	.4163	.4578	2.184	.9092	.4
.7	.4179	.4599	2.174	.9085	.3
.8	.4195	.4621	2.164	.9078	.2
.9	.4210	.4642	2.154	.9070	65.1
25.0	0.4226	0.4663	2.145	0.9063	**65.0**
.1	.4242	.4684	2.135	.9056	64.9
.2	.4258	.4706	2.125	.9048	.8
.3	.4274	.4727	2.116	.9041	.7
.4	.4289	.4748	2.106	.9033	.6
.5	.4305	.4770	2.097	.9026	.5
.6	.4321	.4791	2.087	.9018	.4
.7	.4337	.4813	2.078	.9011	.3
.8	.4352	.4834	2.069	.9003	.2
.9	.4368	.4856	2.059	.8996	64.1
26.0	0.4384	0.4877	2.050	0.8988	**64.0**
.1	.4399	.4899	2.041	.8980	63.9
.2	.4415	.4921	2.032	.8973	.8
.3	.4431	.4942	2.023	.8965	.7
.4	.4446	.4964	2.014	.8957	.6
.5	.4462	.4986	2.006	.8949	.5
.6	.4478	.5008	1.997	.8942	.4
.7	.4493	.5029	1.988	.8934	.3
.8	.4509	.5051	1.980	.8926	.2
.9	.4524	.5073	1.971	.8918	63.1
27.0	0.4540	0.5095	1.963	0.8910	**63.0**
.1	.4555	.5117	1.954	.8902	62.9
.2	.4571	.5139	1.946	.8894	.8
.3	.4586	.5161	1.937	.8886	.7
.4	.4602	.5184	1.929	.8878	.6
.5	.4617	.5206	1.921	.8870	.5
.6	.4633	.5228	1.913	.8862	.4
.7	.4648	.5250	1.905	.8854	.3
.8	.4664	.5272	1.897	.8846	.2
.9	.4679	.5295	1.889	.8838	62.1
28.0	0.4695	0.5317	1.881	0.8829	**62.0**
.1	.4710	.5340	1.873	.8821	61.9
.2	.4726	.5362	1.865	.8813	.8
.3	.4741	.5384	1.857	.8805	.7
.4	.4756	.5407	1.849	.8796	.6
.5	.4772	.5430	1.842	.8788	.5
.6	.4787	.5452	1.834	.8780	.4
.7	.4802	.5475	1.827	.8771	.3
.8	.4818	.5498	1.819	.8763	.2
.9	.4833	.5520	1.811	.8755	61.1
29.0	0.4848	0.5543	1.804	0.8746	**61.0**
Deg.	**Cos**	**Cot**	**Tan**	**Sin**	**Deg.**

Deg.	Sin	Tan	Cot	Cos	Deg.
29.0	0.4848	0.5543	1.804	0.8746	**61.0**
.1	.4863	.5566	1.797	.8738	60.9
.2	.4879	.5589	1.789	.8729	.8
.3	.4894	.5612	1.782	.8721	.7
.4	.4909	.5635	1.775	.8712	.6
.5	.4924	.5658	1.767	.8704	.5
.6	.4939	.5681	1.760	.8695	.4
.7	.4955	.5704	1.753	.8686	.3
.8	.4970	.5727	1.746	.8678	.2
.9	.4985	.5750	1.739	.8669	60.1
30.0	0.5000	0.5774	1.732	0.8660	**60.0**
.1	.5015	.5797	1.7251	.8652	59.9
.2	.5030	.5820	1.7182	.8643	.8
.3	.5045	.5844	1.7113	.8634	.7
.4	.5060	.5867	1.7045	.8625	.6
.5	.5075	.5890	1.6977	.8616	.5
.6	.5090	.5914	1.6909	.8607	.4
.7	.5105	.5938	1.6842	.8599	.3
.8	.5120	.5961	1.6775	.8590	.2
.9	.5135	.5985	1.6709	.8581	59.1
31.0	0.5150	0.6009	1.6643	0.8572	**59.0**
.1	.5165	.6032	1.6577	.8563	58.9
.2	.5180	.6056	1.6512	.8554	.8
.3	.5195	.6080	1.6447	.8545	.7
.4	.5210	.6104	1.6383	.8536	.6
.5	.5225	.6128	1.6319	.8526	.5
.6	.5240	.6152	1.6255	.8517	.4
.7	.5255	.6176	1.6191	.8508	.3
.8	.5270	.6200	1.6128	.8499	.2
.9	.5284	.6224	1.6066	.8490	58.1
32.0	0.5299	0.6249	1.6003	0.8480	**58.0**
.1	.5314	.6273	1.5941	.8471	57.9
.2	.5329	.6297	1.5880	.8462	.8
.3	.5344	.6322	1.5818	.8453	.7
.4	.5358	.6346	1.5757	.8443	.6
.5	.5373	.6371	1.5697	.8434	.5
.6	.5388	.6395	1.5637	.8425	.4
.7	.5402	.6420	1.5577	.8415	.3
.8	.5417	.6445	1.5517	.8406	.2
.9	.5432	.6469	1.5458	.8396	57.1
33.0	0.5446	0.6494	1.5399	0.8387	**57.0**
.1	.5461	.6519	1.5340	.8377	56.9
.2	.5476	.6544	1.5282	.8368	.8
.3	.5490	.6569	1.5224	.8358	.7
.4	.5505	.6594	1.5166	.8348	.6
.5	.5519	.6619	1.5108	.8339	.5
.6	.5534	.6644	1.5051	.8329	.4
.7	.5548	.6669	1.4994	.8320	.3
.8	.5563	.6694	1.4938	.8310	.2
.9	.5577	.6720	1.4882	.8300	56.1
34.0	0.5592	0.6745	1.4826	0.8290	**56.0**
.1	.5606	.6771	1.4770	.8281	55.9
.2	.5621	.6796	1.4715	.8271	.8
.3	.5635	.6822	1.4659	.8261	.7
.4	.5650	.6847	1.4605	.8251	.6
.5	.5664	.6873	1.4550	.8241	.5
.6	.5678	.6899	1.4496	.8231	.4
.7	.5693	.6924	1.4442	.8221	.3
.8	.5707	.6950	1.4388	.8211	.2
.9	.5721	.6976	1.4335	.8202	55.1
35.0	0.5736	0.7002	1.4281	0.8192	**55.0**
.1	.5750	.7028	1.4229	.8181	54.9
.2	.5764	.7054	1.4176	.8171	.8
.3	.5779	.7080	1.4124	.8161	.7
.4	.5793	.7107	1.4071	.8151	.6
.5	.5807	.7133	1.4019	.8141	.5
.6	.5821	.7159	1.3968	.8131	.4
.7	.5835	.7186	1.3916	.8121	.3
.8	.5850	.7212	1.3865	.8111	.2
.9	.5864	.7239	1.3814	.8100	54.1
36.0	0.5878	0.7265	1.3764	0.8090	**54.0**
.1	.5892	.7292	1.3713	.8080	53.9
.2	.5906	.7319	1.3663	.8070	.8
.3	.5920	.7346	1.3613	.8059	.7
.4	.5934	.7373	1.3564	.8049	.6
.5	.5948	.7400	1.3514	.8039	.5
.6	.5962	.7427	1.3465	.8028	.4
.7	.5976	.7454	1.3416	.8018	.3
.8	.5990	.7481	1.3367	.8007	.2
.9	.6004	.7508	1.3319	.7997	53.1
37.0	0.6018	0.7536	1.3270	0.7986	**53.0**
Deg.	**Cos**	**Cot**	**Tan**	**Sin**	**Deg.**

Deg.	Sin	Tan	Cot	Cos	Deg.
37.0	0.6018	0.7536	1.3270	0.7986	**53.0**
.1	.6032	.7563	1.3222	.7976	52.9
.2	.6046	.7590	1.3175	.7965	.8
.3	.6060	.7618	1.3127	.7955	.7
.4	.6074	.7646	1.3079	.7944	.6
.5	.6088	.7673	1.3032	.7934	.5
.6	.6101	.7701	1.2985	.7923	.4
.7	.6115	.7729	1.2938	.7912	.3
.8	.6129	.7757	1.2892	.7902	.2
.9	.6143	.7785	1.2846	.7891	52.1
38.0	0.6157	0.7813	1.2799	0.7880	**52.0**
.1	.6170	.7841	1.2753	.7869	51.9
.2	.6184	.7869	1.2708	.7859	.8
.3	.6198	.7898	1.2662	.7848	.7
.4	.6211	.7926	1.2617	.7837	.6
.5	.6225	.7954	1.2572	.7826	.5
.6	.6239	.7983	1.2527	.7815	.4
.7	.6252	.8012	1.2482	.7804	.3
.8	.6266	.8040	1.2437	.7793	.2
.9	.6280	.8069	1.2393	.7782	51.1
39.0	0.6293	0.8098	1.2349	0.7771	**51.0**
.1	.6307	.8127	1.2305	.7760	50.9
.2	.6320	.8156	1.2261	.7749	.8
.3	.6334	.8185	1.2218	.7738	.7
.4	.6347	.8214	1.2174	.7727	.6
.5	.6361	.8243	1.2131	.7716	.5
.6	.6374	.8273	1.2088	.7705	.4
.7	.6388	.8302	1.2045	.7694	.3
.8	.6401	.8332	1.2002	.7683	.2
.9	.6414	.8361	1.1960	.7672	50.1
40.0	0.6428	0.8391	1.1918	0.7660	**50.0**
.1	.6441	.8421	1.1875	.7649	49.9
.2	.6455	.8451	1.1833	.7638	.8
.3	.6468	.8481	1.1792	.7627	.7
.4	.6481	.8511	1.1750	.7615	.6
40.5	0.6494	0.8541	1.1708	0.7604	**49.5**
.6	.6508	.8571	1.1667	.7593	.4
.7	.6521	.8601	1.1626	.7581	.3
.8	.6534	.8632	1.1585	.7570	.2
.9	.6547	.8662	1.1544	.7559	49.1
41.0	0.6561	0.8693	1.1504	0.7547	**49.0**
.1	.6574	.8724	1.1463	.7536	48.9
.2	.6587	.8754	1.1423	.7524	.8
.3	.6600	.8785	1.1383	.7513	.7
.4	.6613	.8816	1.1343	.7501	.6
.5	.6626	.8847	1.1303	.7490	.5
.6	.6639	.8878	1.1263	.7478	.4
.7	.6652	.8910	1.1224	.7466	.3
.8	.6665	.8941	1.1184	.7455	.2
.9	.6678	.8972	1.1145	.7443	48.1
42.0	0.6691	0.9004	1.1106	0.7431	**48.0**
.1	.6704	.9036	1.1067	.7420	47.9
.2	.6717	.9067	1.1028	.7408	.8
.3	.6730	.9099	1.0990	.7396	.7
.4	.6743	.9131	1.0951	.7385	.6
.5	.6756	.9163	1.0913	.7373	.5
.6	.6769	.9195	1.0875	.7361	.4
.7	.6782	.9228	1.0837	.7349	.3
.8	.6794	.9260	1.0799	.7337	.2
.9	.6807	.9293	1.0761	.7325	47.1
43.0	0.6820	0.9325	1.0724	0.7314	**47.0**
.1	.6833	.9358	1.0686	.7302	46.9
.2	.6845	.9391	1.0649	.7290	.8
.3	.6858	.9424	1.0612	.7278	.7
.4	.6871	.9457	1.0575	.7266	.6
.5	.6884	.9490	1.0538	.7254	.5
.6	.6896	.9523	1.0501	.7242	.4
.7	.6909	.9556	1.0464	.7230	.3
.8	.6921	.9590	1.0428	.7218	.2
.9	.6934	.9623	1.0392	.7206	46.1
44.0	0.6947	0.9657	1.0355	0.7193	**46.0**
.1	.6959	.9691	1.0319	.7181	45.9
.2	.6972	.9725	1.0283	.7169	.8
.3	.6984	.9759	1.0247	.7157	.7
.4	.6997	.9793	1.0212	.7145	.6
.5	.7009	.9827	1.0176	.7133	.5
.6	.7022	.9861	1.0141	.7120	.4
.7	.7034	.9896	1.0105	.7108	.3
.8	.7046	.9930	1.0070	.7096	.2
.9	.7059	.9965	1.0035	.7083	45.1
45.0	0.7071	1.0000	1.0000	0.7071	**45.0**
Deg.	**Cos**	**Cot**	**Tan**	**Sin**	**Deg.**

From CRC Standard Mathematical Tables, 15th edition, 1967.

Courtesy of the Chemical Rubber Company, Cleveland, Ohio.

Solutions to Odd-Numbered Problems

CHAPTER 2

1. (a) 64 C; (b) 1.6 A. **3.** (a) 128.49 mils; (b) 0.12849 in. **5.** 850 V. **7.** 1.5 A. **9.** 160 Ω. **11.** 1.515 Ω. **13.** 274.5°C. **15.** (a) 10 Ω (log scale); (b) 100 Ω ($T \cong 110$°C) (log scale). **17.** 64 mW. **19.** 9.6×10^5 J. **21.** (a) 93.25%; (b) 20 A; (c) 162 W. **23.** (a) 5.714 Ω; (b) 21 A; (c) 6 A; (d) 6 A; (e) 21 V. **25.** (a) 12 Ω; (b) 3 A; (c) $I_1 = 3$ A, $I_2 = 2$ A, $I_3 = 1$ A; (d) 3 W. **27.** $I_S = 4$ A, $R_P = 9$ Ω. **29.** $I = 3.33$ A↓. **31.** 4 V. **33.** (a) 1.714 Ω; (b) 11.81 W; (c) 8.17 W. **35.** 13.5 Ω. **37.** (a) yes; (b) 1.4 A; (c) 0.4 A. **39.** (a) 0.5s; (b) $i_C = 8 \times 10^{-3}e^{-t/\tau}$, $v_C = 40(1 - e^{-t/\tau})$, $v_R = 40e^{-t/\tau}$; (d) $i_C(1\tau) = 2.944$ mA. **41.** (a) 1.2 μF; (b) 24 μF. **43.** (a) $4 \times 10^{-3}s$; (b) $i_L = 16 \times 10^{-3}(1 - e^{-t/\tau})$; $v_L = 8e^{-t/\tau}$, $v_R = 8(1 - e^{-t/\tau})$; (d) $v_R(1\tau) = 5.056$ V. **45.** (a) 0.9 H; (b) 0.2 H. **47.** $I_1 = 12$ mA, $V_1 = 12$ V.

CHAPTER 3

1. (a) Peak value = 120, crosses axis at $\theta = 0$°, $t = 0$; (b) Peak value = 0.04, crosses axis at $\theta = 0$°, $t = 0$. **3.** 0.2s **5.** (a) Peak value = 60, crosses axis at $t = 0$ or $\theta = 0$°, $T = 40$ ms; (b) Peak value = 20×10^{-3}, crosses axis at $t = 0$ or $\theta = 0$°, $T = 31.42$ ms. **7.** (a) Peak value = 100 V, crosses axis 60° to the left of 0°, $T = 16.67$ ms; (b) Peak value = 50 μF, crosses axis 70° to the right of 0°, $T = 15.71$ ms. **9.** (a) 2.82 V, 565.6 mA; (b) Peak value = 311.08 V, crosses axis at $\theta = 0$°. **11.** $v = 20 \sin 377t$. **13.** 2,261.9 Ω. **15.** 530.5 Ω. **17.** (a) $25 + j43.3$; (b) $-60 + j103.92$; (c) $0.02 - j0.0346$; (d) $7.071\underline{/45°}$; (e) $13\underline{/67.38°}$; (f) $98.44\underline{/-63.43°}$. **19.** $v = 150 \sin (\omega t + 60°)$. **21.** $v = 10.6 \sin (377t - 90°)$. **23.** (a) $8.485 \times 10^3\underline{/45°}$; (b) $7.071 + 10^{-3}\underline{/-45°}$; (c) $V_R = 42.427\underline{/-45°}$, $V_L = 141.42\underline{/45°}$, $V_C = 98.994\underline{/-135°}$; (d) $141.42\underline{/45°}$; (e) $0.3w$; (f) 0.7071. **25.**

$12.92 + 10^{-3}/\underline{-21.01°}$. **27.** (a) No. (b) 4 K capacitance to 4 K inductance (or the reverse). **29.** $I_1 = 5 \times 10^{-3}/\underline{0°}$, $R_{p_1} = 4$ K and $I_2 = 20 \times 10^{-3}/\underline{0°}$, $R_{p_2} = 2$ K. **31.** $10.74 \times 10^{-3}/\underline{-16.39°}$. **33.** (a) $2.4 \times 10^3/\underline{-36.9°} = 1.92$ K $- j1.44$ K; (b) 1.2 W. **35.** (a) $V_{\phi_g} = V_{\phi L} = 127.17$ V; (b) $I_{\phi L} = 31.79$ mA; (c) $I_L = 55$ mA; (d) $P = 0$ (totally reactive load). **37.** (a) 3 K; (b) 150; (c) 33.33 Hz; (d) P (at resonance) $= 80$ μW, $P_{\text{HPF}} = 40$ μW; (e) $L = 100$ mH, $C = 0.011$ μF. **39.** (a) band-stop; (b) $f_s = 11.26$ KHz; (c) $Bw = 4139.71$ Hz, $f_1 = 9,190.14$ Hz, $f_2 = 13,329.86$ Hz; (d) 7.69 mV.

CHAPTER 4

1. 800 lines/A·in. **3.** 1.92 A. **5.** (a) $R_{\text{Shunt}} = 0.1$ Ω; (b) $R_s = 180$ K. **7.** (a) 40 Ω; (b) 0.9 W.

CHAPTER 5

1. 4800 rpm. **3.** 180 V. **5.** (a) $I_F = 1.2$ A, $I_A = 4.7$ A; (b) $V_{R_a} = 7.05$ V, $V_{R_s} = 2.35$ V; (c) 129.4 V; (d) 420 W; (e) 608.18 W; (f) 188.18 W. **7.** 0.1062 lbs. **9.** (a) $I_F = 2$ A, $I_A = 3$ A; (b) 114 V; (c) 480 W; (d) 1.88 ft-lbs. **11.** (a) $I_F = 1$ A, $I_A = 6$ A; (b) 7 A; (c) 840 W; (d) 588 W. **13.** 4.5 Ω. **15.** $I_L = 14.45$ A, $I_\phi = 8.35$ A. **17.** 147.85 V. **19.** 1746 rpm. **21.** T drops 50%. **23.** (a) $P_q = 900$ KVARS, $P_a = 1500$ KVA; (b) $P_{a_s} = 900$ KVA.

CHAPTER 6

3. (a) negative portion of input only. (b) 60 V. (c) positive portion of input only with peak $= 75$. **7.** (a) For figure (a) positive portion of input from $0 \rightarrow 10$ V only, for figure (b) V_o fixed at 20 V; (b) For figure (a) that portion of the input not included in V_o, for figure (b) the entire sinusoidal input. **9.** 83 pF. **11.** $0°$C $- 10$ K ohm·cm, $100°$C (log scale) $= 300$ ohm·cm. **13.** 100 fc-R $\cong 2$ MΩ, 1000 fc $-$ R $\cong 80$ K. **15.** (a) $\cong 3$ mA; (b) $\cong 1.8$ mA; (c) $\cong 235$ mV. **17.** (a) $\cong 2.6$mA (b) $\cong 42$ μA; (c) $\cong 3.5$ mA; (d) 130; (e) 0.99. **19.** 6 mA. **23.** 2 mA. **27.** (a) For $V_{BB} = 5$ V; $V_p = 3.5$ V, For $V_{BB} = 20$ V; $V_p = 11$ V; (b) $\cong 7$ mA. **29.** $\cong 4$ to 10 V. **33.** $V_{GK} = -0.5$ V, $I_p \cong 7$ mA; $V_{GK} = -2.5$ V, $I_p \cong 1$ mA.

CHAPTER 8

9. $V_o = 0.8$ V, rms. **11.** $V_o = 130$ mV, rms. **13.** $I_L = 4$ mA, rms. **15.** $I_E = 3.33$ mA, $I_C = 3.26$ mA. **17.** $V_o = 1.5$ V, rms. **19.** $V_{CE_Q} = 4.05$ V. **21.** $A_v = -211$. **23.** $I_C R_C = 8.33$ V, $V_{CE_Q} = 6.67$ V. **25.** $R_E = 1.6$ K. **27.** Yes, 132.7 Hz. **29.** $R_i = 581$ Ω. **31.** $A_v = -333.3$. **33.** $V_{CE_Q} = 3.6$ V. **35.** $R_i = 875$ Ω, $R_o = 2.4$ K. **37.** $g_m = 3.1$ mmhos. **39.** $A_v = -3.96$. **41.** $g_m = 1.26$ mmhos.

CHAPTER 9

1. $A_{v_1} = -16.4$. **3.** $V_o = 2.3$ V, rms. **5.** $I_{c_1} = 1.9$ mA, $I_{c_2} = 2.9$ mA. **7.** $R_{i1} = 2.16$ K, $R_{i2} = 1.47$ K. **9.** $V_{oL} = 1.1$ V, rms, $A_v = 11,000$. **13.** $A_d = 100$. **17.** 28.3 to 1. **19.** $R_o = 15\,\Omega$. **21.** $C_c = 1.6\,\mu$F. **23.** $C_{high} = 371$ pF. **25.** $A_{mid} = -10.8$. **27.** $A_{mid} = -233.8$. **29.** $f_{high} = 26.2$ kHz. **31.** $P_o(ac) = 225$ mW. **33.** $P_i(dc) = 225$ mW.

CHAPTER 10

3. 890 kHz, 910 kHz. **7.** (a) $f_c = 880$ kHz, $f_m = 60$ kHz; (b) $m = 0.5$; (c) $V = 60$ V. **15.** $f_o = 949$ kHz. **17.** $C = 185$ pF. **19.** $C = 0.001\,\mu$F. **21.** $C = 451.5$ pF.

CHAPTER 11

1. $\dfrac{V_o}{V_i} = \dfrac{1}{2}\left(\dfrac{R_2}{R_1 + R_2}\right)$, 0.33.

CHAPTER 12

1. (a) 53_{10}; (b) 28_{10}; (c) 43_{10}. **3.** (a) 35_{10}; (b) 54_{10}; (c) 70_{10}. **5.** (a) 73_{10}; (b) 162_{10}; (c) 1453_{10}. **7.** (a) 111111_2; (b) 260_8; (c) 396_{16}. **9.** 0010 1001 1000 0100. **11.** 1010 0010 1001. **13.** $\bar{A}C + B\bar{C} + A\bar{B}$. **19.** (a) $D = AB + \overline{BC} = A + \bar{B} + \bar{C}$; (b) $T = (\overline{X+W}) \cdot (W + Z) = \bar{X}\bar{W}Z$; (c) $D = ABC + \bar{A} + \bar{C} = \bar{A} + B + \bar{C}$.

CHAPTER 13

7. 31 V. **15.** $\ddot{x} - 20\dot{x} + 4x = 5y$. **17.** 8.39 V.

CHAPTER 14

1. (a) 54.06 V; (b) 94.34 V. **3.** Full wave output, $V_{peak} = 33.33$ V. **5.** (a) 10 V; (b) 0.4 V; (c) 1.38 V; (d) 0.03 F. **7.** $R = 0.8$ K, $C = 6.64\,\mu$F. **9.** 106.7 V. **11.** (a) 181.8 V; (b) 90.9 mA; (c) 1.86%. **13.** 1.7%. **15.** 36.36 V. **17.** $V_L = 17.14$ V, $V_R = 42.86$ V.

CHAPTER 15

1. $V_P = 11$ V, $V_{rms} = 7.8$ V. **3.** $V_{P-P} = 275$ mV, $V_{rms} = 97.2$ mV. **5.** (a) 58 msec; (b) 2.9 msec; (c) 29 μsec. **7.** $f = 5.56$ kHz. **9.** (a) $V_{P-P} = 300$ mV, $V_{rms} = 106.1$ mV; (b) $t = 60$ msec; (c) $f = 16.7$ Hz. **11.** 15°.

Index

A

ac generator, 72, 191–97
Admittance, 100–6
Air capacitor, 52–53
Alpha, 233
Alternators, 72, 191–97
Ammeter, 5, 62, 159
AMP-CLAMP, 169–70
Ampere, 4–6
Ampere's circuital law, 155–56
Ampere-turns, 152
Amplifiers:
    differential, 322, 453–54
    large signal, 311, 337
    phase shift, 333
    transistor, 282–94
Amplitude modulation (AM):
    detection, 352
    sidebands, 350
    transmission, 348
Analog:
    divider, 439
    multiplier, 438
Analog-digital conversion, 442–48
    dc voltage, 443
    shaft position, 447
    successive approximation, 446
Analyzer:
    spectrum, 495
    wave, 493

B

Band frequencies, 132–33, 377
Bandwidth, 132–34, 136–39
Battery, 10–12
Beta, 233
B-H curve, 154, 155
Bias circuits:
    beta independent, 290
    fixed bias, 287
Bistable multivibrator, 403
BJT (*see also* transistors):
    amplifier circuit, 282
    input resistance, 289
    small-signal amplifier, 282–94
Block diagram, 371
Bode plot, 374
Boolean algebra, 397

Angular velocity, 75–76
Apparent power, 121–23, 129, 202–3
Armature reaction, 182, 186
Astable multivibrator, 402
Asymptotic curve:
    gain, 377
    phase shift, 378
Audio frequency range, 348
Automatic gain control, 355
Average value, 80–81
AWG number, 7–9

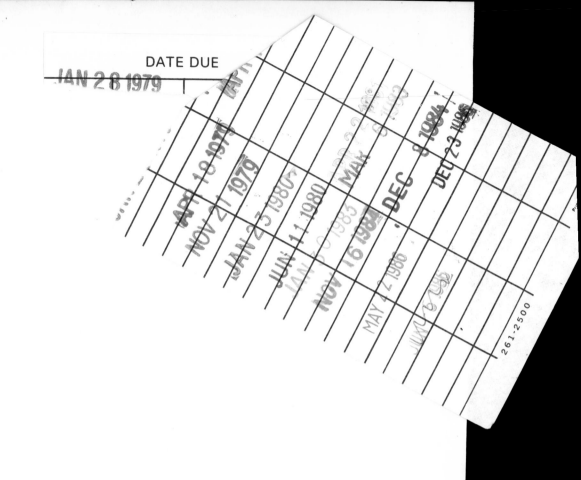